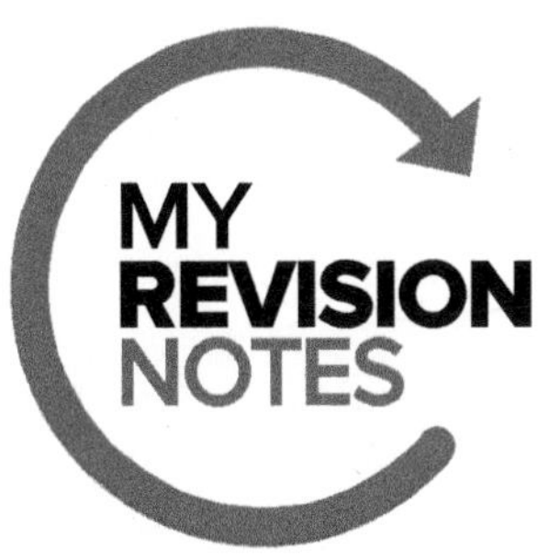

Pearson Edexcel

A-level

GEOGRAPHY

THIRD EDITION

Michael Chiles
Philly Simmons
Michael Witherick
Dan Cowling

The Publishers would like to thank the following for permission to reproduce copyright material.

Photo credits

p.45 John Eveson/Alamy Stock Photo; **p.53** *l* Victor Lucas/Adobe Stock, *c* Stepo/Adobe Stock, *r* Paul Richardson/Alamy Stock Photo; **p.98** Anthony/Adobe Stock, **p.115** Richard Jolley/Cartoonstock.

Every effort has been made to trace all copyright holders, but if any have been inadvertently overlooked, the Publishers will be pleased to make the necessary arrangements at the first opportunity.

Although every effort has been made to ensure that website addresses are correct at time of going to press, Hodder Education cannot be held responsible for the content of any website mentioned in this book. It is sometimes possible to find a relocated web page by typing in the address of the home page for a website in the URL window of your browser.

Hachette UK's policy is to use papers that are natural, renewable and recyclable products and made from wood grown in well-managed forests and other controlled sources. The logging and manufacturing processes are expected to conform to the environmental regulations of the country of origin.

Orders: please contact Hachette UK Distribution, Hely Hutchinson Centre, Milton Road, Didcot, Oxfordshire, OX11 7HH. Telephone: +44 (0)1235 827827.
Email education@hachette.co.uk. Lines are open from 9 a.m. to 5 p.m., Monday to Friday.
You can also order through our website: www.hoddereducation.co.uk

ISBN: 978 1 3983 2549 4

First published in 2017.
This edition published in 2021 by
Hodder Education,
An Hachette UK Company
Carmelite House
50 Victoria Embankment
London EC4Y 0DZ

www.hoddereducation.co.uk

Impression number 10 9 8 7 6 5 4 3 2

Year 2025 2024 2023

Cover photo © Nightman1965 - stock.adobe.com

Illustrations by Aptara, Inc.

Typeset in India by Aptara, Inc.

Printed and bound by CPI Group (UK) Ltd, Croydon, CR0 4YY

A catalogue record for this title is available from the British Library

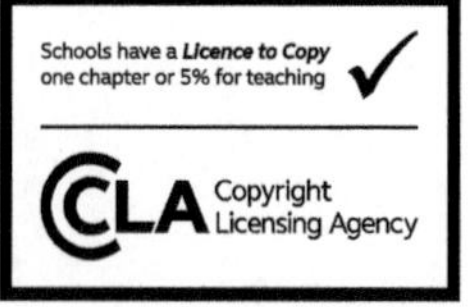

Get the most from this book

Everyone has to decide his or her own revision strategy, but it is essential to review your work, learn it and test your understanding. These Revision Notes will help you to do that in a planned way, topic by topic. Use this book as the cornerstone of your revision and don't hesitate to write in it — personalise your notes and check your progress by ticking off each section as you revise.

Tick to track your progress 

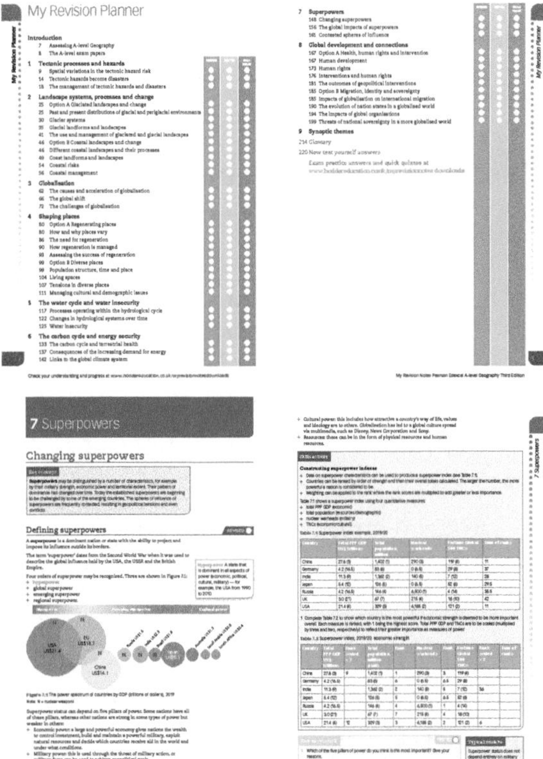

Use the revision planner on pages 4 and 5 to plan your revision, topic by topic. Tick each box when you have:

- revised and understood a topic
- tested yourself
- practised the exam questions and gone online to check your answers and complete the quick quizzes.

You can also keep track of your revision by ticking off each topic heading in the book. You may find it helpful to add your own notes as you work through each topic.

Features to help you succeed

Key concepts

Essential concepts are explained more fully to aid understanding.

Exam tips and summaries

Expert tips are given throughout the book to help you polish your exam technique in order to maximise your chances in the exam. The summaries provide a quick-check bullet list for each topic.

Typical mistakes

The authors identify the typical mistakes candidates make and explain how you can avoid them.

Now test yourself

These short, knowledge-based questions provide the first step in testing your learning. Answers are at the back of the book.

Definitions and key words

Clear, concise definitions of essential key terms are provided where they first appear. Key words from the specification are highlighted in colour throughout the book.

Skills activities

Skills activities are provided for each topic. Use them to consolidate your revision and practise your exam skills.

Revision activities

These activities will help you to understand each topic in an interactive way.

Exam practice

Practice exam questions are provided for each topic. Use them to consolidate your revision and practise your exam skills.

Making links

Overarching themes help you to make links between different geographical ideas and concepts.

Online

Go online to check your answers to the exam questions and try out the extra quick quizzes at **www.hoddereducation.co.uk/myrevisionnotesdownloads**

My Revision Planner

Exam practice answers and quick quizzes at www.hoddereducation.co.uk/myrevisionnotesdownloads

Countdown to my exams

6–8 weeks to go

- Start by looking at the specification — make sure you know exactly what material you need to revise and the style of the examination. Use the revision planner on pages 4 and 5 to familiarise yourself with the topics.
- Organise your notes, making sure you have covered everything on the specification. The revision planner will help you to group your notes into topics.
- Work out a realistic revision plan that will allow you time for relaxation. Set aside days and times for all the subjects that you need to study, and stick to your timetable.
- Set yourself sensible targets. Break your revision down into focused sessions of around 40 minutes, divided by breaks. These Revision Notes organise the basic facts into short, memorable sections to make revising easier.

REVISED

2–6 weeks to go

- Read through the relevant sections of this book and refer to the exam tips, summaries, typical mistakes and key terms. Tick off the topics as you feel confident about them. Highlight those topics you find difficult and look at them again in detail.
- Test your understanding of each topic by working through the 'Now test yourself' questions in the book. Look up the answers at the back of the book.
- Make a note of any problem areas as you revise, and ask your teacher to go over these in class.
- Look at past papers. They are one of the best ways to revise and practise your exam skills. Write or prepare planned answers to the exam practice questions provided in this book. Check your answers online and try out the extra quick quizzes at **www.hoddereducation.co.uk/myrevisionnotesdownloads**
- Use the revision activities to try out different revision methods. For example, you can make notes using mind maps, spider diagrams or flash cards.
- Track your progress using the revision planner and give yourself a reward when you have achieved your target.

REVISED

One week to go

- Try to fit in at least one more timed practice of an entire past paper and seek feedback from your teacher, comparing your work closely with the mark scheme.
- Check the revision planner to make sure you haven't missed out any topics. Brush up on any areas of difficulty by talking them over with a friend or getting help from your teacher.
- Attend any revision classes put on by your teacher. Remember, he or she is an expert at preparing people for examinations.

REVISED

The day before the examination

- Flick through these Revision Notes for useful reminders, for example the examiners' tips, examiners' summaries, typical mistakes and key terms.
- Check the time and place of your examination.
- Make sure you have everything you need — extra pens and pencils, tissues, a watch, bottled water, sweets.
- Allow some time to relax and have an early night to ensure you are fresh and alert for the examinations.

REVISED 

My exams

A-level Geography Paper 1

Date: ....

Time: ....

Location: ....

A-level Geography Paper 2

Date: ....

Time: ....

Location: ....

A-level Geography Paper 3

Date: ....

Time: ....

Location: ....

Introduction

Assessing A-level Geography

As a student of Geography it is important that you understand three things in relation to your A-level examination:

1 The key assessment objectives
2 The main command words and their meanings
3 The structure of the course

Assessment objectives

- **AO1:** Demonstrate knowledge and understanding of places, environments, concepts, processes, interactions and change, at a variety of scales.
- **AO2:** Apply knowledge and understanding in different contexts to interpret, analyse and evaluate geographical information and issues.
- **AO3:** Use a variety of relevant quantitative, qualitative and fieldwork skills to:
  - investigate geographical questions and issues
  - interpret, analyse and evaluate data and evidence
  - construct arguments and draw conclusions.

Command words

It is crucial you have a clear understanding of the command words that you will expect to see in your examination papers. This will help you to focus your response so that it meets the expectation of the command word used in the exam question. Prior to answering the question it is a good idea to **BUG** the question — **BOX** the command word, **UNDERLINE** the key words and **GLANCE** back at the question.

As a Geographer, the core command words that you will expect to find in your examination papers are as follows:

Command word	Mark tariff	Assessment objective	Definition
Calculate	4	AO1	Produce a numerical answer, showing relevant working.
Draw/Plot	4	AO1	Create a graphical representation of geographical information.
Suggest	6	AO1	For an unfamiliar scenario, provide a reasoned explanation of how or why something may occur. A suggested explanation requires a justification/exemplification of a point that has been identified.
Explain	8	AO1	Provide a reasoned explanation of how or why something occurs. An explanation requires understanding to be demonstrated through the justification or exemplification of points that have been identified.
Analyse	8	AO1/AO3	Use geographical skills to investigate an issue by systematically breaking it down into individual components and making logical, evidence-based connections on the causes and effects or interrelationships between the components.
Assess	12	AO1/AO2	Use evidence to determine the relative significance of something. Give balanced consideration to all factors and identify which are the most important.
Evaluate	16/18/20/24	AO1/AO2/AO3	Measure the value or success of something and ultimately provide a balanced and substantiated judgement/conclusion. Review information and then bring it together to form a conclusion, drawing on evidence such as strengths, weaknesses, alternatives and relevant data.

The A-level exam papers

Your Geography A-level is split into four core components, which are **three written papers and one NEA (non-examined assessment)**.

Paper 1 Physical Geography

Duration: 2 hours 15 minutes

Structure: Split into three sections

Core content

Section A relates to Topic 1: Tectonic processes and hazards.

Section B relates to Topic 2: Landscape systems, processes and change. You must answer questions on either Topic 2A: Glaciated landscapes and change or Topic 2B: Coastal landscapes and change.

Section C relates to Topic 5: The water cycle and water insecurity and **Topic 6**: The carbon cycle and energy security.

Paper 2 Human Geography

Duration: 2 hours 15 minutes

Structure: Split into three sections

Core content

Section A relates to Topics 3 and 7: Globalisation/Superpowers.

Section B relates to Topic 4: Shaping places. Students answer questions on either Topic 4A: Regenerating places or Topic 4B: Diverse places.

Section C relates to Topic 8: Global development and connections. You must answer questions on either Topic 8A: Health, human rights and intervention or Topic 8B: Migration, identity and sovereignty.

Paper 3 Synoptic Paper

Duration: 2 hours 15 minutes

Structure: A series of questions that build in demand using a resource booklet and links to prior knowledge.

Core content

This synoptic paper will require you to demonstrate knowledge and understanding based on a geographical issue that is linked to the synoptic themes from two or more of the compulsory topics of study.

NEA Independent Investigation

An internally assessed and externally moderated geographical investigation based on a topic of your choosing that is linked to the specification.

1 Tectonic processes and hazards

A natural hazard is a naturally occurring event that has the potential to create risk and vulnerability for people. Earthquakes, volcanoes and tsunamis are examples of tectonic hazards.

Spatial variations in the tectonic hazard risk

The global distribution of tectonic hazards

REVISED 

Earthquakes

The global distribution of tectonic hazards is far from random. The main earthquake zones occur along **plate boundaries**, particularly **convergent** and **conservative** ones (see Figure 1.1). **Divergent** boundaries lead to the formation of new crustal material. **Collision** zones occur when two tectonic plates move towards each other and collide. Occasionally earthquakes occur in the middle of tectonic plates (intra-plate earthquakes).

Intra-plate earthquake
Occurs when there is a release in strain energy away from the plate boundaries. The causes of these events are still not fully understood.

Key concepts

A **plate boundary** is the point at which two plates meet. The movement of these plates in relation to each other leads to the formation of distinctive landforms. There are three types:

+ **Convergent:** occur where two tectonic plates are moving together. Where a dense oceanic plate collides with a less dense continental plate, the former is thrust underneath the latter, forming a subduction zone. Mountain building and volcanic eruptions are the outcomes.
+ **Divergent:** the moving apart of the plates create rifts filled with new crustal material from volcanic eruptions.
+ **Conservative:** two plates slide past each other. The friction often triggers earthquakes.

A **collision** occurs when two continental plates collide and crush against each other, pushing up mountains.

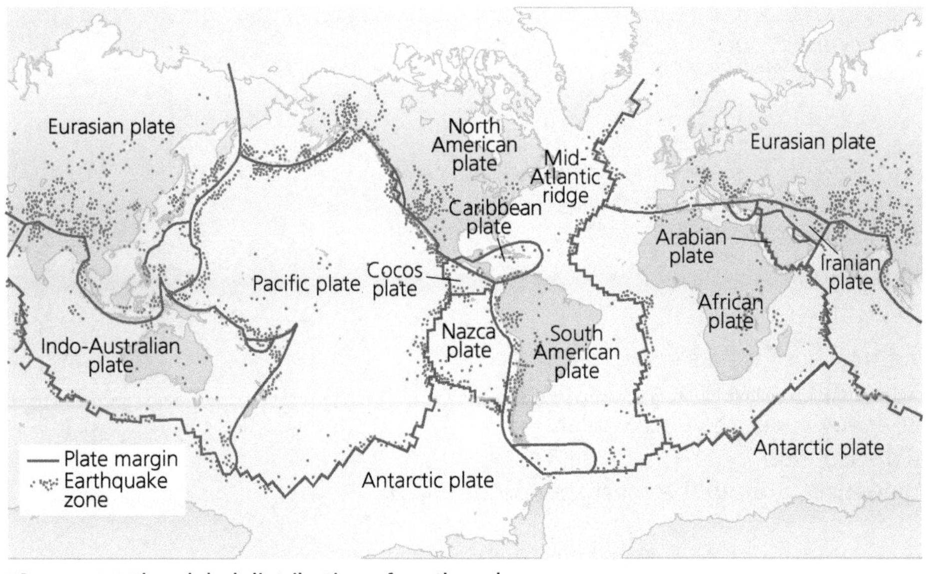

Figure 1.1 The global distribution of earthquakes

Exam tip

Remember to know each of the plate boundaries and have a clear understanding of the different processes and subsequent features created.

Now test yourself TESTED

1 Summarise what happens at convergent (destructive) plate boundaries.

Answer on p. 220

Revision activity

Create a flash card for each of the plate boundaries. Write the name of the plate boundary on one side and bullet point notes of what processes and features are created on the back.

Hotspot volcanoes
Volcanoes found in the middle of tectonic plates and thought to be fed from the underlying mantle (a thick layer of high-density rocks lying between the Earth's crust and its molten core). These volcanoes occur where the mantle is unusually thin and hot. The summits of the Hawaiian Islands are classic examples.

Volcanoes

There are around 500 active volcanoes across the world. Figure 1.2 shows that a significant number of these are located in the 'Ring of Fire' around the Pacific Ocean. Most volcanoes occur near plate boundaries, but there are hotspot volcanoes.

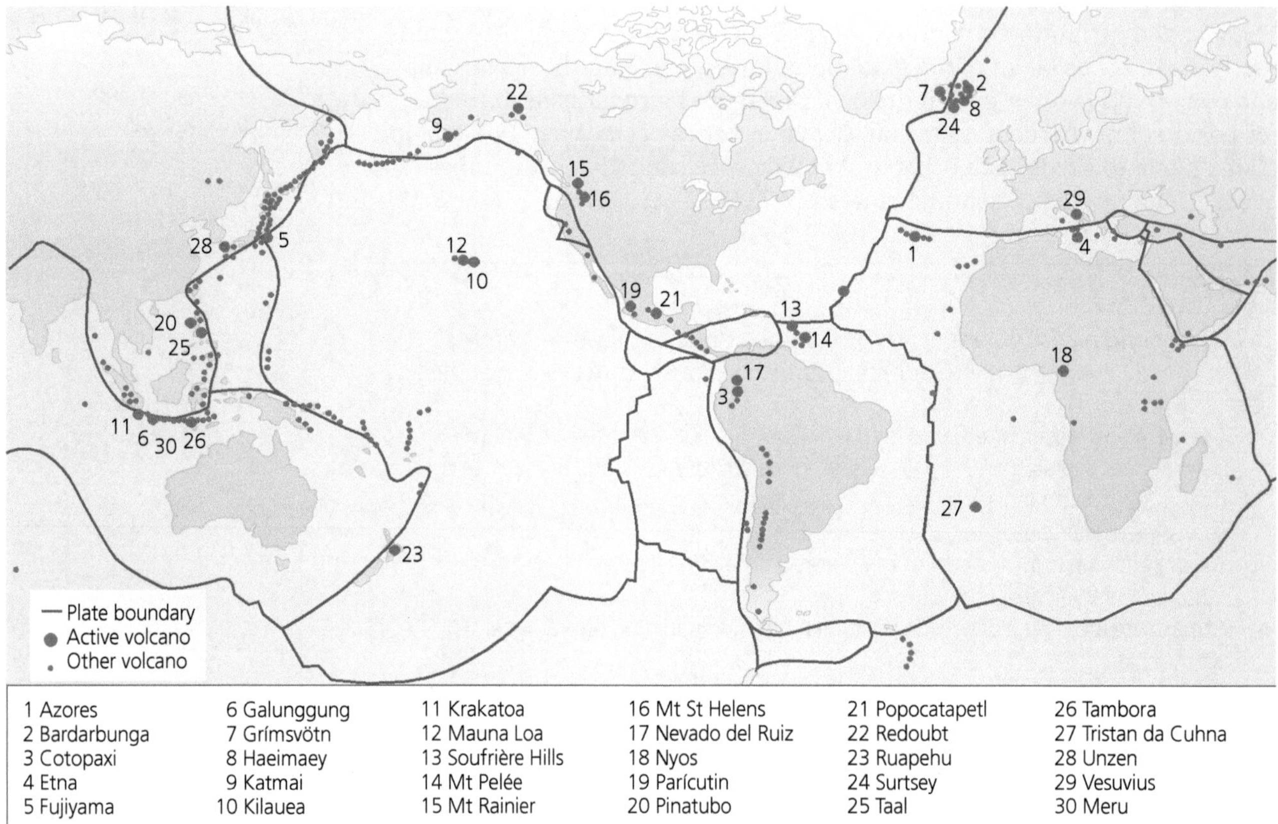

Figure 1.2 The global distribution of active volcanoes

Revision activity

Take time to explore different maps and practise describing the distribution. You can use the analogy of **TEA** to support your structure — **T**rend, **E**xamples, **A**nomalies.

Tsunamis

- A tsunami is caused by submarine shock waves generated by an earthquake (usually greater than magnitude 7) or a volcanic eruption. The earthquake/volcano displaces enough water to form waves.
- Tsunamis have a wide global distribution.
- They are most commonly experienced around the coastlines of the Pacific Ocean.
- They are potentially most devastating where a gently sloping continental shelf allows tsunamis to build to great heights.

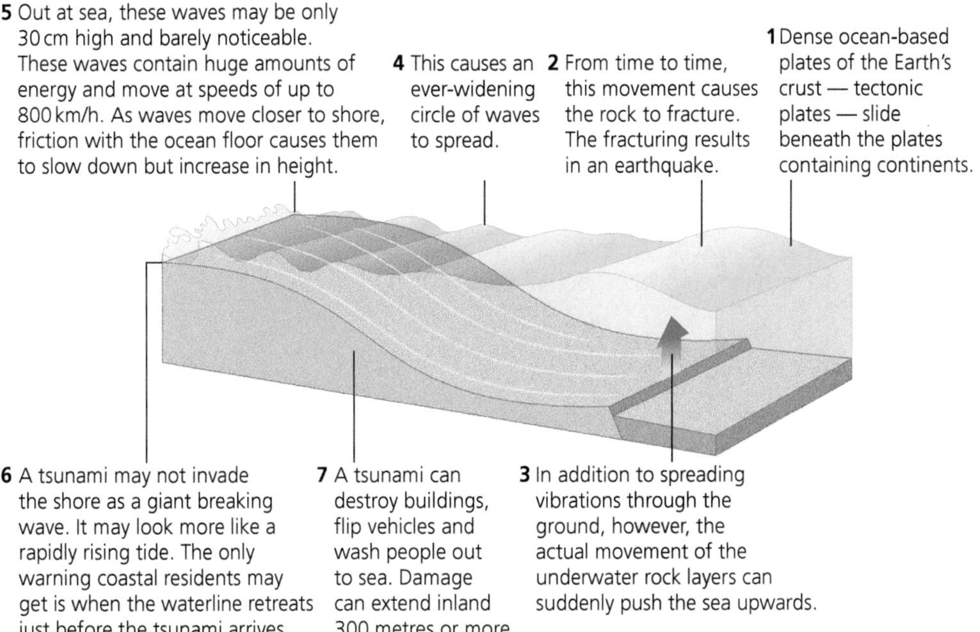

5 Out at sea, these waves may be only 30 cm high and barely noticeable. These waves contain huge amounts of energy and move at speeds of up to 800 km/h. As waves move closer to shore, friction with the ocean floor causes them to slow down but increase in height.
4 This causes an ever-widening circle of waves to spread.
2 From time to time, this movement causes the rock to fracture. The fracturing results in an earthquake.
1 Dense ocean-based plates of the Earth's crust — tectonic plates — slide beneath the plates containing continents.
6 A tsunami may not invade the shore as a giant breaking wave. It may look more like a rapidly rising tide. The only warning coastal residents may get is when the waterline retreats just before the tsunami arrives.
7 A tsunami can destroy buildings, flip vehicles and wash people out to sea. Damage can extend inland 300 metres or more.
3 In addition to spreading vibrations through the ground, however, the actual movement of the underwater rock layers can suddenly push the sea upwards.

Figure 1.3 The causes of a tsunami

Source: US National Oceanographic and Atmospheric Administration

Now test yourself TESTED

2 Using the annotated diagram in Figure 1.3, summarise the causes of a tsunami.

Answer on p. 220

Theoretical frameworks

REVISED

The theory

Over a long period of geological time, as the plates move relative to each other, they cause:

- the continents to drift apart
- the ocean basins to change in size and form
- the formation of major landforms such as mountain chains and mid-ocean ridges
- earthquakes, volcanic eruptions and tsunamis.

The Earth's structure is a relatively thin crust broken up into plates and wrapped around a thick and largely molten mantle.

Convection within the mantle causes crustal plates to move and was believed to be the main driver of plate movement. This theory, known as **plate tectonics**, is now considered less viable.

Subduction zones Broad areas where two plates are moving together, often with the thinner, more dense oceanic plate descending beneath a continental plate. Fold mountains form at the edge of the overriding plate, with associated volcanic activity. Stress between the two plates also triggers earthquakes.

Paleomagnetism Results from magma locking in the Earth's magnetic polarity when it cools. Scientists can use this to reconstruct past plate movements.

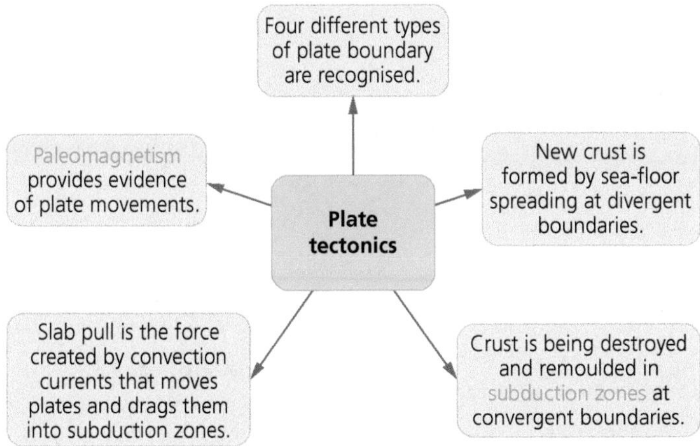

Figure 1.4 Important elements in the theory of plate tectonics

Now test yourself

3 What is paleomagnetism?

Answer on p. 220

TESTED

Key concept

Plate tectonics theory views the Earth's crust as consisting of a number of mobile yet rigid elements (plates). These plates are of two different types:

- thin crust underlying the ocean basins
- thicker crust underlying the continents.

The low density of the thick continental crust allows it to 'float' on the much higher-density mantle below. Heat derived from the Earth's molten core rises within the mantle to create convection currents which, in turn, move the tectonic plates.

Now test yourself

4 Only earthquakes occur at convergent plate boundaries. True or false?

Answer on p. 220

TESTED

The type and magnitude of event

- The convergent boundary is the most productive of both earthquakes and volcanic eruptions, followed by the divergent margin. The conservative margin produces only earthquakes.
- Science has still to discover what determines the magnitude of a tectonic event.
- The Benioff Zone is thought to be important in the case of some earthquakes. This is the boundary between an oceanic plate that is undergoing subduction and an overriding continental plate. It is a sloping plane, and stresses are built up as the cold oceanic plate sinks into the hot mantle.

Physical processes behind tectonic hazards

Earthquakes

Earthquakes are caused by sudden movements of the Earth's crust relatively close to the surface, usually along a pre-existing fault. The movement is the outcome of a gradual build-up of tectonic pressure and then its sudden release. The sudden movement creates vibrations (seismic waves) of three different kinds:

- P (fast): the first to reach the surface and can travel through both liquids and solids.
- S (slower): only travel through solids and do more damage than P waves.
- L (surface love) waves: the slowest type of waves but cause the most damage.

Exam tip

Make sure you understand the differences between the various types of seismic waves and the impact they have.

The hypocentre of an earthquake, sometimes referred to as the focus, is the point of origin within the Earth's crust where the pressure is released — the point of rupture. The epicentre is the point on the Earth's surface directly above the hypocentre. It is the surface location where the shock waves are likely to be strongest.

Hypocentre (focus) The point of origin where the pressure is released inside the Earth.

Epicentre The point directly above the hypocentre (focus) on the Earth's surface.

The overall severity of an earthquake is determined by the amplitude and frequency of these waves. The S and L waves are more destructive than the P waves. They cause crustal fracturing, ground shaking and three secondary hazards:

- Liquefaction: this affects loose rock and sediment. The seismic waves trigger the ground to lose its load-bearing capacity, causing large buildings to settle into the ground, tilt and possibly collapse.
- Landslides: these occur where slopes are weakened by seismic waves and slide under the influence of gravity.
- Tsunamis (see below).

Now test yourself

5 What is the difference between the epicentre and the hypocentre of an earthquake?

Answer on p. 220

TESTED

Tsunamis

These waves are potentially the most lethal of the secondary earthquake hazards. Out at sea they do not represent a hazard since they are low in height and generally go unnoticed. It is only as they approach a coastline and the sea becomes shallower that they grow in height.

Figure 1.5 shows the physical and human factors affecting the impact of tsunamis.

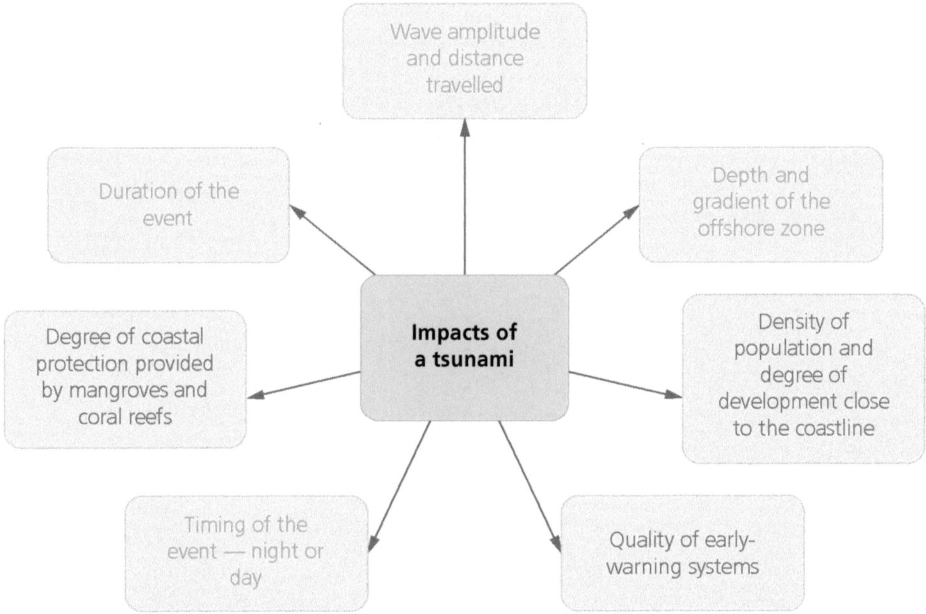

Figure 1.5 Physical and human factors affecting the impact of tsunamis

Primary effect Occurs as a direct result of a hazard.

Secondary effect Occurs as a result of the primary effect.

Volcanoes

Table 1.1 lists the primary and secondary hazard effects of volcanoes.

Table 1.1 The primary and secondary hazard effects of volcanoes

Primary effects	Secondary effects
Pyroclastic flows: the frothing of magma at the vent produces bubbles that burst explosively to eject hot and poisonous gases as well as hot, fine materials. Clouds formed of these gases and materials are most lethal when they roll down the sides of a volcano. Tephra (ash falls): rock fragments ejected into the atmosphere and ranging in size from 'bombs' to fine dust. The accumulation of tephra on roofs starts fires and causes buildings to collapse. Lava flows: flows of molten rock, often fast moving and lethal. Volcanic gases: mixed gases emitted during explosive eruptions. The carbon dioxide is particularly dangerous.	Lahars: mudflows created by the combination of heavy rain on slopes covered by fine volcanic material. Jökulhlaups: catastrophic floods caused by volcanic eruptions beneath glaciers.

Exam tip

Remember that compared with other hazards, such as earthquakes and tsunamis, volcanoes have historically killed far fewer people. An important factor is that there is often some form of advanced warning of an eruption.

Revision activity

Make sure that you have a located, recent example of each of the three tectonic hazards, together with the date of the event and some indications of the scale of the human impact.

Now test yourself TESTED

6 Which of the primary hazards of a volcanic eruption is potentially the most lethal? Give your reasons.

Answer on p. 220

Tectonic hazards become disasters

Vulnerability, risk, resilience and disaster

REVISED

Vulnerability and risk are key factors in turning hazards into disasters.

Key concepts

Vulnerability relates to the ability of a community to cope with the impacts of a hazard. That ability is determined by a range of factors, from the quality of warning systems and emergency responses to the level of development and settlement density. So it is argued that a developed country, with good governance and access to technology and relevant resources, is less vulnerable to the same hazard than a developing country. The likelihood of that hazard becoming a disaster is reduced.

Risk is the exposure of people to a hazardous event. It relates to the probability of a hazard leading to a loss of life and/or livelihoods. The assessment of risk is complicated by many factors, including:

- the perceptions of individuals and communities
- the unpredictability of hazards, with people being caught out by the timing or magnitude of a tectonic event
- the lack of alternatives — people continue to live in hazardous areas because they have no other options
- the fact that the benefits of a hazardous location may outweigh the risks involved in staying there
- acceptance of the risk that something might happen.

Hazards Natural events that threaten or actually cause injury and death, as well as damage and destruction to property.

Disasters Occur when hazards have a significant impact on vulnerable populations. Officially, a hazard becomes a disaster when 100 or more people are killed and/or 100 or more people are affected.

Typical mistake

The terms hazard and disaster are often taken to mean the same thing. In fact, they mean very different things.

The hazard–risk formula involves the components that influence the amount of **risk** a community is taking with a particular type of hazard:

Risk = hazard × exposure × vulnerability/manageability

The pressure and release (PAR) model adopts a slightly different approach to the assessment of the risk of a hazard becoming a disaster. It sees disaster as occurring at the intersection of two processes:

- those generating **vulnerability**
- those of the natural hazard event (see Figure 1.6).

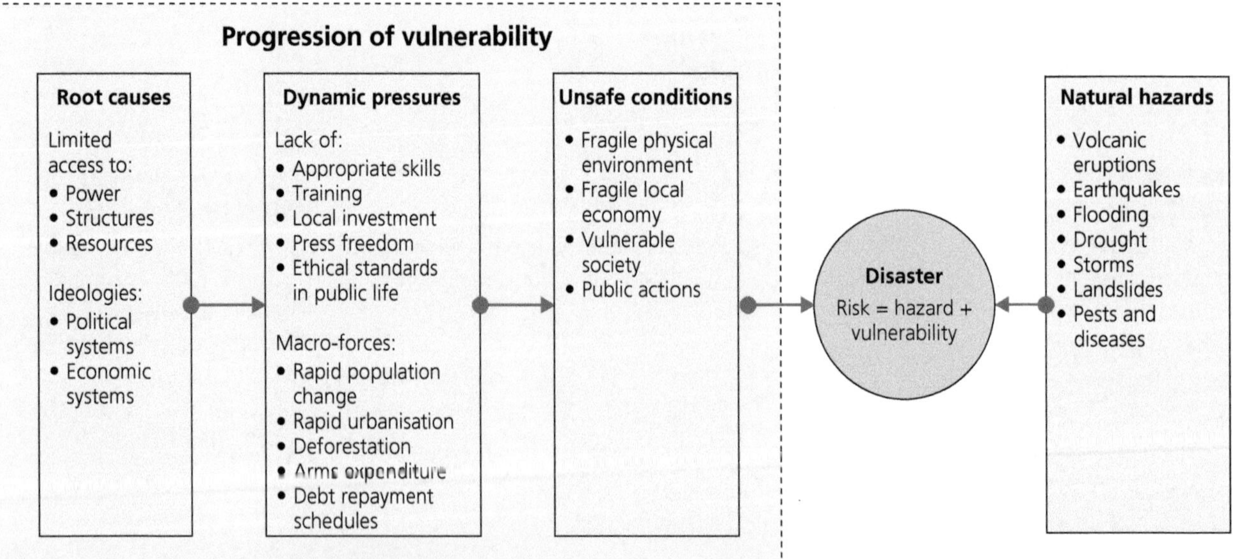

Figure 1.6 The pressure and release model

Root causes create vulnerability through different pressures such as inadequate training and poor government. Dynamic pressures produce unsafe conditions (environmentally and socially) for the most vulnerable people.

One other component that needs to be taken into account when weighing up the risk of a hazard becoming a disaster is **resilience**.

Now test yourself TESTED

7 What is the difference between a hazard and a disaster?
8 What is the difference between vulnerability and resilience?

Answers on p. 220

Typical mistake

Tectonic events by themselves are not hazards. They become hazardous only when they adversely impact on people, their settlements and livelihoods.

Key concept

Resilience is the ability of a community or country exposed to hazards to resist, absorb and recover from the impacts of a hazard. Resilience can often help prevent a hazard from becoming a disaster.

Tectonic hazard impacts

+ The economic and social impacts of tectonic hazards vary considerably: over time, from place to place, and from minor nuisances to major disasters.
+ The impacts of earthquakes and their secondary hazards are generally much greater than those of volcanic eruptions.
+ The concentration of active volcanoes in relatively narrow belts means that only a small land area lies in close proximity to them.
+ It is estimated that less than 1 per cent of the world's population is likely to suffer the impacts of a volcanic eruption, whereas the estimate for earthquakes is 5 per cent.

The economic impacts of a tectonic hazard are roughly proportional to the land area exposed to the particular hazard. But there are other factors involved, such as:

+ level of development and per capita GDP
+ total number of people affected
+ speed of recovery from the hazardous event (resilience)
+ degree of urbanisation
+ amount of uninsured losses.

Now test yourself

9 Explain why the global impacts of earthquakes are greater than those of volcanic eruptions.

Answer on p. 220

TESTED

Tectonic hazard profiles

Magnitude and intensity

Magnitude and **intensity** are important aspects of tectonic hazards. Observations and measurements are converted to mathematical scales. Of four widely used scales, three relate to earthquakes and one to volcanic activity (see Table 1.2).

Table 1.2 Scales used for two different types of tectonic hazard

	Hazard	Scale	Overview
Richter Scale	Earthquake	0–9	A measurement of the height (amplitude) of the waves produced by an earthquake. The Richter Scale is an absolute scale — wherever an earthquake is recorded, it will measure the same on the Richter Scale.
Mercalli Scale (modified)	Earthquake	I–XII	Measures the experienced impacts of an earthquake. It is a relative scale because people experience different amounts of shaking in different places. It is based on a series of key responses, such as people waking up, the movement of furniture and damage to structures.
Moment Magnitude Scale (MMS)	Earthquake	0–9	A modern measure used by seismologists to describe earthquakes in terms of energy released. The magnitude is based on the 'seismic moment' of the earthquake, which is calculated from the amount of slip on the fault, the area affected and an Earth-rigidity factor. The US Geological Survey (USGS) uses the MMS to estimate magnitudes for all large earthquakes.
Volcanic Explosivity Index (VEI)	Volcanic eruption	0–8	A relative measure of the explosiveness of a volcanic eruption, which is calculated from the volume of products (ejecta), height of the eruption cloud and qualitative observations. Like the Richter Scale and the MMS, the VEI is logarithmic: an increase of one index indicates an eruption that is ten times as powerful.

None of the scales shown in Table 1.2 is perfect. For example, they do not take into account the duration of the hazard, the physical exposure, or the vulnerability and resilience of the affected communities.

Hazard profiles

Given a set of criteria, it is possible to compile a tectonic hazard profile, which can then be compared with the profiles of other events. Figure 1.7 shows one style in which the characteristics of earthquakes at two different plate boundaries are compared.

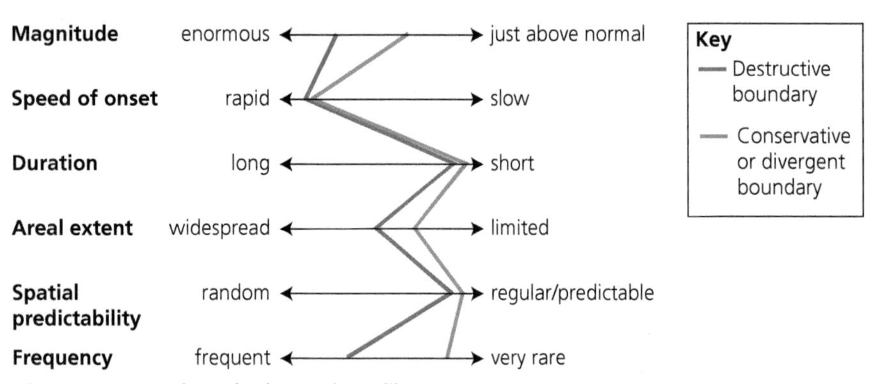

Figure 1.7 Earthquake hazard profiles

The hazard profile is not the only factor that determines the social and economic impacts of an event. For example, it is generally assumed that the impacts of tectonic hazards are likely to be greater in developing countries because of higher levels of vulnerability and lower levels of resilience. The significance of development and other factors is explored in a little more detail in the following section.

Exam tip

Any assessment of the risks posed by tectonic hazards must identify the nature and magnitude of the hazard, the number of people at risk, the amount of economic investment, the vulnerability of the society, and the society's ability to respond to and mitigate the impacts of the hazard.

Now test yourself TESTED

10 What is the value of compiling hazard profiles?

Answer on p. 220

Typical mistake

Often the media will incorrectly refer to the Richter scale when reporting on earthquakes. In fact, the numbers are measured using the MMS.

Tectonic hazard profile A technique used to try to understand the physical characteristics of different types of hazards, such as earthquakes, volcanic eruptions and tsunamis.

Revision activities

Create a series of revision cards for each of the scales used to measure the magnitude and intensity of tectonic hazards.

Make sure you have notes detailing the impacts of one specific tectonic hazard event in each of three different geographical locations: a developing country, an emerging country and a developed country.

The importance of development and governance

REVISED

Economic development gives communities and countries access to the resources, organisations and technology needed to cope with hazard events. With increasing income, people are better able to ensure their own safety by living in 'safe' locations and in 'hazard-proofed' property.

However, non-economic aspects of development are also significant:

- Access to education: education means that people can be made more aware of the hazard risks of living where they do and of what to do in the event of a hazard.
- Access to healthcare: the better people's health, the better they are at withstanding the health and food risks resulting from the hazard.
- Housing: poorly built housing is usually unable to withstand earthquake shock waves, leading to serious injury and death.
- Governance: the quality of governance can be critical (see below).

Governance

Poor **governance** in the form of corrupt local and national government and weak political organisation increases hazard vulnerability in two ways:

- By failing to invest properly in infrastructure that might mitigate the impacts of a tectonic hazard, for example failing to invest in warning systems, 'hazard-proofing' buildings, etc.
- By being ill-prepared to deal with the emergency situation immediately following a hazard.

Key concept

Governance is the way a country, city, community, company, etc. is run by the people in control. It is based on three concepts: authority, decision making and accountability. Good governance embodies the recognition and practice of a range of principles, such as transparency, the rule of law, equity, consensus and participation.

Making links

Clearly, when it comes to governance, national and local governments are the top players — their transparency and efficiency are paramount.

But governance is not only about political authority. There are other people and organisations (stakeholders), both public and private, that have a role to play in good governance, largely by observing the key principles of good governance, such as accountability and participation in responsible decision making in the present context of a community's readiness and ability to cope with the hazard risk.

Stakeholders Individuals, communities, organisations, businesses and government with a specific interest in a situation — in this instance, in hazard risk and hazard mitigation.

Now test yourself TESTED

11 Why is good governance so important in the context of tectonic hazards?

Answer on p. 220

Geographical factors

Finally, various geographical factors can increase hazard vulnerability. These include the following:

- Population density: the higher the density, the more people are at risk.
- Urbanisation: the more people and businesses concentrated in cities, the higher the risk and vulnerability.
- Isolation and inaccessibility: this is particularly critical in the immediate aftermath of a hazard event when there is an urgent need to provide rapid emergency aid.
- Community spirit: a strong spirit can certainly help boost morale and the collective wish to survive the hazard.

Exam tip

Remember that the significance of these factors is always conditioned by the magnitude and intensity of the tectonic hazard event.

Contrasting locations

The specification requires you to make a comparative study of tectonic hazard events in **three different geographical contexts** (a developed country, an emergent country and a developing country), which focus on the significance of development.

The student book published by Hodder Education contains studies of three earthquake events, in Japan (2016), Iran (2003) and Nepal (2015). Despite considerable damage and destruction, only in Japan did the earthquake event not become a disaster.

In the case of Iran (Bam) and Nepal, two factors turned the earthquakes into disasters:

- The poorly constructed and vulnerable housing and buildings.
- A poor emergency response in terms of lack of equipment and specialised medical and rescue training. In Bam's case, the situation was not helped by the destruction of the three main hospitals during the event. In Nepal, the problem was made worse by the inaccessibility of the stricken areas — roads in remote mountainous locations were rendered impassable by huge landslides.

Revision activity

It is important that you keep notes about the impacts of the same type of hazard in the three different contexts. Did you study those place studies in your student book or did your teacher introduce you to different examples? Whichever is the case, be sure to brush up on the details.

The management of tectonic hazards and disasters

Trends and patterns

REVISED

Compared with other natural hazards, few tectonic hazards develop into disasters. Tectonic hazards cannot be prevented. Neither can their spatial occurrence be changed.

Figure 1.8 shows that the annual number of tectonic (geophysical) hazard events involving losses (life, property, etc.) has remained fairly stable compared with meteorological and hydrological hazard events. In other words, such events seem to occur regularly.

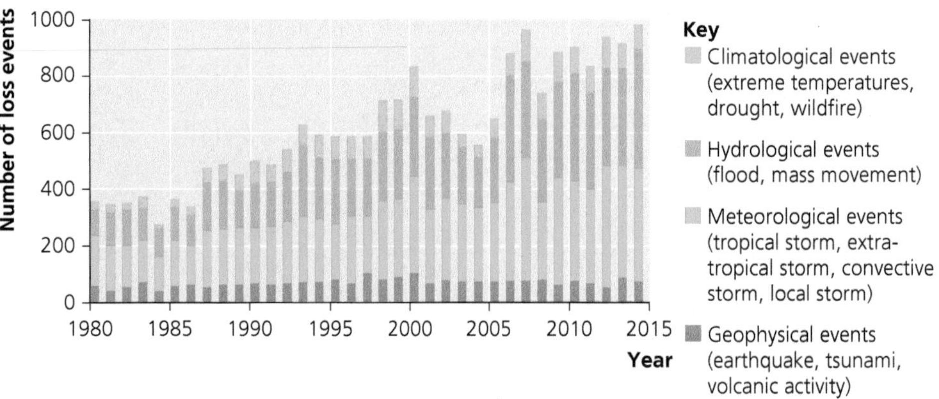

Figure 1.8 The number of hazard loss events, by type (1980–2014)

Table 1.3 confirms that volcanic eruptions cause far fewer disasters than earthquakes. But the smaller numbers also reflect the fact that there are many more earthquakes than volcanic eruptions occurring during the course of a year. The table does not show the expected simple pattern when the number of tectonic disasters is analysed in terms of the Human Development Index (HDI) of the countries in which they occurred. Is the fact that most tectonic disasters occurred in medium HDI countries (emerging countries) really explained by the level of development? Could it be that their incidence reflects the fact that these countries happen to be located in unstable parts of the Earth's crust?

All hazard data needs to be treated with some caution for a number of reasons:

- There is no universally agreed definition of a disaster.
- Smaller events in remote locations are often under-recorded.
- Disaster deaths and damage are sometimes under-recorded for political reasons.

Table 1.3 Total number of tectonic disasters grouped by country's level of development (2004–13)

Hazard	Very high HDI	High HDI	Medium HDI	Low HDI	Total
Earthquakes and tsunamis	41	71	121	36	269
Volcanic eruptions	5	12	30	10	57

Skills activity

Study Table 1.3, which illustrates the number of tectonic disasters grouped by countries' level of development.

Calculate the percentage of earthquakes and tsunamis that occur in very high HDI (Human Development Index) countries.

Remember — to calculate a percentage, divide the value by the total and times by 100.

Exam tip

It is recommended in the specification that you know about one tectonic mega-disaster, such as the 2004 Asian tsunami, the 2010 Eyjafjallajökull eruption or the 2011 Japanese tsunami.

Mega-disasters Result from tectonic hazards and show several diagnostic features:
- They are large-scale in terms of the area involved and their economic and human impacts.
- They pose huge challenges, particularly at the emergency stage.
- They usually require substantial amounts of international disaster aid.

Hazard hotspots Locations that are extremely disaster prone for a number of reasons. Notable is the fact that they experience more than one type of natural hazard.

It needs to be understood that there are parts of the world at risk from more than one type of hazard. So it is appropriate to think in terms of multiple-hazard zones. Figure 1.9 plots zones regularly experiencing three different types of hazard. Locations where the three distributions overlap can be identified as hazard hotspots.

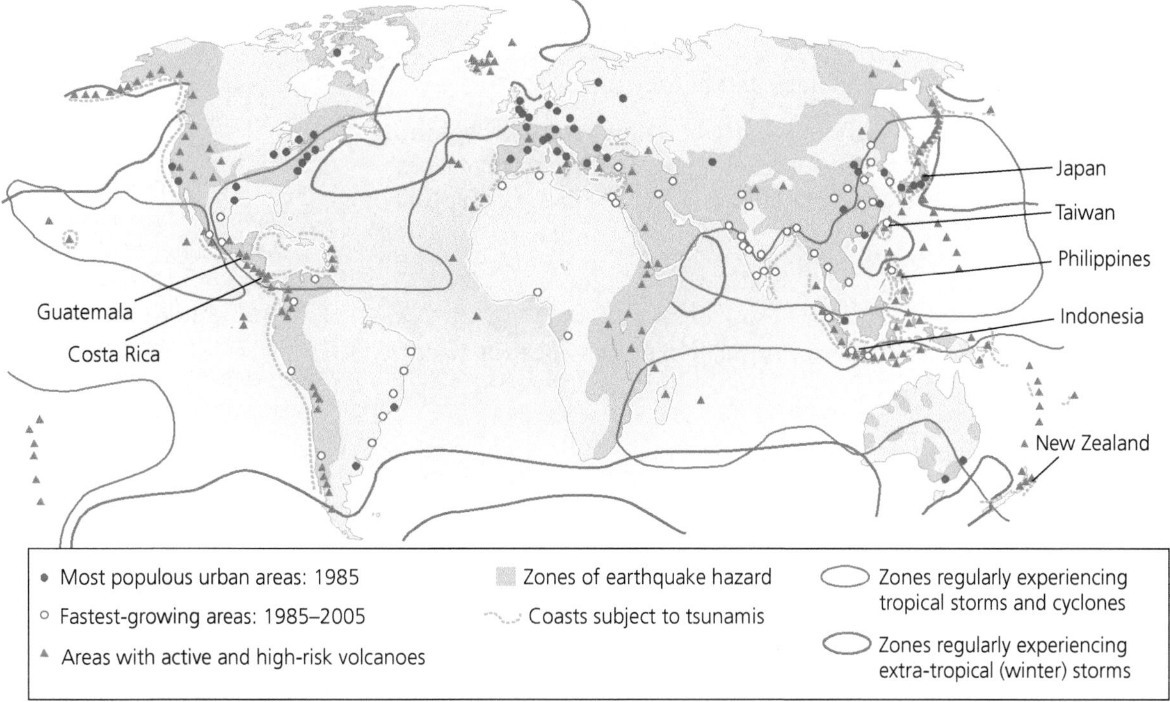

Figure 1.9 The global pattern of multiple hazards

Table 1.4 shows the eight countries most exposed to multiple hazards. In the case of tectonic hazards, it should be noted that their impacts are often aggravated by hydro-meteorological events, which encourage liquefaction and landslides on slopes weakened by earthquakes.

Table 1.4 The countries most exposed to multiple hazards

Country	Total area exposed (%)	Population exposed (%)	Number of different hazards the country is exposed to
Taiwan	73.1	73.1	4
Costa Rica	36.8	41.1	4
Vanuatu	28.8	20.5	3
Philippines	22.3	36.4	5
Guatemala	21.3	40.8	5
Ecuador	13.9	23.9	5
Chile	12.9	54.0	4
Japan	10.5	15.3	4

Skills activity

Study Table 1.4, which illustrates the countries most exposed to multiple hazards.

Calculate the mean population exposed.

Remember — to calculate a mean, add all the values together and divide by the number of values.

Now test yourself

TESTED

12 What makes a hazard into a disaster and a disaster into a mega-disaster?

Answer on p. 220

Prediction and management

REVISED 

Prediction

Predicting the occurrence of tectonic hazards is an obvious starting point in any attempt to reduce the deaths and destruction they cause.

Earthquakes are altogether more difficult to predict. However, it is beginning to look as if there might be some early warning signs. The key to success is being able to detect those areas of particular stress in the Earth's crust that trigger earthquakes.

Hazard management cycle

The hazard management cycle involves a number of stages once the hazard has struck:

1 Emergency response
2 Initial recovery (rehabilitation)
3 Reconstruction (including mitigation)
4 Return to normality
5 Appraisal of the lessons learned during the hazard event and implementation of remedial actions
6 Improving preparedness

The choice of response depends on a complex series of interlinked physical and human factors, such as those shown in Figure 1.10. This figure does not show the critical stages 5 and 6 in the 'disaster-free period'.

Mitigation Any action taken to reduce or eliminate the long-term risk to human life and property from natural hazards. Those actions are largely the outcome of stage 5 in the hazard management cycle (see Figure 1.11). However, they are also likely to be taken during stage 3, sometimes referred to as adjustment or adaptation.

Preparedness Educating people about what they should do in that emergency (where to seek shelter, how to assist others) as well as improving warning systems and training, and equipping rescue teams. It focuses on the emergency stage immediately following a hazard.

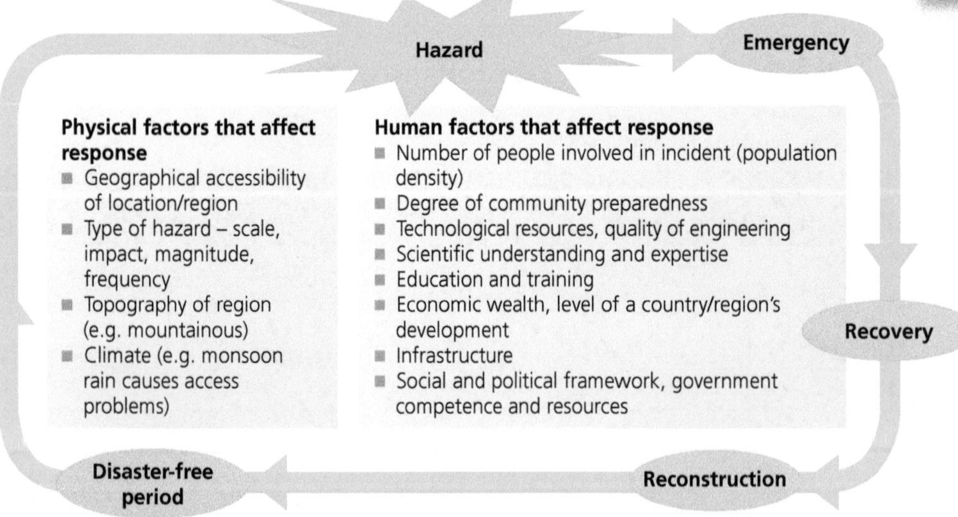

Figure 1.10 The range of factors affecting the response to hazards

Revision activity

Which of the following factors affecting the response to hazards are either physical or human?

A The degree of community preparedness.
B The topography of a region.
C The climate of the place.
D The technological resources available.

Now test yourself

13 What is the difference between mitigation and preparedness?

Answer on p. 220

TESTED

Making links

Scientists are key players in the prediction of tectonic disasters. But the accuracy of any forecasting depends on the type and location of the tectonic hazard.

Park's disaster response curve

Park's disaster response curve is a model that can be used to help analyse the timeline between when a hazard strikes and when a place or community returns to normal life (see Figure 1.11). The model recognises five stages, which are a near match with the stages in the hazard management cycle. The model allows the response curves of different hazard events to be compared.

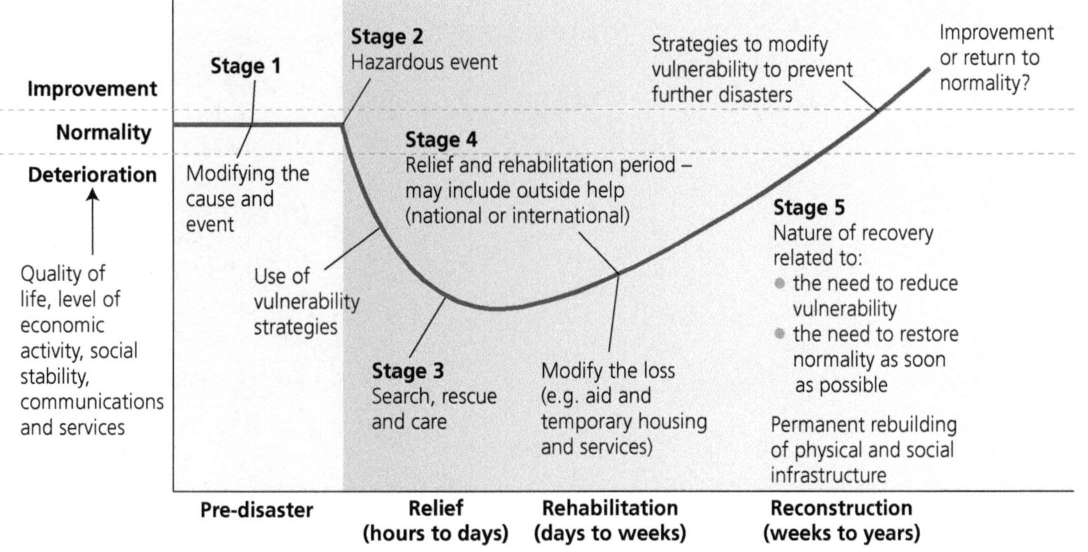

Figure 1.11 Park's model: the disaster response curve

Making links

The actions involved in planned emergency procedures are particularly critical in the mitigation of hazard events. All too often, the planning of such actions has been found to be wanting, with unnecessary delays and confusion as to who does what.

Exam tip

It is tempting to think that the shorter the timeline of the first five stages of the hazard management cycle, the greater the resilience of the strike community. It is also tempting to think that response curves will generally be shorter in developed countries compared with those in developing countries. However, remember that the nature of the response curve is also greatly affected by the magnitude of the hazard event.

Mitigation and adaptation strategies

REVISED

There are three actions that can be undertaken to mitigate the impacts of a tectonic hazard:

- Modify the hazard event.
- Modify both vulnerability and resilience.
- Modify the potential financial loss.

Modifying the hazard event

No technologies are yet capable of preventing tectonic disturbances. The best that can be achieved is to modify, i.e. reduce the hazard impacts by mitigating or adaptive actions (see below). However, it is possible to:

- strengthen coastal defences against tsunamis
- divert or chill lava flows
- increase the stability of slopes where there is a high risk of landslides.

Now test yourself

14 What are the three ways to mitigate the impacts of a tectonic hazard?

Answer on p. 220

TESTED

Modifying vulnerability and resilience

This involves reducing vulnerability and improving resilience. There are a number of different approaches or focuses:

- Improving prediction, forecasting and warning systems — for example, scientific research is constantly seeking to become more proficient in hazard prediction and forecasting, while modern technology is providing us with more efficient warning systems.
- Improving community preparedness — for example, enforcing building codes aimed at 'hazard proofing' structures, particularly public buildings such as hospitals, police stations and pipelines, which need to be fully operational immediately after the hazard event.
- Changing behaviours that reduce the hazard risk — for example, moving people away from high-risk areas.

Making links

Key players in hazard mitigation and adaptation are planners (avoiding developments in hazardous locations) and engineers (hazard-proofing buildings and locations, possibly modifying hazard events).

Modifying losses

- Insurance is one approach to reducing the losses associated with a hazard event.
- Insurance is expensive, but in most instances the actual costs of repair and reconstruction are significantly more.
- Insurance companies consider the following when issuing premiums:
  1. The level of risk in a particular location.
  2. The probability of a hazard of a certain magnitude happening.
  3. The market value of the properties to be insured.
  4. The likely costs of repair or reconstruction.
- For earthquakes, seismologists work with risk analysts to help the insurance industry calculate premiums and risk.
- Computer simulations are used to estimate the probability of damage from different scales of earthquake events.
- However, with volcanic eruptions there is a greater confidence in the assessments of risk and the potential scale of damage.

Now test yourself

TESTED

15 Insurance companies consider which two of the following when issuing premiums?

A The likely cost of repairs or reconstruction.
B The available land around the property.
C The market value of the properties to be insured.
D The number of residents in the property.

Answer on p. 220

Disaster aid is another way in which hazard losses might be reduced, particularly during the emergency and early recovery stages. The funding for disaster aid has two sources:

- Donations by governments to intergovernmental organisations like the United Nations.
- Private donations to voluntary organisations and charities, such as the Red Cross, Oxfam and Médecins Sans Frontières.

Disaster aid is often criticised, largely on the grounds that national and local distribution systems are often inefficient or corrupt, it does not encourage self-help and it does not encourage a more bottom-up management of disasters at a local level.

The Sendai Framework (2015) set out four priorities for disaster management:

1. Understand the disaster risk.
2. Ensure a strengthening of governance to manage the hazard risk.
3. Invest in improving resilience and disaster preparedness.
4. 'Build back better' in the recovery, rehabilitation and reconstruction stages.

It is also recognised today that:

+ the Millennium Development Goals (2000) gave insufficient prominence to risk reduction and resilience
+ the distribution of international disaster relief is too complex, fragmented and disorganised, and needs to be properly co-ordinated.

Now test yourself

TESTED

16 Why is disaster aid often criticised?

Answer on p. 220

Making links

Key players in seeking to reduce the burden of hazard losses are non-governmental organisations (NGOs) — through appropriate aid and advice — and the insurance industry. But how many developing countries can afford the necessary insurance premiums?

Revision activity

Create a table summarising the ways in which people attempt to cope before, during and after one type of tectonic hazard.

Exam skills

You should be familiar with the following skills and techniques used in the geographical investigation of tectonic hazards:

+ analysis of global and regional distribution maps
+ use of block diagrams to identify key features of plate boundaries
+ analysis of time–distance maps to predict the spatial impact of tsunamis
+ use of correlation techniques to analyse links between the magnitude of events and deaths and damage
+ statistical analysis to compare hazard profiles
+ interrogation of large data sets to assess data reliability and identify trends
+ use of geographic information systems (GIS) to identify hazard risk zones and the degree of risk.

Exam practice

A-level

1 Study Figure 1. Explain why plate movement is the key to understanding what the map shows. (4)

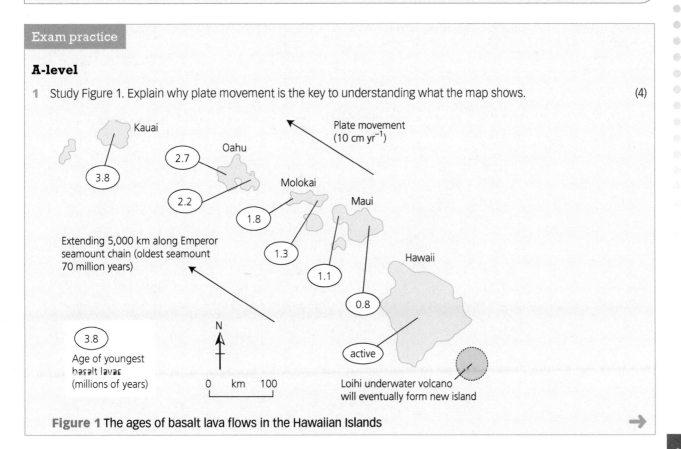

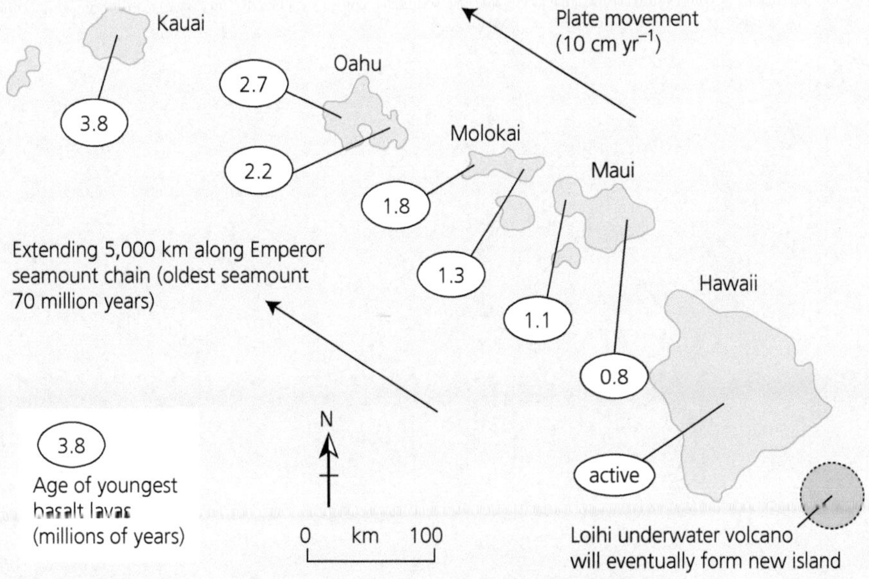

Figure 1 The ages of basalt lava flows in the Hawaiian Islands

2 Study Table 1. Calculate the mean number of countries exposed to multiple hazards. (1)

Table 1 The countries most exposed to multiple hazards

Country	Total area exposed (%)	Population exposed (%)	Number of different hazards the country is exposed to
Taiwan	73.1	73.1	4
Costa Rica	36.8	41.1	4
Vanuatu	28.8	20.5	3
Philippines	22.3	36.4	5
Guatemala	21.3	40.8	5
Ecuador	13.9	23.9	5
Chile	12.9	54.0	4
Japan	10.5	15.3	4

3 Assess the factors affecting the response to tectonic hazards. (12)

4 Assess the reasons why some places are more vulnerable to hazards. (12)

5 Assess the extent to which strategies can modify the vulnerability of tectonic hazards. (12)

Answers and quick quiz 1 online

Summary

You should now have an understanding of:

- the global distributions and causes of earthquakes, tsunamis and volcanic eruptions
- the distinction between divergent, convergent and conservative plate boundaries
- the distributions and associated hazards of different plate boundaries
- intra-plate earthquakes and hotspot volcanoes
- the key elements of the theory of plate tectonics
- tectonic processes at different plate boundaries
- factors affecting earthquake magnitude and the type of volcanic eruption
- earthquake shock waves and secondary hazards
- volcanic emissions and secondary hazards
- factors affecting tsunamis
- the difference between a hazard and a disaster
- mega-disasters and multiple hazard zones
- disaster trends and differential impacts
- predicting and forecasting tectonic hazards
- the hazard management cycle
- Park's response curve model
- modifying tectonic events, vulnerability, resilience and losses.

2 Landscape systems, processes and change

2A Glaciated landscapes and change

The landscapes covered in this topic may be collectively referred to as **cold environments**. They have all been affected, over considerable periods of time, by sub-zero temperatures and associated glacial and periglacial processes. The topic takes into account not only those parts of the world currently experiencing these processes but also locations that have been glaciated and periglaciated in the past.

Glaciation The modification of landscapes while covered by ice sheets or glaciers.

Periglaciation The modification of landscapes located adjacent to the margins of ice sheets and glaciers.

Past and present distributions of glacial and periglacial environments

Exam tip

You need to be aware of the distinction between areas that are currently undergoing glaciation or periglaciation and those areas that have experienced those conditions in the past but no longer do so.

Causes of climate change

REVISED

Climatic oscillations

The Quaternary period in which we live is divided into two geological epochs:

- The Pleistocene: from the beginning of the Quaternary to 11,500 years ago when the most recent continental glaciation ended.
- The Holocene: the interglacial period of today.

During the Pleistocene, the Earth's climate fluctuated between colder and warmer conditions, between **ice-house** and **greenhouse conditions**. The ice-house or glacial phases have left evidence of erosional and depositional features created by glaciers, ice sheets and their meltwaters. However, the landforms created by one glacial phase have usually been reworked, reshaped and even destroyed by later glacial phases.

Today, the features produced by the most recent glacial phases are being modified by post-glacial processes.

Causes of short-term climatic oscillations

Long-term changes in the Earth's orbit around the Sun are currently seen as the primary causes of the oscillations between glacial and non-glacial conditions. The Milankovitch theory attributes the oscillations to three main characteristics of the Earth's orbit:

- Eccentricity: the orbit changes from elliptical to circular and back over a period of around 100,000 years.
- Axis tilt: this varies between 21.8° and 24.4° over a period of around 41,000 years.

+ Wobble: like a spinning top, the Earth wobbles on its axis and this changes the distance from the Sun over a 21,000-year cycle.

These three different orbital cycles can combine to minimise the amount of solar energy reaching the Northern Hemisphere. When this happens, the climate cools and ice-house conditions return.

Evidence indicates that even within glacial and non-glacial periods, there are short-term fluctuations with frequent warming and cooling periods (**stadials**). Two main factors are thought to be responsible for these oscillations:

+ Fluctuations in the amount of energy emitted by the Sun (related to sunspot activity). High levels of sunspots increase the Earth's temperature — these changes are generally small, between –0.5°C and +0.5°C.
+ Volcanic activity. Eruptions with a high **volcanic explosivity index (VEI)** eject huge volumes of ash, sulphur dioxide, carbon dioxide and water vapour into the atmosphere. These ejected substances are distributed around the globe by high-level winds. It is thought that such eruptions can cool the Earth's temperature as they reduce the amount of solar energy reaching the Earth.

Examples of these short-term oscillations in climate in the UK are the Loch Lomond Stadial (between 12,500 and 11,500 years ago) and the Little Ice Age (between 1550 and 1750).

Revision activity

Just be aware of these three characteristics of the Milankovitch theory. It is not necessary to understand the physics behind them.

Typical mistake

Do not think that global warming is a recent phenomenon for which people are responsible. The history of the Earth has been one of recurring warming and cooling.

Now test yourself TESTED

1 What is the difference between the causes of long-term and short-term climate change?
2 What evidence is there that climate has changed in the past 450,000 years?
3 Why is it important to understand past climate change?

Answers on p. 220

Revision activity

Create a mind map of the long-term and short-term causes of climate change.

Present and past distributions of ice cover

REVISED 

Classification of ice masses

Ice masses can be classified by their morphological characteristics, size and location, as seen in Table 2.1.

Table 2.1 The characteristics of ice masses

Ice mass	Characteristics
Ice sheet	Complete submergence of topography beneath ice up to several kilometres deep. Found in areas of high latitude. Antarctica is the largest of the two ice sheets — it covers 14 million square kilometres and is up to 4 kilometres thick.
Ice cap	Smaller version of ice sheet burying upland topography. Found in mountain ranges, e.g. Vatnajökull in Iceland.
Ice field	Area of less than 500,000 square kilometres that is not thick enough to bury an upland area. Found in high-altitude regions, e.g. the Southern Patagonian Ice Field.
Valley glacier	Glacier confined within valley walls.
Piedmont glacier	Valley glacier that spreads out beyond the valley end.
Cirque glacier	Glacier occupying hollow on a mountainside.
Ice shelf	Large area of floating glacier ice that extends beyond the coast.

Key concept

Cryosphere comprises those parts of the Earth's crust and atmosphere subject to temperatures below 0°C for at least part of each year, for example ice sheets and glaciers, sea and lake ice, ground ice (permafrost) and snow cover.

Key concepts

In terms of glacier behaviour and impact on the landscape, an important distinction is made between **warm-based** and **cold-based glaciers**:

- Warm-based glaciers, also known as 'wet' glaciers, occur in high-altitude areas outside the polar region. From the surface to the base, temperatures are close to 0°C. During the summer they generate large amounts of meltwater, which acts as a lubricant, allowing the glacier ice to slide over the bedrock.
- Cold-based glaciers, also known as 'polar' glaciers, occur in high latitudes, particularly in Antarctica and Greenland. Temperatures are well below freezing, so there is no basal sliding.

The significance of this distinction will be made evident in the sections 'Glacier systems' and 'Glacial landforms and landscapes' (see pages 30–40).

Global ice cover

Table 2.2 shows that the global ice cover today is one-third of what it was in the Pleistocene. Outside Antarctica and Greenland, the contraction has been so great that ice cover has almost disappeared. This applies particularly to the great mountain ranges, such as the Alps, Andes and Himalayas.

Table 2.2 Estimates of present and past global ice cover

Region	Present area (estimated thousand km²)	Past area (estimated thousand km²)	Reduction in ice cover (%)
Antarctica	1,350	1,450	6.9
Greenland	180	235	31.9
Arctic Basin	32	1,600	98.0
Andes	3	88	96.6
European Alps	0.4	4	90.0
Scandinavia	0.4	660	99.9
Asia	12	390	95.4
Rest of the world	0.2	104	99.8
Total	**1,578**	**4,531**	**65.2**

About 85 per cent of current ice cover is located in Antarctica. Next, but a long way back, comes Greenland, which accounts for just over 10 per cent.

A number of factors influence the distribution of ice cover:

- Latitude: areas of permanent ice sheets found at high latitudes, e.g. inside the Arctic and Antarctic circles.
- Altitude: areas of high latitudes create cold conditions, which results in the formation of glaciers and glacial landscapes, e.g. the European Alps, Himalayas, Andes and Rockies.
- Aspect.
- Relief.

The two most important factors are latitude and altitude.

Typical mistake

The data in Table 2.2 relates to the area covered by ice — it does not take into account the thickness of ice cover. In this respect, the volumes of the Antarctic and Greenland ice sheets may have suffered much higher percentage losses.

Evidence of Pleistocene ice sheet in the UK

- Erosional evidence, e.g. erosional landforms including arêtes, corries and glacial troughs in the Cairngorms, Scotland, Snowdonia, Wales, and the Lake District, England.
- Depositional evidence, e.g. depositional landforms including moraines in the Cairngorms and erratics in the Lake District.
- Meltwater evidence, e.g. meltwater channels in North Yorkshire and glacial till along the Holderness coast.

Now test yourself

TESTED

4 Why has there been such a great reduction in the ice cover of the major mountain ranges?

5 How does aspect affect the distribution of ice cover?

Answers on p. 221

Revision activity

On a blank world map, use an atlas to locate present-day high-latitude ice sheets and high-altitude glaciers. Make sure you can describe their distribution.

Periglacial processes and their landforms and landscapes

REVISED

A key feature of periglacial areas is the climate:

- daily temperatures below freezing for at least nine months
- low precipitation — less than 600 mm per year
- frequent cycles of freezing and thawing
- intense frosts throughout the year.

This climate means that the ground surface is frozen for much of the year. Indeed, the ground below the surface layer is permanently frozen (permafrost). The extent of the permafrost is usually taken as indicating the distribution of periglacial environments.

In summer, overlying snow and ice melt away to produce a seasonally unfrozen zone above the permafrost called the active layer. This may vary in thickness from a few centimetres to as much as 3 metres.

Because of the global distribution of land, most **permafrost** areas are found in the Northern Hemisphere around the Arctic Ocean. Figure 2.1 shows the distribution of permafrost around the world currently.

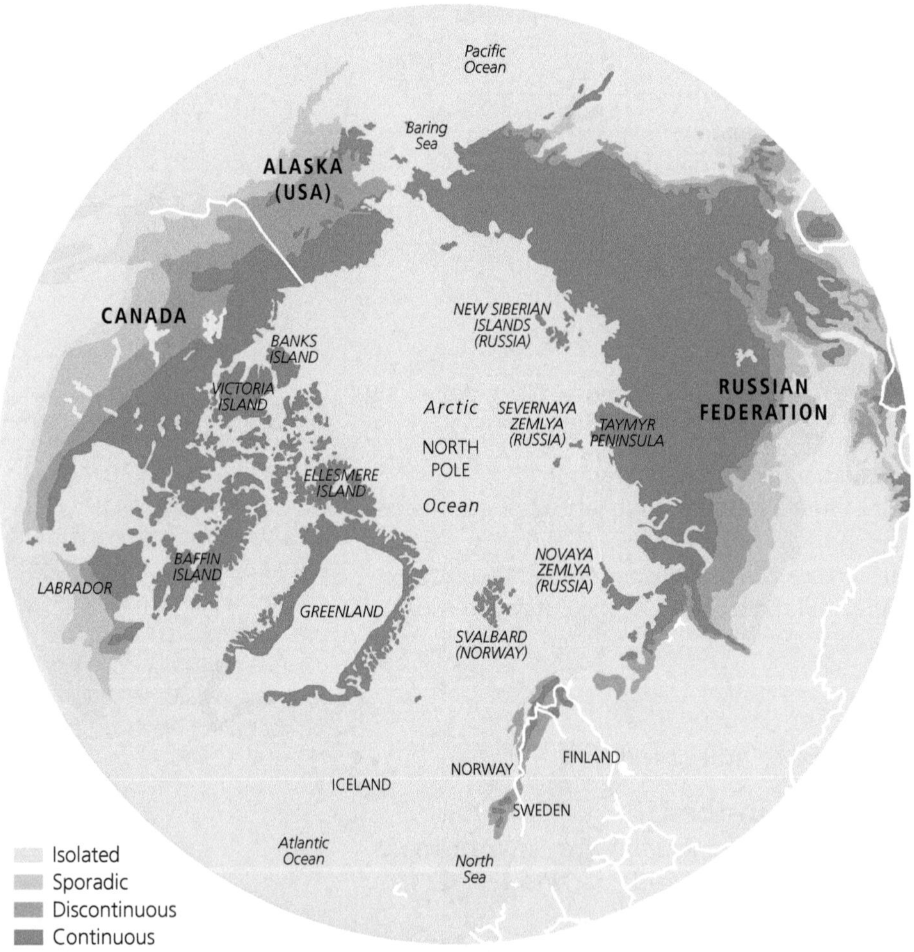

Figure 2.1 Present distribution of permafrost

Source: International Permafrost Association

Key concept

Permafrost is soil and rock that remain frozen as long as temperatures do not exceed 0°C in the summer months for at least 2 years. Three types of permafrost are recognised:

- continuous: occurs in the coldest areas of the world where mean annual air temperatures are –6°C. It can be hundreds of metres deep
- discontinuous: is more fragmented and less deep
- sporadic: occurs at the margins of periglacial areas and is usually fragmented and very thin.

Periglacial processes

The distinctive climate means that periglacial areas are distinguished by a particular variety of landform processes, as summarised in Table 2.3.

Table 2.3 Landform processes in periglacial areas

Process	Definition
Freeze–thaw weathering	Shattering of rock as a result of water in its joints and pores freezing and expanding. Particularly active when temperatures fluctuate around freezing point.
Frost heave	Upward movement of rock or soil particles as a result of the pressures generated by the formation of ice segregations in the ground.
Nivation	Sometimes known as snow-patch erosion, occurs when both weathering and erosion take place around and beneath a snow patch.
Solifluction	Mass movement of the active layer downslope.
Wind action	The lack of vegetation cover due to the low temperature plus the prevailing aridity mean there are plenty of opportunities for the strong wind to pick up and transport fine sediment. When deposited this is known as loess.
Meltwater action	This takes place during the short summer period only, when temperatures are above freezing.

Weathering The breakdown or disintegration of rock in situ.

Mass movement The large-scale downward movement of material under the influence of gravity.

Periglacial landforms

The distinctive mix of processes is responsible for the creation of often unique periglacial landforms, as summarised in Table 2.4 (see also Figure 2.2).

Table 2.4 Periglacial landforms

Periglacial features	Description
Ground ice features	Mostly caused by frost heaving and include ice-wedge polygons, patterned ground and pingos.
Frost-shattering features	Mostly caused by freeze–thaw and include block fields, scree slopes, rock glaciers and tors.
Mass-movement features	Mostly formed by the downslope movement of weathered materials and include asymmetric valleys and solifluction terraces.
Wind action	A lack of vegetation enables strong winds to pick up large amounts of fine, loose material and redeposit it as loess.
Meltwater action	As saturated soil slumps downhill due to solifluction during the summer, this can create solifluction lobes. Also forms braided streams/rivers as the meltwater flows across glacial outwash plains.

Although periglacial conditions no longer prevail in the UK, relict periglacial features are still to be found. Most have been subject to modification by the processes of a warmer climate.

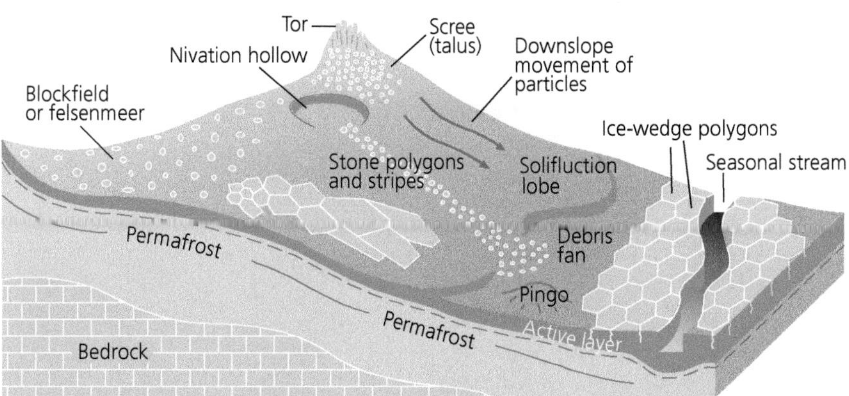

Figure 2.2 Periglacial landforms

Exam tip

It is helpful to classify all the distinctive periglacial landforms under the five headings listed in Table 2.4. Make sure you know an example of where periglacial landscapes form, e.g. tundra environment of Northern Canada.

Now test yourself

TESTED 

6 What is the difference between periglacial and glacial?
7 How have periglacial processes contributed to upland landscapes?
8 What are the typical characteristics of a periglacial landscape?

Answers on p. 221

Revision activity

Check that you have brief notes about the periglacial processes and at least one landform associated with each process.

Glacier systems

Mass balance

REVISED

The glacier system is made up of two zones:

- accumulation zone of direct snowfall, precipitation, wind deposition and debris avalanching from slopes above the glacier
- ablation zone where there is a loss in the amount of ice as a result of melting, evaporation (sublimation), the calving of icebergs and ice blocks, and the deposition of rock debris.

The most critical feature of the glacier system is its mass balance:

- This is the balance between the inputs and the outputs (see Figure 2.3).
- Where accumulation is greater than ablation (inputs exceed outputs), a zone of excess will form and the mass balance will be positive. The glacier will grow.
- When the situation is reversed, with ablation greater than accumulation (output exceeding inputs), a zone of deficiency will form and the mass balance will be negative. The glacier will shrink.

Key concept

A **systems view** of glaciers is helpful in understanding how they behave. Systems are made up of three components:

- inputs: precipitation, rock debris, energy (kinetic and solar)
- processes or throughputs: ice movement, erosion, transport, deposition
- outputs: debris, meltwater, calving.

Sublimation The change from the solid state (ice) to gas (water vapour) with no intermediate liquid stage (water).

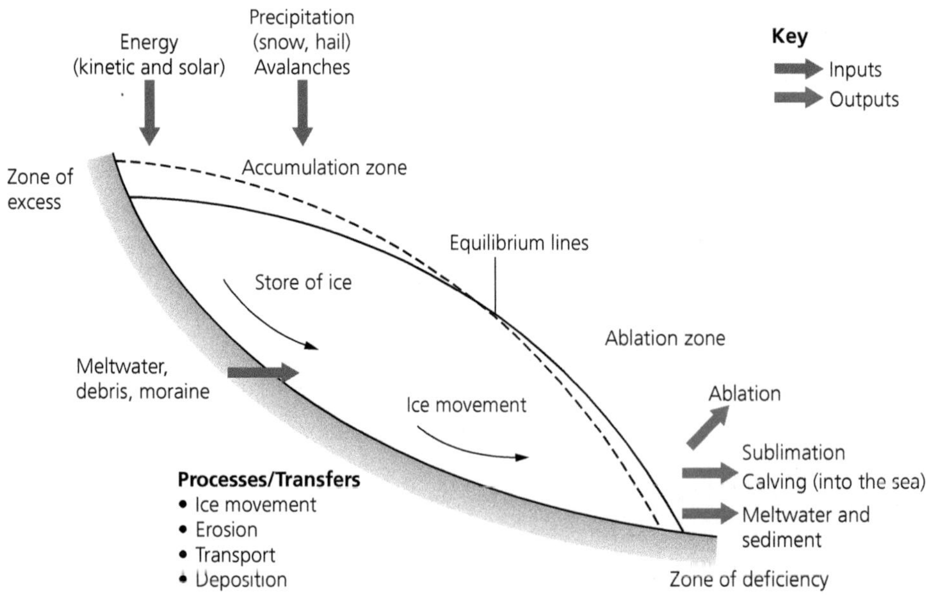

Figure 2.3 Glacier mass balance

Now test yourself

TESTED

9 What is the mass balance of a glacier?

10 Why do you get positive mass balance values in the winter and negative mass balance values in the summer?

Answers on p. 221

The mass balances of glaciers vary over different time scales:

- This is because of variations in the rates of accumulation and ablation.
- Figure 2.4 shows how the mass balance varies during the course of a year.
- The same principles apply when dealing with fluctuating mass balances on longer time scales, as between stadials and interstadials, or between glacials and interglacials.

Stadials and **interstadials** Short-term fluctuations within glacial periods. Stadials are colder phases that lead to ice advances, while interstadials are slightly warmer phases during which ice sheets and glaciers retreat.

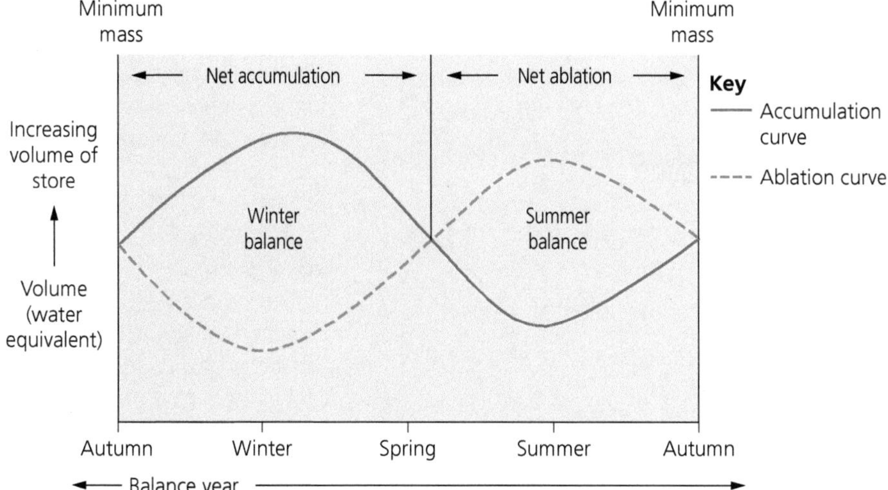

Figure 2.4 Annual mass balance of a typical glacier

Now test yourself

TESTED

11 Is the glacier a closed or an open system, and what are its components?

12 What is the difference between positive and negative feedback?

Answers on p. 221

Key concept

Glacier feedback comprises those effects that can either amplify a small change and make it larger (positive feedback) or diminish the change and make it smaller (negative feedback). An example of positive feedback in a glacier is increase in temperature > glacier ice begins to melt > glacier moves faster > total amount of ice in glacier decreases > glacier gets stretched out and thins.

Exam tip

It is recommended that you have some information about fluctuating glaciers, as for example glaciers in South Georgia, the Alps or the Rockies.

Glacial movement

REVISED

When it comes to glacier movement, the earlier distinction between warm-based (temperate) and cold-based (polar) glaciers becomes significant.

Warm-based glaciers move much faster than cold-based ones because of:

- the imbalance between accumulation and ablation, with the former being greater than the latter
- the availability of summer meltwater acting as a lubricant to encourage basal sliding
- regelation slipping
- internal deformation of the basal ice.

The rate of glacier movement is also controlled by other factors, as shown in Table 2.5.

Table 2.5 Factors affecting the rate of glacier movement

Factor	Description
Altitude	Affects temperatures (decreases with altitude and therefore reduces speed of movement) and precipitation inputs (snow rather than rain).
Slope	The steeper the slope, the faster the velocity as gravity causes ice to move.
Lithology	Friction with hard, resistant rock tends to restrain movement at the base and sides of the glacier.
Size	The greater the glacier mass, the greater the potential velocity.
Mass balance	The nature of this determines not just the velocity but also whether the glacier is retreating or advancing.
Meltwater	This lubricates the base of the glacier, which enables it to move downhill at a faster rate.

Because temperate glaciers are more mobile, they are capable of much more erosion, transportation and deposition than polar glaciers.

It is important to understand that rates of movement vary within the individual glacier, both laterally and vertically. The critical factor is friction between the glacier ice and the valley floor and sides. As a consequence, the part of the glacier moving fastest is its surface in the middle of the valley (see Figure 2.5).

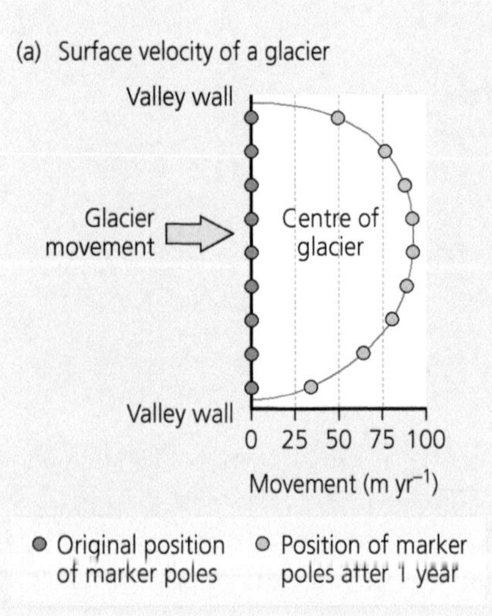

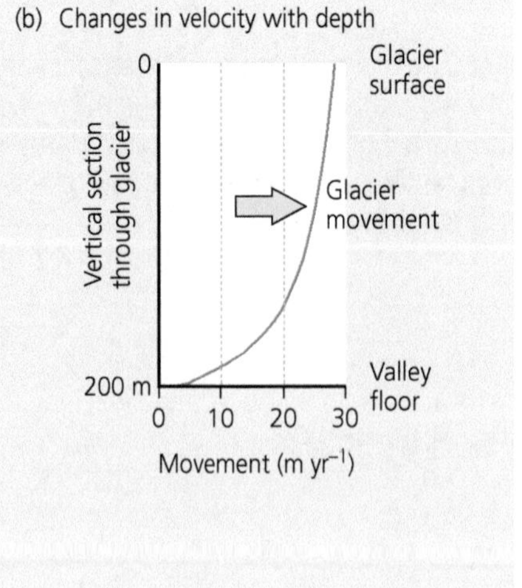

Figure 2.5 Glacier velocities

Basal sliding Occurs where ice temperatures are at or close to 0°C and a layer of basal meltwater forms between the ice and the bedrock.

Regelation Occurs where basal ice is forced against a rock obstacle — it melts and then refreezes on the down-glacier side. The temporary meltwater acts as a lubricant.

Internal deformation A plastic-like quality caused when, under pressure, ice crystals are affected by recrystallisation.

Exam tip

Glacier movement is very dependent on the thawing of ice to provide meltwater, which acts as a lubricant between the glacier ice and the bedrock.

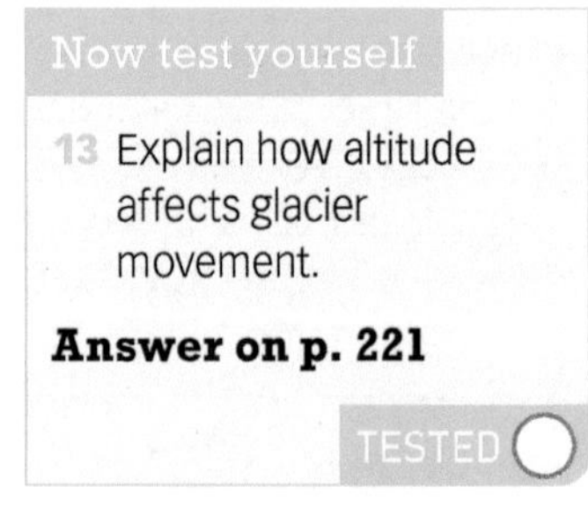

Now test yourself

13 Explain how altitude affects glacier movement.

Answer on p. 221

TESTED

Revision activity

Create a flow diagram to show how glaciers move, advance and retreat.

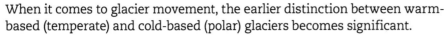

Skills activity

Table 2.6 Two different glaciers and their rate of movement

Glacier 1 day number	Metres per day	Glacier 2 day number	Metres per day
1	18.2	1	0.12
2	18.5	2	0.11
3	17.1	3	0.09
4	17.3	4	0.13
5	17.5	5	0.13
6	17.9	6	0.12
7	17.6	7	0.14
8	17.8	8	0.14
9	17.5	9	0.15
10	17.5	10	0.16
11	17.9	11	0.16
12	18.3	12	0.18
13	18.2	13	0.16
14	18.3	14	0.16
15	18.3	15	0.14
16	18.6	16	0.14
17	18.8	17	0.13
18	19.1	18	0.12
19	19.3	19	0.12
20	19.3	20	0.11

Table 2.6 shows two different glaciers and their rate of movement.

1 Calculate the mean daily rates of flow for both Glacier 1 and Glacier 2.
2 Calculate the median daily rates of flow for both Glacier 1 and Glacier 2.
3 Calculate the modal daily rates of flow for both Glacier 1 and Glacier 2.

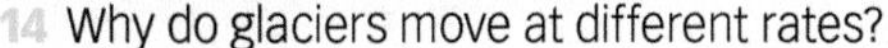

Now test yourself

TESTED

14 Why do glaciers move at different rates?
15 Why do polar and temperate glaciers move at different rates?

Answers on p. 221

Revision activity

Create a mind map to show the factors that affect the rate at which glaciers move.

The glacier landform system

REVISED

Glacier processes

Glaciers alter landscapes and produce distinctive landforms by a number of processes.

Erosion

+ This process principally occurs by abrasion (the scraping, scouring, rubbing and grinding action of debris being carried along by the glacier) and plucking (whereby the glacier freezes around rocks on the valley sides and floor, which are then pulled away by the movement of the glacier).

- A crucial aspect of erosion is entrainment, which is the incorporation of debris onto and into the glacier from subglacial and supraglacial sources.

Entrainment Process by which surface sediment is incorporated into a fluid flow (e.g. air, water or ice) as part of the process of erosion.

Transport

This process takes place at three levels:

- supraglacially: debris that has fallen onto the surface of the glacier
- englacially: debris that has worked its way into the heart of the glacier
- subglacially: debris picked up in the basal layer from the bedrock.

Deposition

- This process occurs when material is released from the glacier ice either at its margins or at the base of the glacier.
- Deposition may take place directly, on the ground (ice contact), or indirectly, when sediments are released into meltwater and are subsequently laid down.

For more on glacial processes, see 'Glacial landforms and landscapes' (see pages 35–40).

Glacial landforms

The distinctive glacier processes are responsible for creating a wealth of distinctive landforms. Although the same landform can often be found in many different sizes, all the glacial landforms may be broadly classified into three size groups (see Table 2.7).

Table 2.7 Selected glacial landforms broadly classified by size

Size	Landforms
Macro-scale	Ice sheet eroded knock and lochan landscapes; cirques, arêtes and pyramidal peaks; glacial trough and ribbon lakes; till plains, terminal moraines and sandurs
Meso-scale	Crag and tail and roches moutonnées; drumlins, kames, eskers and kame terraces; kettle holes
Micro-scale	Glacial striations, grooves and chatter marks; erratics

These landforms are produced in different parts of the glacier system. Some are subglacial in origin (e.g. drumlins); others are either at or just beyond the margins of the glacier (e.g. terminal moraines). Some geographers distinguish between proglacial, periglacial and paraglacial.

Subglacial Locations that lie beneath the base of a glacier or ice sheet.

Proglacial Locations that lie close to the ice front of a glacier or ice sheet.

Periglacial Locations are those where frost action and permafrost processes dominate.

Paraglacial Locations are those recovering from the disturbance of glaciation.

For more on glacial landforms, see 'Glacial landforms and landscapes' (see pages 35–40).

Glacial landscapes

Figure 2.6 shows the sequence of links in the glacier system that produces the individual landforms that together make up the glacial landscape. Those landscapes can be of a different character depending on the dominant glacial processes (erosion or deposition). Erosional landscapes are characteristically found in upland areas, depositional ones in lowland areas.

A complication with glaciated landscapes is that they are polycyclic. They are the product of several periods of glaciation and most likely will have been modified in between by periglacial or paraglacial conditions.

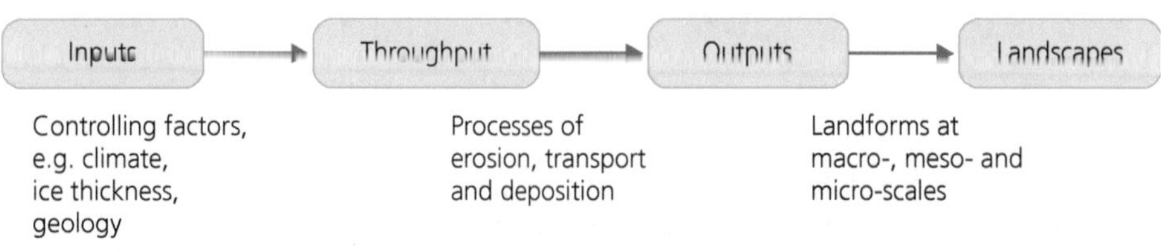

Figure 2.6 The glacier landscape system

Now test yourself TESTED

16 How do glaciers transport material?

17 What glaciated landforms would distinguish an upland glaciated landscape from a lowland one?

Answers on p. 221

Revision activity

Categorise the following features and landforms into subglacial, marginal and proglacial:

- Esker
- Medial moraine
- Erratic
- Drumlin
- Braided stream
- Terminal moraine
- Meltwater
- Kame terrace
- Steep valley side
- Kettle hole
- Ribbon lake
- Till
- Wasting ice sheet
- Outwash plain

Glacial landforms and landscapes

Glacial erosion

REVISED

Processes

The rate, intensity and effectiveness of glacial erosion vary with time and from place to place within the world's cold environments. Figure 2.7 summarises the main factors influencing the rates of abrasion and plucking.

Abrasion The process by which solid rock is eroded by rock fragments being transported by glaciers.

Plucking The detachment of joint-bounded blocks by glaciers. It was thought that the ice froze onto the rock and wrenched blocks of it away. However, it is now believed that the breaking away of blocks is due to the movement of the glacier.

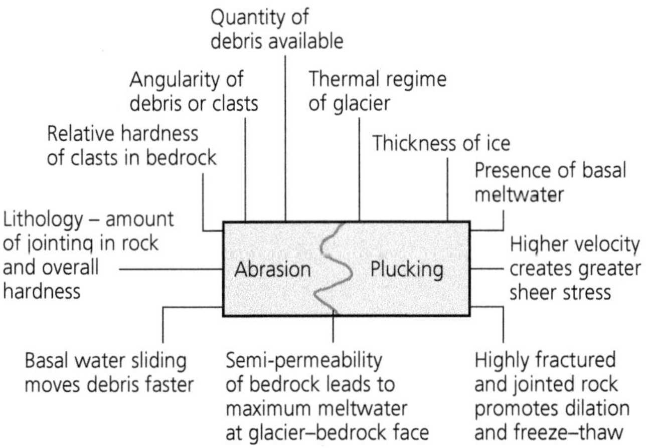

Figure 2.7 Factors affecting abrasion and plucking rates

Revision activity

Make revision cards about three factors affecting each of the two processes of abrasion and plucking rates.

Exam tip

Remember that there are other, less significant erosional processes at work. These are quarrying, crushing, basal melting, freeze–thaw and mass movement.

Macro landforms

Macro features (see Figure 2.8) include U-shaped valleys with their truncated spurs and tributary hanging valleys. Also conspicuous are the three related landforms of cirques, arêtes and pyramidal peaks (see Table 2.8).

Table 2.8 Landforms associated with cirque and valley glaciers

Landform	Description
Corries	Enlarged, deep hollow on a mountainside. Characteristic features include a steep, cliff-like back wall.
Arêtes	When two neighbouring glaciers cut back into the mountainside. The narrow ridge that forms between the two corries is known as an arête.
Pyramidal peaks	Where three or more corries erode back-to-back. The ridge becomes an isolated peak known as a pyramidal peak.
Glacial troughs	Steep-sided, flat-bottomed and deep valley. Created as a glacier moves through and erodes the valley.
Truncated spurs	As a glacier moves through an upper river valley it cuts off the top of the interlocking spurs as it moves downhill. This leaves behind steep cliffs.
Hanging valleys	Smaller glacier is a tributary and erodes less than a much larger glacier. When the ice melts away the smaller valley is left 'hanging' above the main valley.
Ribbon lakes	Sometimes found in glacial troughs. Formed by localised overdeepening due to enhanced erosion.

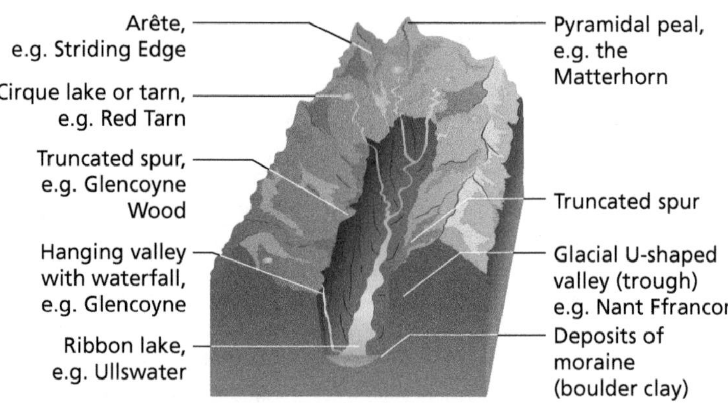

Figure 2.8 Macro features of a glaciated upland area

Meso landforms

Some meso landforms, such as roches moutonnées and whalebacks, are found within macro features such as glacial troughs. However, meso landforms are more commonly found where ice sheets and glaciers spread out over large areas of lower relief (see Figure 2.9). They are largely the product of ice sheet scouring (see Table 2.9).

Table 2.9 Landforms associated with ice sheet scouring

Landform	Description
Roches moutonnées	Bare outcrops of rock on the valley floor, sculpted by moving ice.
Crag and tail	Formed when a very large resistant object (crag) obstructs the flow of a glacier. The ice is forced around the obstruction, eroding the weaker rock. Material in the lee of the obstruction is protected by the crag, which leads to the formation of a gently sloping tail of deposited material.

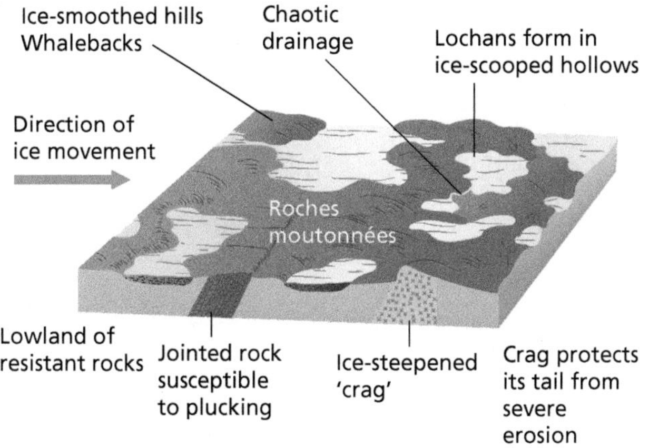

Figure 2.9 Lowland glacial erosion

Now test yourself TESTED

18 How are corries, arêtes and pyramidal peaks related?

19 What is the difference between abrasion and scouring?

Answers on p. 221

Glacial deposition

REVISED

All accumulations of glacial debris are referred to as moraines (see Table 2.10). There are two broad types:

- Subglacial: moraines deposited beneath the glacier. These are made up largely of lodgement till and create extensive flat areas that cover pre-existing topography. In places, the till is moulded into elongated, streamlined mounds (drumlins).
- Ice-marginal: moraines deposited along the edges of glaciers. They include lateral moraines (along the edge of the valley floor), medial moraines (along the middle of the valley floor), terminal moraines (a ridge of moraine extending across the valley at the furthest point reached by the glacier) and recessional moraines (a series of cross-valley ridges behind the terminal moraines).

Table 2.10 Glacial depositional features

Glacial depositional feature	Description
Medial moraines	Formed when lateral moraines from two merging glaciers join up. A line of debris is left in the centre of the combined glacier flow.
Lateral moraines	Formed along the outer edge of a glacier. Can be several metres high.
Recessional moraines	When a glacier retreats it can experience periods of stability. A second ridge of sediment forms at the snout. It has the same characteristics as terminal moraines.
Terminal moraines	A ridge of sediment piled at the furthest extent of an advancing glacier.
Drumlins	An oval hill made of glacial till. They are aligned in the direction of the glacial ice flow. Usually occur in clusters on flat valley floors.

Exam tip

Be aware that not all drumlins are thought to have been formed in the same way. There are two main theories, one emphasising the deposition of moraine in the lee of a solid obstacle and the other the impact of subglacial meltwater.

Table 2.11 summarises lowland depositional features associated with glaciers.

Table 2.11 Lowland depositional features

Lowland depositional feature	Description
Till plains	An extensive plain created by the melting of a large ice sheet that detached from a glacier. Angular, unsorted till is divided into lodgement till (dropped by moving glaciers) and ablation till (dropped by stagnant or retreating ice).
Erratics	A boulder or rock fragment deposited far from its origin. Help to recreate paleo-environments.

Now test yourself

TESTED 

20 Distinguish between (a) terminal and recessional moraines and (b) lateral and medial moraines.

21 How can different glacial landforms be used to reconstruct formed ice extent?

Answers on p. 221

Glacial meltwater

REVISED 

Meltwater from glaciers plays a vital role in the fluvio-glacial processes of erosion, entrainment, transport and deposition. The two main sources of meltwater are:

- surface melting produced mainly during the short summer
- basal melting caused by the temperature of the ice being at **pressure melting point**. Meltwater beneath a glacier facilitates basal sliding, subglacial bed formation and erosion.

Key concept

Pressure melting point is the temperature at which ice melts at a given pressure. It is normally 0°C at the surface of a glacier, but inside, the melting point is fractionally lowered by the raised pressure within the ice.

Fluvio-glacial deposits differ from glacial ones in that they are rounded rather than angular, sorted by size, with the heaviest load deposited first (glacial unsorted), and stratified into distinct layers by the meltwater (glacial unstratified).

Table 2.12 summarises the main features resulting from subglacial meltwater.

Table 2.12 The main features resulting from subglacial meltwater

Ice-contact features	Description
Eskers	Long, sinuous ridges of the valley floor. Up to 30 m high. Formed by subglacial river deposition during glacial retreat.
Kames	Mounds of stratified material deposited beneath the ice of a decaying glacier or ice sheet.
Kame terraces	Ridges of material deposited along the edge of a valley floor.

Table 2.13 summarises the main features created by meltwater as it exits the snout of the glacier.

Table 2.13 The main proglacial features created by meltwater as it exits the snout of a glacier

Proglacial feature	Description
Outwash plains (sandurs)	Formed when meltwater spreads out after issuing from the glacier snout and its load is deposited to create a gently sloping surface.
Proglacial lakes	Created when meltwater becomes impounded between the glacier snout and high ground.
Overflow channels	Spillways cut when proglacial lakes overflow their confines.
Kettle holes	Created when large blocks of ice are covered by deposits from meltwater streams. When the ice melts, a depression is left behind, which then fills with water.

Skills activity

Student's *t*-test

- When collecting fieldwork data you can use statistical analysis to test the reliability of your data.
- Where the sample size is lower than 30, the Student's *t*-test can be used.
- Table 2.14 shows data collected on sediment size and shape in outwash plains. Use the Student's *t*-test to analyse changes in sediment size.

Table 2.14 Data collected on sediment size and shape in outwash plains

Site	Distance in metres from snout	Mean sediment size (cm)	Average sediment shape (Caileux index). Greater R values equal greater roundness
1	1	16.3	159
2	15	15.2	201
3	50	12.1	170
4	65	6.0	245
5	100	3.2	260
6	136	2.9	345
7	243	3.5	502
8	365	2.4	688
9	498	1.8	760
10	698	1.4	802

To calculate the coefficient *t* for the Student's *t*-test, use the following formula:

$$t = \frac{x - y}{\sqrt{\left(\Sigma \times \frac{2}{n_x}\right)}} - \frac{x}{(n_x - 1)} + \frac{\left(\Sigma \times \frac{2}{n_y}\right) - y}{(n_y - 1)}$$

Where:

$x - y$ are the means of each sample.

n_x and n_y are the number of individuals in each sample.

Σ means the 'sum of'.

To work out the answer, use Table 2.15, adding two extra columns and three extra rows to carry out the calculations.

Table 2.15 Answers table

Site	Mean sediment size (cm) (x)	Average sediment shape (Caileux index). Greater R values equal greater roundness (y)	x_2	x_2
1	16.3	159	265.69	25281
2	15.2	201	231.04	40401
3	12.1	170	146.41	28900
4	6.0	245	36	60025
5	3.2	260	10.24	67600
6	2.9	345	8.41	119025
7	3.5	502	12.25	252004
8	2.4	688	5.76	473344
9	1.8	760	3.24	577600
10	1.4	802	1.96	643204
	$\sum x = 64.8$	$\sum y = 4132$	$\sum x^2 = 721$	$\sum y^2 = 2287384$
	$\bar{x} = 6.48$	$\bar{y} = 413.2$		
	$n_x = 10$	$n_y = 10$		

1 $x - y = 6.48 - 413.2 = -406.72$ (This number must be a positive so the minus sign is removed.)

2 $\frac{\sum x^2}{n_x} = \frac{721}{10} = 72.1$

3 $\left(\frac{\sum x^2}{n_x}\right) - x$ (insert macron)^2 $= 72.1 - 41.9904 = 30.1096$

4 $n_x - 1 = 9$

5 $\left(\frac{\sum x^2}{n_x}\right) - \frac{x^2}{n_x} - 1 = \frac{30.10906}{9} = 3.346$

6 $\frac{\sum y^2}{ny} = \frac{2287384}{10} = 228738.4$

7 $\left(\frac{\sum y^2}{ny}\right) - \bar{y}^2 = 228738.4 - 170734.24 = 58004.16$

8 $ny - 1 = 9$

9 $\left(\frac{\sum y^2}{ny}\right) - \frac{\bar{y}^2}{(ny-1)} = \frac{58004.16}{9} = 6444.907$

10 $\left(\frac{\sum x^2}{n_x}\right) - \frac{x^2}{(n_x-1)} + \left(\frac{\sum y^2}{n_y}\right) - \frac{\bar{y}^2}{(n_y-1)} = 3.346 + 6444.907 = 6448.253$

11 $\sqrt{6448.253} = 80.301$

12 $\frac{x-y}{\sqrt{\left(\frac{\sum x^2}{n_x}\right)}} - \frac{x^2}{(n_x-1)} + \left(\frac{\sum y^2}{n_y}\right) - \frac{\bar{y}^2}{(n_y-1)} = \frac{406.72}{80.301} = 5.06$

To test the significance of this number, we do the following:

1. Calculate the degree of freedom, which is $n - 1$. As there are two data sets the value must be calculated twice and added together. $n - 1$ for x and $y = 9$, which becomes 18 when added together.
2. If the value is greater than the critical value then the result is significant; if it is less than the critical value then the result is not significant.
3. In this case, the t value is higher than the critical value so there is a significant difference between mean sediment size and average sediment shape.

Exam tip

You will always be given the formula in an exam and you will only be asked to complete part of the equation.

Revision activity

Create flashcards on glacial landforms created by erosion, deposition and meltwater. Include diagrams to help remember how the landforms are created.

Now test yourself

22 What is the significance of the pressure melting point?

23 What are the similarities and differences between eskers and kames?

Answers on p. 221

TESTED

The use and management of glaciated and glacial landscapes

Intrinsic value

REVISED

People value relict glaciated and active glacial and periglacial landscapes for a variety of reasons, including:

- **wilderness** and the opportunity to escape from the hustle and bustle of modern living
- the spectacular scenery they provide
- their unique biodiversity, ecology and wildlife
- the challenge they present to scientists, explorers and travellers
- their role in the global water and carbon cycles
- glaciers — an important source of water
- the economic opportunities they present, such as farming, forestry, tourism, mining and quarrying, and hydroelectric power (HEP)
- the artistic inspiration they have provided to generations of artists and writers
- the cultural heritage created by indigenous peoples.

Clearly, wilderness has much greater value beyond its economic potential:

- The quality of **wilderness** offered by these landscapes varies from place to place and is inversely proportional to the degree to which they have been modified by people (see Figure 2.10).
- The **wilderness continuum** runs from the almost pristine and active Antarctic and Arctic to the well-trampled and commercially exploited relict landscapes of Snowdonia and the Cairngorms.

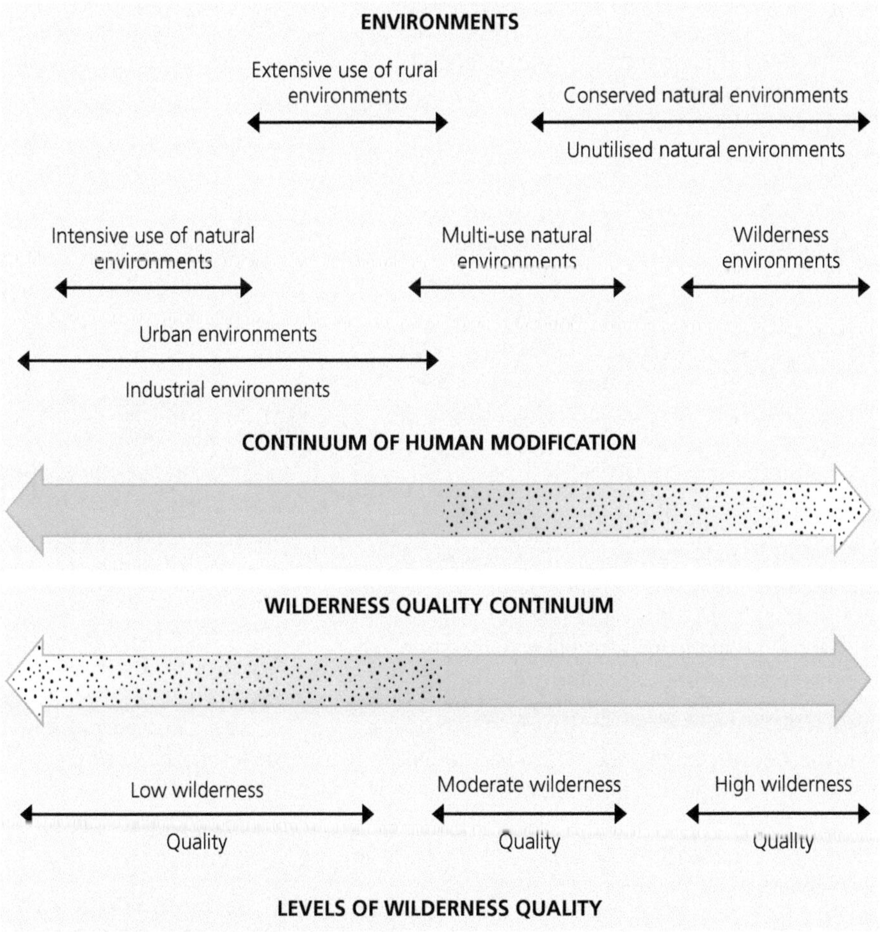

Figure 2.10 The wilderness continuum

Key concept

Wilderness is any area of the world that has remained relatively untouched by human activity (for example, Antarctica), but may be home to small numbers of indigenous people (for example, Amazonia). The overriding characteristics of wilderness are remoteness and detachment from modern living. The value of wilderness can be spiritual, scientific, artistic and therapeutic. But wilderness is threatened: by its economic potential as a source of resources, and its appeal as a tourist destination.

Typical mistake

Do not think that all wildernesses are vast in extent.

Making links

Unfortunately, not all players involved in glacial and periglacial environments share the same attitudes. What they seek ranges from outright exploitation of resources to resource preservation.

Revision activity

Make sure you have a definition of wilderness and that you understand the meaning. Does it have a value, and what is that value?

Now test yourself

TESTED

24 Why are glacial landscapes valued? Make sure you know the value of one glacial landscape that you could write about, e.g. the Lake District, UK.

Answer on p. 221

Exam tip

Be aware of the distinction between present and past glaciated landscapes. The former are more in focus in this part of the topic.

Threats

REVISED

The threats facing active and relict glaciated landscapes fall into three categories:

- the natural hazards that pose a risk to humans
- the impacts of human activity
- global warming.

Natural hazards

It is only the human presence in these environments that converts the following natural events into possible hazards (see Table 2.16).

Table 2.16 Natural events that are possible human hazards

Natural event	Possible human hazard
Avalanche	A rapid descent of a large mass of rock, ice and snow down a mountain slope. Avalanches are frequently triggered by human activity, such as skiing.
Lahar	A rapid flow of mud and debris resulting from meltwater overflowing a glacial lake.
Glacial outburst flood (jökulhlaup)	A short-lived but sometimes catastrophic flood resulting from the sudden release of meltwater stored within or on the surface of a glacier or ice sheet. The build-up of meltwater is often triggered by the escape of geothermal heat beneath the ice.

Human impacts

Clearly, it is human activity that truly threatens to degrade and damage fragile glaciated landscapes. The activities in Table 2.17 may be singled out as causing the most concern.

Table 2.17 Human activities that threaten fragile glaciated landscapes

Activity	Concerns
Polar tourism	A booming but very expensive business. To date it is fairly well policed by the International Association of Antarctica Tour Operators (IAATO).
Mountain tourism	The modern tourist infrastructure and the sheer pressure of tourist numbers are inflicting considerable environmental damage. In particular, there is the wear and tear on fragile ecosystems, soil erosion and upsetting water budgets that are critical to the supply of water to rivers and nearby populated areas.
Mining and quarrying	Possibly the biggest threat here comes from the oil industry. The Alaskan experience serves as a reminder of the immense damage that this industry can inflict on the environment — roads and pipelines cutting swathes across the tundra, pollution around oil wells and oil spills in coastal waters.

One particular threat concerns the indigenous peoples of the Northern Hemisphere who still make their living in these environments, be it by fishing and sealing or herding caribou. It is important that they, their traditional livelihoods and cultural heritage be protected.

Making links

It is often the case that the actions of players have indirect and unforeseen impacts on the natural systems of glaciated landscapes. For example, IAATO, by encouraging responsible polar tourism, is also encouraging more people to visit Antarctica. That means more people damaging fragile natural systems.

The very presence of people, let alone economic players, has a debilitating impact on the resilience of fragile landscapes.

Global warming

- This is undoubtedly the greatest threat to glacial environments. Already most of the world's glaciers are in retreat.
- Data from satellite surveys of Greenland show a huge decrease in the ice-covered area. Surveys of Antarctica show alarming losses of shelf ice.
- These changes will have serious consequences for the global population: the disruption of global water and carbon cycles, as well as a rise in sea level that threatens coastal cities and millions of inhabitants.

Now test yourself TESTED

25 Which of the human impacts on glaciated environments poses the greatest threat? Give your reasons.

26 How could global warming impact glacial mass balances?

Answers on p. 221–222

Revision activity

Make sure you have an example of a glacial environment that is facing threats, e.g. Alpine valleys.

Management approaches

There are various stakeholders with an interest in the future of glaciated landscapes. They range from conservationists to oil companies, from scientists to tourist operators, from indigenous people to investors.

The way ahead

There are a number of possible approaches to the management of cold environments (see Figure 2.11). They range from 'do nothing' and 'business as usual' to 'comprehensive conservation' and 'total protection'.

Revision activity

Make a list of these stakeholders and identify their particular visions of the glaciated landscapes.

Figure 2.11 The spectrum of management options for cold environments

Which strategy is most appropriate depends on the area and the views of the players involved. In the event that unanimity rarely prevails among stakeholders with different values and viewpoints, compromise is the only way ahead. Possible strategies here include:

- sustainable exploitation
- sustainable management
- zoning — a useful middle way, with the high-value environments given full protection and buffer zones created between these areas and zones where sustainable activities are permitted.

Making links

The attitudes and actions of players in glaciated landscapes range from all-out exploitation to preservation.

Legislation

If it is to be successful, any management of glacial environments requires the support of legislation, both international and national.

Examples on the international scale include:

- The Antarctic Treaty (1959), which established that all areas south of 60°S were to be without any national claims and protected for peace and science.
- Since then, a large number of recommendations and international agreements have been adopted to form the Antarctic Treaty System (ATS).
- While the Antarctic now looks reasonably 'future proof', the same cannot be said for the Arctic, an oceanic area largely surrounded by the two superpowers of Russia and the USA.

Effective legislation on the national scale has led to the creation of a range of conservation areas, from national parks to more local nature reserves, particularly in parts of the Arctic and in high mountainous regions such as the Alps, Himalayas and Rockies.

The problem with well-intentioned national legislation is the lack of overarching international legislation. Without it, conservation and protection become piecemeal.

The crucial point is that the future of the world's cold environments is a global issue requiring the backing of international treaties to give real 'teeth' to both conservation and protection.

Revision activity

Find out more about why and how global warming, resource exploitation and geopolitics are threatening the future of the Arctic.

Climate change

- Climate warming is by far the most serious threat to the world's glaciated areas. Currently, their extent is being reduced at an alarming rate. What, if anything, can be done to halt this?
- Much attention is focused today on carbon emissions as a causal factor and on the need for concerted international actions to reduce these emissions.
- However, what if today's global warming is simply one of those natural changes in climate that have characterised the Earth during both geological and historic times?
- If this is the case, then there is little that humans can do, on any spatial scale, to change the forces and courses of nature.

Now test yourself TESTED

27 What are the options when it comes to the possible management of cold environments?

28 Why do different stakeholders have different attitudes to the management of cold environments?

Answers on p. 222

Making links

There are uncertainties about the future, and different people approach these questions in different ways:

- 'Business as usual', e.g. carry on as usual.
- 'We need more sustainable strategies', e.g. landscape protection and management of alpine areas.
- 'We must take radical action', e.g. adapting or mitigating threats to climate change.

The risks associated with climate warming are creating uncertainty. This is creating a need to devise appropriate mitigation and adaptation strategies.

Exam skills

You should be familiar with the geographical skills used in the investigation of the following aspects of glaciated landscapes:

- representing reconstructed past climates by means of graphs
- comparing past and present distributions of glacial and glaciated landscapes using global and regional maps
- calculating the mass balance and the position of the equilibrium line using data from global and regional maps
- identifying the main features of glaciers using GIS
- comparing rates of glacier movement using measures of central tendency
- analysing and correlating cirque orientation, height and size using data from large-scale maps
- representing till fabric analysis by means of rose diagrams
- reconstructing past ice extent and ice-flow direction from Ordnance Survey and British Geological Survey (drift) maps
- analysing changes in sediment size and shape in glacial and fluvio-glacial deposits using the Student's *t*-test and measures of central tendency
- analysing data relating to mean rates of glacial recession in different regions
- comparing drumlin morphometry and orientation in two contrasting locations using field measurements.

Revision activity

For further information relating to the issue of climate change, see Topic 6.

Exam practice

A-level

1 Study Figure 1.

Figure 1 The Lake District

a) Explain the economic importance of glacial landscapes. (6)

b) Explain how human activity can threaten glacial environments. (6)

c) Explain the role of erosional processes in the formation of glacial landscapes. (8)

d) Evaluate the importance of different stakeholders in managing the threats to glacial landscapes. (20)

2 a) Study Figure 2. Explain the occurrence of high erosional intensity. (6)

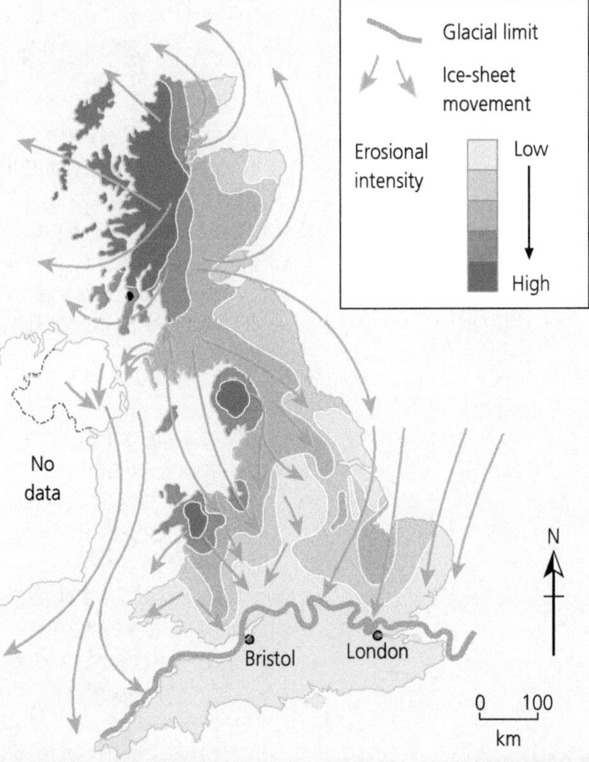

Figure 2 Ice sheet movement and erosional intensity in the UK during the last glacial advance

b) Explain how meltwater contributes to the glaciated landscape. (6)

c) Explain how the balance between net accumulation and net ablation is the key to understanding glacier behaviour. (8)

d) Evaluate the extent to which glacial landscapes are more threatened by climate change than periglacial landscapes. (20)

Answers and quick quiz 2A online

Summary

You should now have an understanding of:

- the causes of long- and short-term changes in climate and the creation of icehouse and greenhouse conditions
- the present distribution of ice cover and its role in global systems
- evidence for the extent of ice cover during the Pleistocene
- periglacial processes and their distinctive landscapes
- glacial dynamics and glacial systems at work
- glacial movement and variations in its rates
- the glacier landform system
- glacial erosion and its contribution to the glaciated landscape
- glacial deposition and its contribution to the glaciated landscape
- glacial meltwater and its contribution to the glaciated landscape
- the value of glacial and periglacial landscapes
- the threats facing glaciated landscapes
- managing the threats to glaciated landscapes.

2B Coastal landscapes and change

Coasts are dynamic landscapes and an important distinction between land and sea, subject to both terrestrial and marine processes, and they experience extreme events, such as tropical cyclones, tsunamis and storm surges.

Different coastal landscapes and their processes

The coast and littoral zone

REVISED

All coastlines show the same littoral sub-zones (see Figure 2.12), but not all coastlines have similar landscapes. The littoral zone can be a dynamic one of rapid change.

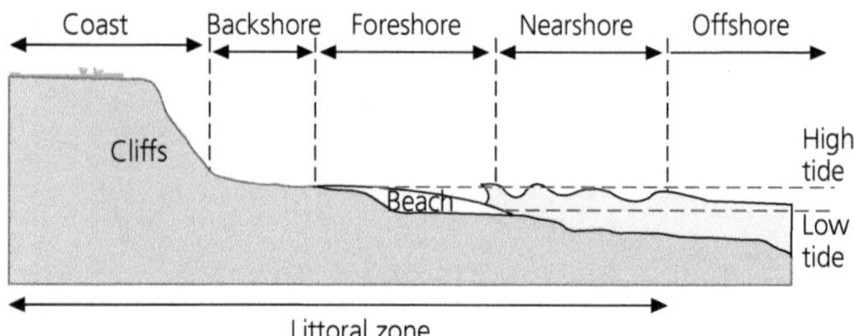

Figure 2.12 The littoral zone and its sub-zones

Littoral zone The wider coastal zone, which includes adjacent land areas, the shore and the shallow part of the sea just offshore. It comprises four sub-zones: coast, backshore, foreshore and nearshore.

Dynamic equilibrium The balanced state of a system when inputs and outputs balance over time. If one of the inputs changes, then the internal equilibrium of the system is upset. By a process of feedback, the system adjusts to the change and the equilibrium is regained.

Classifying coasts

Coasts may be classified in various ways, as for example by:

- geological characteristics (lithology and structure)
- the impacts of sea-level changes (rising or falling)
- the dominant coastal process (erosion or deposition).

A common but simple coastal classification distinguishes between:

- rocky or cliffed: where there is a clear distinction between land and sea, mainly because of the height of the cliffs. Exposure to the erosive forces of the sea, rain and wind creates a high-energy coastline.
- coastal plains: where the land slopes gently towards the sea and there is an almost imperceptible transition from one to the other. These are often maintained in a state of dynamic equilibrium between the deposition of sediment by river stems entering the sea and sediment from offshore sources and marine erosion. They are typically low-energy coastlines.

Exam tip

Be sure you are able to name actual stretches of coastline that exemplify both cliffed and coastal plain types. These will help add authenticity to your answers.

Rocky coasts and coastal plains

Rocky coasts result from a geology that is resistant to the erosive forces of the sea and weather in high-energy environments. Coastal plains are found in areas of low relief and depend on the supply of terrestrial sediment.

Now test yourself

1 Define dynamic equilibrium.

Answer on p. 222

TESTED

Geological structure and the development of coastal landscapes

REVISED

Key concept

Geological structure refers to the arrangement of rocks in three dimensions and involves three key elements:

- strata: the different layers of rock in a location and how they relate to each other
- deformation: the degree to which rock units have been tilted or folded by tectonic activity
- faulting: the presence of fractures along which rocks have moved.

All three elements affect coastal landscapes and the development of coastal landforms.

Typical mistake

Do not confuse geological structure and lithology. The former is about the disposition of rocks, while the latter is about the composition of rocks, whether they are limestones, sandstones, marls, etc.

Concordant and discordant coasts

Geological structure produces two main types of coast (see Figure 2.13):

- concordant: formed when rock strata run parallel to the coastline. The typical coastline is generally a smooth or slightly indented one, also known as Dalmatian type.
- discordant: formed when different rock strata intersect the coast at an angle, so that lithology varies along the coastline. The typical coastline is one of bays and headlands, also known as Atlantic type.

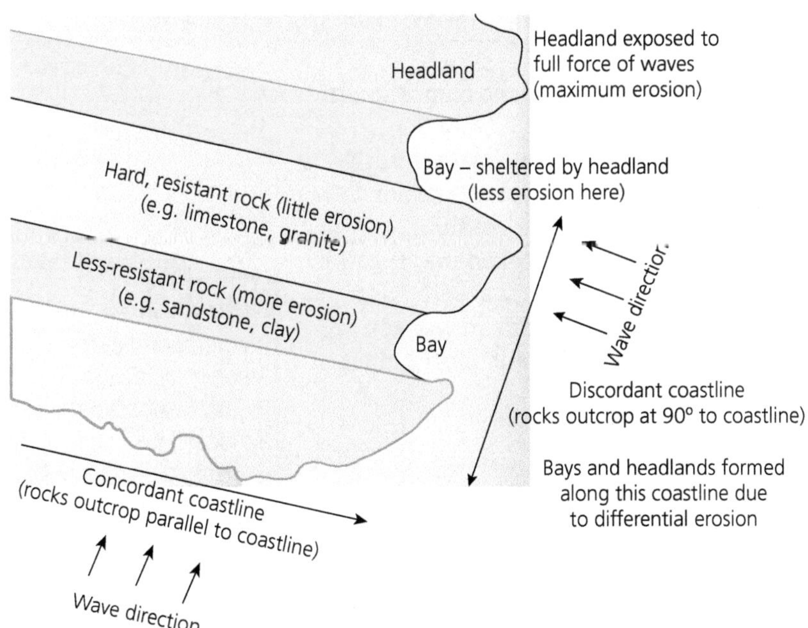

Figure 2.13 The impacts of lithology and rock structure on coastlines

Cliff profiles

Cliff profiles are influenced by two different aspects of geology: the resistance of the rock to erosion and the dip or angle of rock strata in relation to the coastline.

In addition to the dip of strata, other geological features influence cliff profiles and rates of erosion:

- faults: rocks are fractured and therefore weakened on either side of a fault line
- joints: occur in most rocks and are potential lines of weakness
- fissures: small cracks in rocks also represent weakness that erosion can exploit.

Exam tip

The Isle of Purbeck (Dorset) is part of the Jurassic Coast and provides excellent examples of the impact of geology on the coast landscape.

Rates of coastal recession

REVISED 

Geological factors

The rate of coastal erosion and in turn the rates of recession at a coastline are determined by a number of interrelating factors, which include the lithology of the coastline. Coastlines are made up of three key rock types, illustrated in Table 2.18.

Table 2.18 Rates of erosion for different rock types

Rock type	Examples	Erosion rate
Igneous	Granite Basalt Dolerite	Very slow Rocks are resistant to erosion because they are crystalline and usually have few joints.
Metamorphic	Slate Schist Marble	Slow Crystalline rocks are resistant to erosion. Metamorphic rocks are often folded and fractured and therefore vulnerable to erosion.
Sedimentary	Sandstone Limestone Shale	Moderate to fast Most sedimentary rocks erode faster than the other two types. Younger rocks tend to be softer and weaker. Rocks with bedding planes and fractures are more vulnerable to erosion.

- Differential erosion of alternating strata in cliffs can produce complex cliff profiles.
- The lithology of the strata and their geological structures, together with their degree of exposure to the erosive forces of the sea, are the major factors influencing the rate of cliff retreat and of coastal recession.
- Permeability is one other geological factor that should be taken into account.
- Permeable rocks, such as sandstone and limestone, allow water to pass through them. Groundwater flow through permeable rock can weaken rocks by removing the cement that binds the rock sediment. Slumping is a common outcome.

Exam tip

Remember that the geological factors affecting coastal landscapes include lithology, folding, faulting, jointing, dip and the orientation of rock outcrops relative to the alignment of the coastline.

Now test yourself

TESTED 

2 What distinguishes metamorphic rock from igneous and sedimentary?

3 How is the resistance of rocks to erosion affected by their geological structure?

Answers on p. 222

Coastal vegetation

While lithology plays an important role in how coastlines are shaped, the role of vegetation is equally important. Unconsolidated sediment can be protected from coastal processes by the action of plants acting as stabilisers.

There are a number of examples of plants in action on the coastline, including salt marshes, mangrove swamps and sand dunes.

Many of the plants that grow in coastal environments are halophytes; some are xerophytes.

Halophytes Plants that can tolerate salt water, be it around their roots, being submerged at high tide or being exposed to sea spray.

Xerophytes Plants that can tolerate very dry conditions, such as those found in coastal sand dunes.

Sand dunes are particularly effective in encouraging coastal accretion. Through plant succession, sand dunes can convert a supply of sediment into land (see Figure 2.14).

- The succession starts with specialised halophytic plants capable of growing in salty, bare sand.
- Once established, they trap more sand and this leads to the formation of embryo dunes.
- The embryo dunes, in their turn, alter the environmental conditions to an environment in which xerophytic plants can flourish.
- The dunes gradually become fixed and the plant cover develops into a climax community of heath or woodland.

A similar process of plant succession occurs on bare mud deposited in estuaries.

- Estuaries are ideal for the development of salt marshes because of the sheltered conditions and the supply of mud and silt provided by the river.
- The succession starts with algae, followed by various halophytic grasses, then sea thrift and lavender and ending with a climax community of rush and sedge.

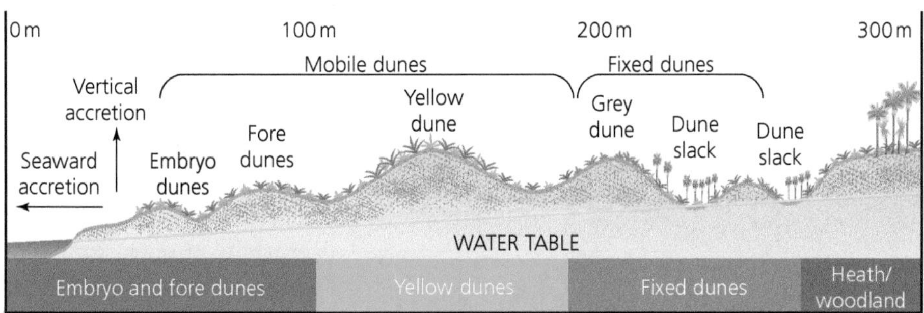

Figure 2.14 Cross-section across sand dunes showing plant succession

Plant succession The sequential development of vegetation from its initial establishment on bare ground through to the ultimate vegetation cover or climax plant community.

Revision activity

Draw a cross-section of sand dune succession and annotate it to describe how vegetation changes.

Now test yourself

4 What is the difference between halophytes and xerophytes?

Answer on p. 222

TESTED 

Coast landforms and landscapes

Marine erosion

REVISED

Waves

- Waves are caused by friction between wind and water; they directly influence the three marine processes of erosion, transport and deposition.
- Wave size and strength depend on the strength of the wind, the length of time the wind blows for, water depth and wave fetch.

Fetch The uninterrupted distance across water over which the wind blows. It is the distance over which waves are able to grow in size.

Typical mistake

Waves and tides are frequently confused. Tides are formed by the gravitational pull of the Moon on water on the Earth's surface. This causes sea level (the tide) to rise and fall twice a day. Tidal ranges vary from place to place. Waves are more localised disturbances of the sea caused mainly by wind.

Now test yourself

TESTED 

5 Identify the factors that influence wave size and strength.

Answer on p. 222

There are two types of waves, as illustrated in Figure 2.15.

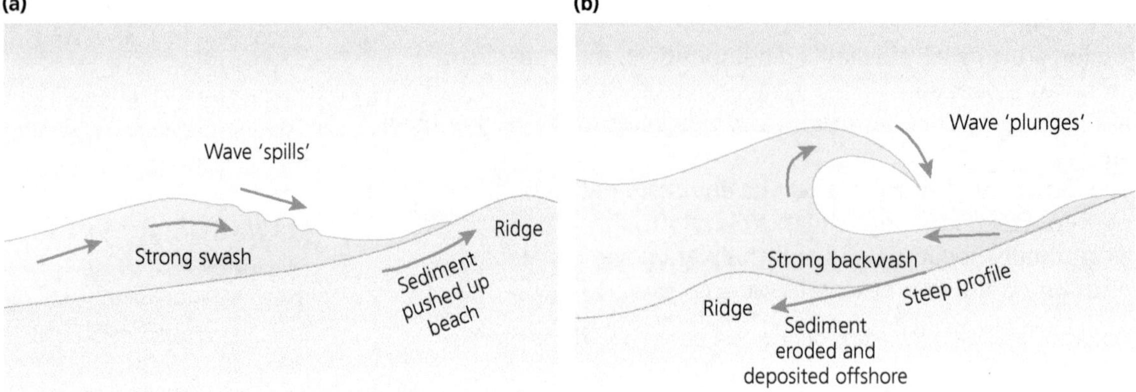

Figure 2.15 Constructive (a) and destructive (b) waves

Table 2.19 lists the main characteristics of constructive and destructive waves.

Table 2.19 Constructive and destructive waves

Constructive waves	Destructive waves
Low in height and long in length	High and short in wavelength
Spilling waves	Plunging waves
Strong **swash** and weak **backwash**	Weak swash and strong backwash
Deposits sediment on the beach	Erodes the beach

Swash The flow of seawater up a beach as a wave breaks.

Backwash The return flow of seawater back down the beach to meet the next incoming wave.

Beach morphology The shape of a beach, including its width and slope (the beach profile) and features such as berms, ridges and runnels. Also includes the type of sediment (shingle, sand, mud) forming the beach.

Beach morphology is strongly conditioned by the nature of the prevailing waves. Wave conditions can fluctuate over time and bring with them changes to beach morphology.

There are four diagnostic beach features of prevailing wave conditions (see Table 2.20).

Table 2.20 Diagnostic beach features of prevailing wave conditions

Storm beach	Berms	Cusps	Offshore bars
Caused by constructive waves during stormy weather.	Small ridges created by constructive waves during relatively calm weather.	The product of gentle destructive waves eroding berms.	Formed by persistent destructive waves.

Now test yourself TESTED

6 Which of the following statements are true about the characteristics of destructive waves?

A Destructive waves have a strong swash and a weak backwash.
B Destructive waves have a more circular motion.
C Destructive waves have a weak swash and a strong backwash.
D Destructive waves have a long wavelength.

Answer on p. 222

Erosion processes

Waves cause erosion. Erosion is not necessarily a continuous process. It mostly occurs during storms and when:

- waves approach the coast at right angles
- the tide is high
- heavy rainfall has weakened the rocks of the cliff
- debris at the foot of the cliff has been removed and no longer protects this critical point.

Figure 2.16 illustrates the four main types of marine erosion.

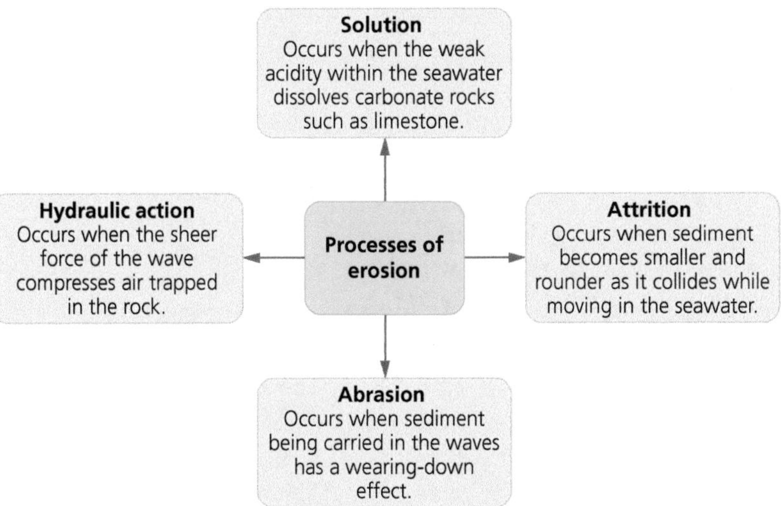

Figure 2.16 The four main types of marine erosion

Erosional landforms

The interaction of physical processes causes the formation of distinctive landforms over time, as shown in Figure 2.17.

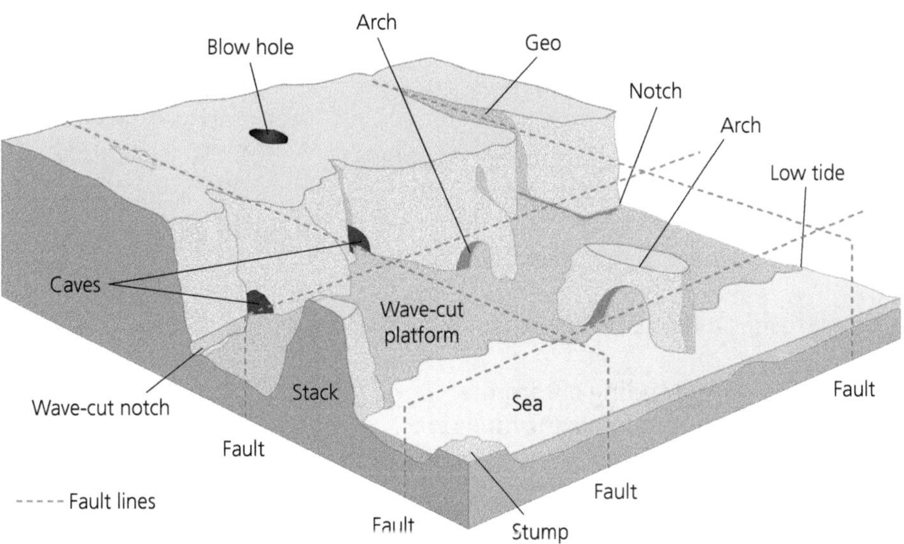

Figure 2.17 Erosional landforms in resistant sedimentary rocks

One of the most critical erosional features is the wave-cut notch formed by the processes of hydraulic action and abrasion. As the notch becomes deeper, the rocks overhanging it become unstable and eventually collapse. Repeated cycles of notch cutting and collapse cause cliffs to recede inland.

Revision activity

Create a series of summary revision cards which outline the sequence of events that lead to the formation of the erosion landforms shown in Figure 2.17.

Now test yourself TESTED

8 Explain the sequence of erosional landforms that starts with a cave and finishes with a stack.

Answer on p. 222

Now test yourself

7 Explain how hydraulic action erodes the coastline.

Answer on p. 222

TESTED

Marine transport and deposition

REVISED

Sediment transportation

Sediment is transported by the sea in four different ways:

- traction: heavier sediment (pebbles, boulders) rolls along the sea floor, pushed by waves and currents
- saltation: sediment (mainly sand particles) bounces along the floor
- suspension: fine sediment (silt, clay) is carried within the body of water
- solution: dissolved sediment (calcium carbonate) is carried in the water as a solution.

Currents Flows of seawater in a particular direction driven by wind, tides and differences in density, salinity and temperature.

In most cases the transportation of sediment occurs into and away from the shore through the process of **longshore drift**.

- This process occurs as the waves approach the coast at an acute angle, directed by the prevailing wind.
- This leads to the swash of the wave pushing sediment up on to the beach.
- The sediment is then dragged back into the sea by the wave's backwash.
- The process repeats, causing sediment to be slowly transported along the coast in the direction of the prevailing wind.

Exam tip

Remember that wave direction is determined by the direction of the wind, which means the direction of sediment transported by longshore drift is reflected in the direction of the prevailing wind.

Depositional features

The main features produced by the deposition of transported sediment are shown in Figure 2.18. They include:

- bayhead beach: an accumulation of sand at the head of a sheltered stretch of water between two headlands
- spit: a sand or shingle beach ridge extending beyond a turn in the coastline
- recurved hooked spit: a spit built out into a bay or across an estuary, the end of which curves landward into shallower water
- bar: a sand or shingle beach extending across a coastal indentation with a lagoon behind
- tombolo: a sand or shingle bar that attaches a former offshore island to the coast
- cuspate foreland: a triangular area of shingle extending out from a shoreline, possibly formed by longshore drifts from opposing directions.

Now test yourself

9 Explain the difference between traction and saltation.

Answer on p. 222

TESTED

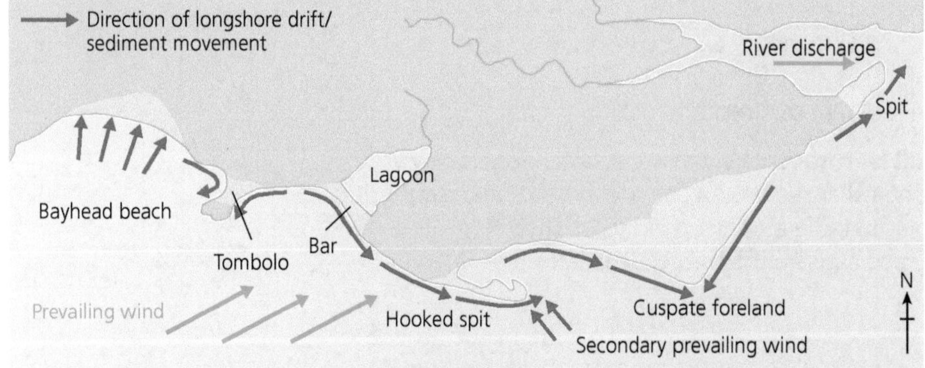

Figure 2.18 Depositional landforms of the coast

Sediment cells

Recognising the existence of **sediment cells** and understanding how they work is of fundamental importance to the management of the coast, particularly at a time of global warming and when there is much human pressure on the coast.

Revision activity

Make notes about one of the sediment cells around the UK's coast, preferably one on a stretch of the coast that you know about firsthand.

Sub-aerial processes

REVISED

Weathering and mass movement are processes that affect most coastlines.

Weathering

There are three types of weathering, which are influenced by the climatic conditions of the location (see Table 2.21).

Table 2.21 Three types of weathering

Mechanical (freeze-thaw/salt crystallisation)	Chemical (carbonation/oxidation)	Biological (plant roots/rock boring)
Freeze-thaw: the repeated force applied to cracks in the rock from the expansion of water as it freezes **Salt crystallisation:** when salt crystals grow and exert pressure within the cracks of rocks	**Carbonation:** the dissolution of limestone from the weak carbonic acid in rainwater **Oxidation:** the reaction of oxygen to minerals, which produce iron oxide	**Plant roots:** the action of tree and plant roots forcing rocks apart as they grow into the cracks **Rock boring:** when species such as clams bore holes into the rock

Mass movement

The role of mass movement can be significant on some coastlines, causing large-scale cliff collapse. There are several different types of mass movement (see Figure 2.19) that you should be able to explain.

Weathering The disintegration and decomposition of rocks in situ by the combined actions of the weather, plants and animals.

Mass movement A collective term for the processes responsible for the downslope movement of weathered material under the influence of gravity.

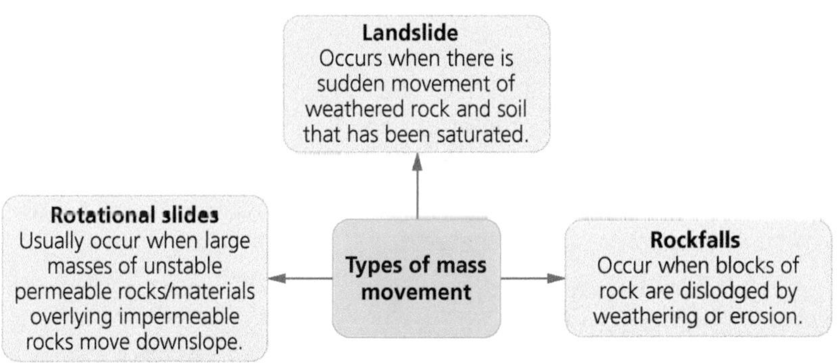

Figure 2.19 Types of mass movement

The distinctive coastal landforms created by mass movement include screes (the product of mechanical weathering of an exposed rock face), rotational scars (the outcome of slumping) and cliff terraces (the result of rotational sliding).

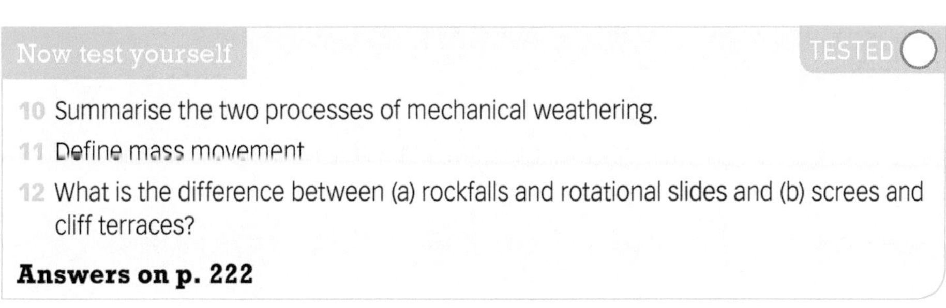

Now test yourself

TESTED

10 Summarise the two processes of mechanical weathering.

11 Define mass movement.

12 What is the difference between (a) rockfalls and rotational slides and (b) screes and cliff terraces?

Answers on p. 222

Coastal risks

Sea-level change

REVISED

- Sea levels change on a day-to-day basis as a result of tides, changes in atmospheric air pressure and winds.
- Over long time scales, sea-level changes can be more permanent and the outcome of complex factors.
- Part of the complexity is that a change in sea level can be brought about by a change either in land level (**isostatic change**) or in the volume of the sea (**eustatic change**).

Long-term changes

A marine regression results from a eustatic fall in sea level (as during glacial periods, when water becomes locked up in ice and snow) and an isostatic fall in sea level (when ice sheets melt and the land rises). Both movements expose the seabed and produce an emergent coast (see Figure 2.20).

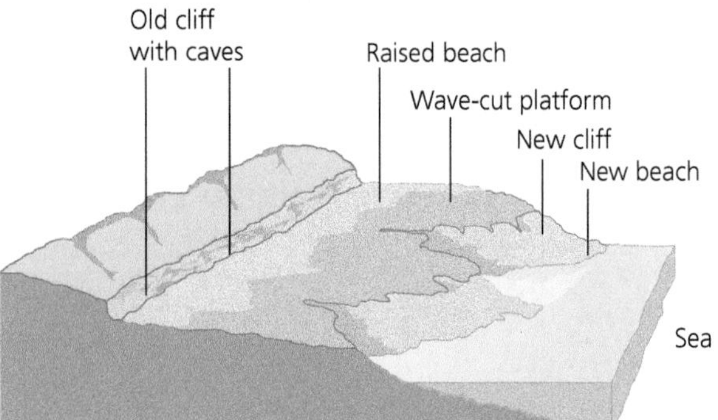

Figure 2.20 Features of an emergent coastline

Emergent and submergent coastlines

A marine transgression results from a eustatic rise in sea level (at the end of a glacial period) and an isostatic rise in sea level (when land sinks under the weight of accumulated snow and ice). In both cases, large areas of land are submerged beneath the sea, producing a submergent coast. The ria coastline is one of the best examples of such a coast (see Figure 2.21).

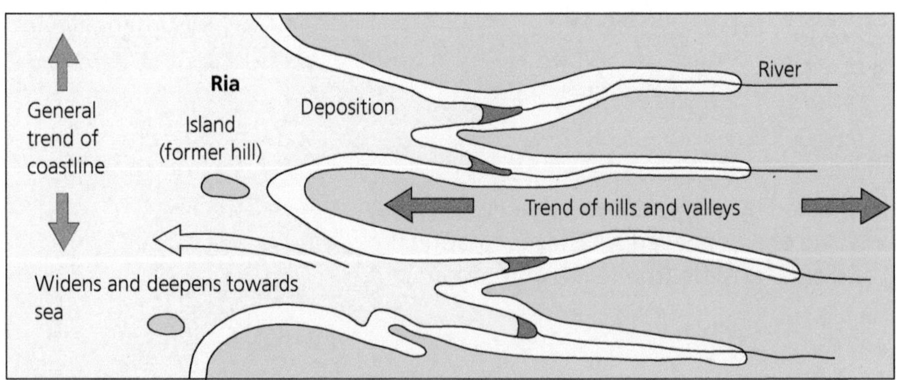

Figure 2.21 A ria or submergent coastline

Now test yourself

TESTED

13 Barrier islands are another characteristic feature of submergent coasts. Check that you understand how they are formed.

Answer on p. 222

Contemporary sea-level change

Sea levels are rising globally and most scientists attribute this to the impact of global warming. The current rate of rise is about 2 mm per year. There are two components:

- thermal expansion of the oceans as they are warmed by the change in climate
- the melting of ice sheets and glaciers increasing the water volume of the oceans.

In some locations, shorelines are being either lifted or dropped by earthquakes. For example, the earthquake responsible for the 2004 Indian Ocean tsunami caused the coastline of part of Sumatra to drop by 1 m, while some offshore islands were raised by up to 2 m.

Coastal recession

REVISED

Rapid recession

Rapid coastal recession threatens people, their property and their livelihoods. It is caused by physical factors, such as:

- long wave fetch and large destructive waves
- strong longshore drift
- soft or unconsolidated geology
- cliffs with structural weaknesses and vulnerable to weathering and mass movement.

It can also be accelerated by human actions, such as:

- dredging the offshore seabed for sand and gravel
- river dams reducing the supply of sediment to the coast (e.g. the impact of the Aswan High Dam on the Nile Delta)
- coastal management (construction of groynes).

Coastal erosion and recession are not constant, even on coasts, such as the Holderness coast of Yorkshire and that of California, which are receding rapidly. Changes in the weather account for much of this variation in the rate of erosion.

Making links

The actions of players may alter the natural systems of the coast, either deliberately (by reclaiming coastal marshes and mudflats) or inadvertently (by constructing groynes to protect a stretch of coastline and so disrupting longshore drift).

Typical mistake

It is wrong to attribute the present rapid erosion of some coastlines to climate change. In most locations, erosion has been going on for centuries, even millennia. Remember that climate change will simply accelerate what is already underway.

Now test yourself TESTED

14 How is it that river dams constructed a long way from the coast can contribute to coastal erosion?

Answer on p. 222

Coastal flooding

REVISED

Factors

Coastal flooding is another significant and increasing risk along some low-lying coastlines. High-risk areas include coastal plains, estuaries and deltas. The risk is being increased by:

- the rising sea level associated with global warming
- human actions, such as the removal of coastal vegetation (e.g. mangroves), the building of coastal tourist resorts and the general pressure of population drawn, for example, to deltas by the fertile soils and farm productivity. Asia's six mega deltas are home to more than 100 million 'at risk' people.

Storm surges, short-term rises in sea level caused by low air pressure, are a particular hazard in low-lying coastal areas and are capable of causing immense damage and destruction. Classic examples include the North Sea storm surges of 1953 and 2013 (see Figure 2.22).

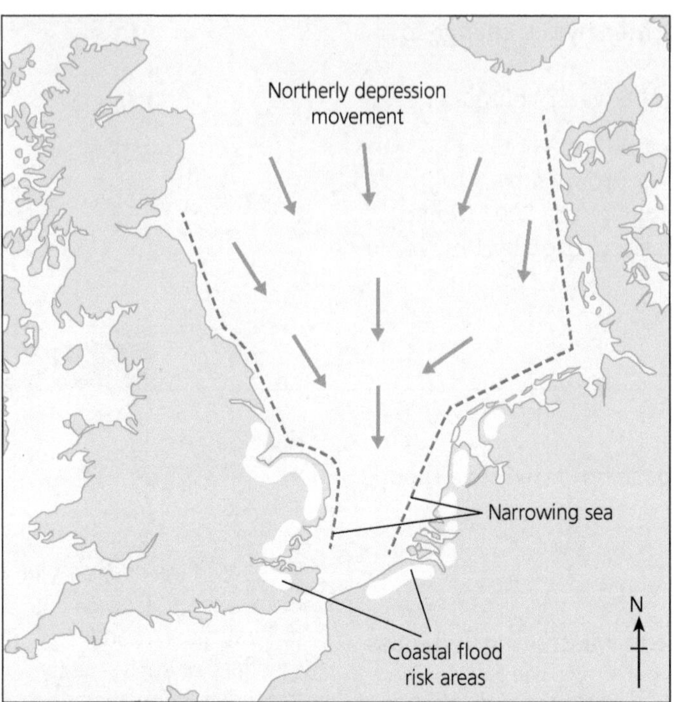

Figure 2.22 North Sea coastal topography and storm surges

Climate change

It is important to remember that storm surges, cyclones and depressions have always been a part of weather in many locations around the world.

No doubt they will continue to be hazards, but the concern is that global warming will intensify the frequency and magnitude of such events.

In short, many believe that coastal flooding will become still more of a threat to human lives and livelihoods.

Revision activity

Be sure you have some examples of islands (e.g. the Maldives, Tuvalu and Vanuatu) and deltaic areas (e.g. Bangladesh) that are particularly exposed to coastal flooding.

Making links

The risks associated with climate warming are resulting in uncertainty. This is creating a need to devise appropriate mitigation and adaptation strategies. Think back to the work you have done on the water and carbon cycle.

Now test yourself

15 What causes storm surges?

Answer on p. 222

TESTED

Coastal management

Risks and their consequences for communities

REVISED

The risks of coastal recession and coastal flooding look set to increase in the near future. These will clearly affect coastal communities. The costs of both changes fall into three broad categories, as seen in Table 2.22.

Table 2.22 The costs of coastal recession and coastal flooding

Economic costs	Social costs	Environmental costs
These include the loss of property in the form of homes and businesses, the loss of transport lines, as well as the loss of farmland and other means of livelihood.	These include the impacts on people, such as the costs of relocation and community disruption, as well as the impacts on health and well-being, levels of stress and hardship.	These include the loss of coastal habitats and ecosystems.

Of course, the higher the population density and the greater the concentration of economic wealth, the greater the consequences and the higher the losses involved.

While in many parts of the world rising sea levels will be managed by building higher flood walls and stronger sea defences, there will be locations where the situation becomes unmanageable.

Most at risk are islands such as the Maldives, Tuvalu and Barbados. Here, land will simply have to be abandoned and left to disappear beneath the sea, creating a growing number of environmental refugees.

The major issue will be one of deciding where they should go to start a new life.

Environmental refugees People and communities forced to abandon their homes due to natural processes. These processes may be sudden, as with landslides or volcanic eruptions, or gradual, such as coastal erosion and rising sea levels.

Now test yourself TESTED

16 Do you agree that the consequences of coastal change for communities are directly proportional to the density of population in the coastal zone?

Answer on p. 222

Different approaches to managing coastal risks

REVISED

Hard engineering

The traditional approach to dealing with the threat of coastal recession or flooding has been the hard-engineering approach (see Table 2.23) — that is, to make extensive use of concrete, stone and steel to provide the required protection.

Table 2.23 The main types of hard-engineering sea defence

Type	Description
Sea walls	Made of reinforced concrete.
Rip-rap (rock armour)	Huge rock boulders piled up at the base of a sea wall.
Rock breakwaters	Usually built of huge boulders, but offshore.
Revetments	Stone, timber or interlocking concrete structures on dune faces and mud banks.
Groynes	Vertical stone or timber fences built at 90 degrees to the coast and spaced along a beach.

Only the first of these, the sea wall, is widely used in the context of combatting coastal flooding. All other approaches are employed to deal with coastal erosion.

The hard-engineering approach has both advantages and disadvantages:

- Advantages:
  - It is obvious to those at risk that something is being done to protect them.
  - It can be a one-off action that protects for decades.
- Disadvantages:
  - Construction and ongoing maintenance costs are high.
  - Even very carefully designed engineering solutions can fail.
  - The hard engineering is not visually attractive.
  - Coastal ecosystems can be badly affected.
  - Defences in one location can have adverse effects further along the coast in the direction of longshore drift.

Soft engineering

Many coastal planners consider soft engineering (see Table 2.24) as a less invasive and more cost-effective approach to managing the coastline by working with the natural processes.

Table 2.24 Types of soft engineering

Type	Description
Beach nourishment	This is achieved by topping up beaches with sediment transported from elsewhere.
Cliff stabilisation	Stabilisation can be achieved by planting vegetation through a tough, flexible membrane that holds soil and often rock in place. It can also be achieved by regrading and reducing cliff slopes.
Dune stabilisation	Dunes are an effective form of coastal defence, but they are easily degraded. They can be stabilised by planting marram grass and by constructing relatively cheap dune fencing.

Exam tip

You should have an awareness of the implications when applying the different types of hard and soft engineering on both the physical and human environment.

Making links

Remember that any human intervention or action in the coastal system is likely to have an impact elsewhere. These impacts may be indirect and may have unforeseen consequences, such as accelerating erosion on other parts of the coast.

Sustainable coastal management

Coastal landscapes are dynamic and constantly changing due to rising sea levels and increased stormy weather, but they are equally desirable locations for different stakeholders.

This means that coastal landscapes require careful consideration on the most effective approach to their management, which has sustainability at the forefront (see Figure 2.23).

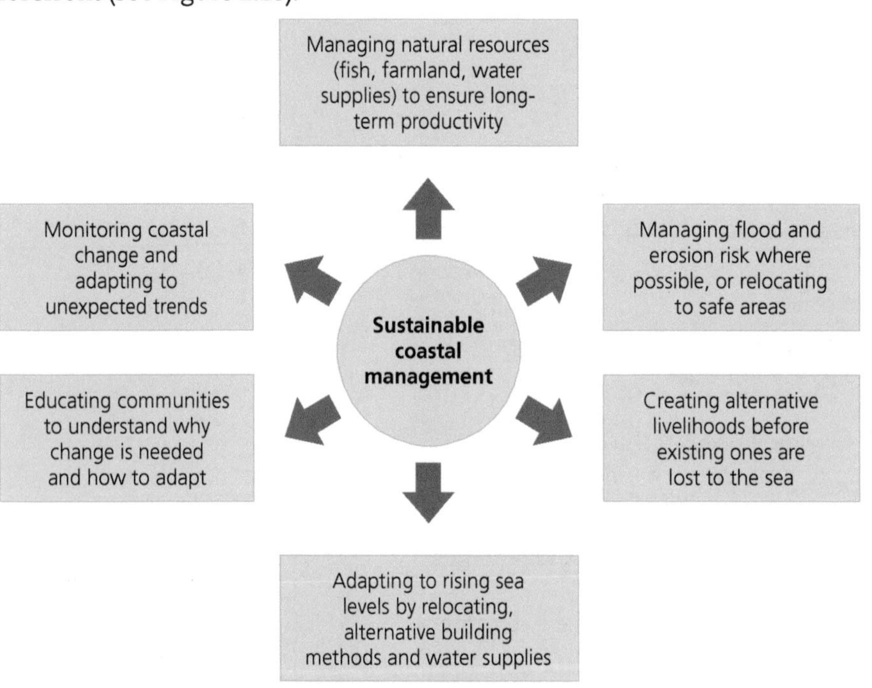

Figure 2.23 Sustainable coastal management

Making links

Coping with increased storm events and rising sea levels calls for imaginative mitigation and adaptation strategies to ensure the future of the present coast. Think back to the work you did when studying the water and carbon cycle on pages 117–147. Some of these strategies could link here.

Revision activity

Be sure you note the six different strands of sustainable coastal management shown in Figure 2.23.

Integrated coastal zone management

REVISED

Key concept

Integrated coastal zone management (ICZM) dates from the Rio Earth Summit in 1992. The concept has three key features. It recognises:

- that the entire coastal zone needs to be managed, not just the zone where the breaking waves are causing erosion or flooding
- the importance of the coastal zone to people's livelihoods and well-being
- the need to make the management of the coast sustainable.

ICZM is a joined-up or holistic approach to coastal management, which must:

- plan for the long term
- involve all stakeholders and ensure they have a say in any policy decisions
- follow an 'adaptive' approach to unforeseen changes
- try to work with natural processes rather than against them.

ICZM works well with the concept of littoral cells (also referred to as sediment cells):

- These are natural subdivisions of the coastline containing sediment sources, transport paths and sinks.
- Each cell is isolated from adjacent cells and can be managed as a holistic unit.
- The coastline can be divided up into littoral cells and each managed as an integrated unit.

The coastline of England and Wales involves 11 cells. Each cell is managed by a shoreline management plan (SMP), which may be subdivided into sub-cells.

The hard part of **ICZM** is the decision-making process to decide what actions to take. In the UK, the critical decision concerns which one of four different management options to follow (see Table 2.25).

Table 2.25 Four different coastline management options

Option	Description
No active intervention	The coast is left to erode or flood.
Hold the line	Coastal defences are built so that the shoreline remains the same over time.
Managed realignment	The coastline is allowed to change but in a controlled way (sometimes known as 'managed retreat').
Advance the line	New coastal defences are built on the seaward side of the existing coastlines. This usually involves land reclamation.

The choice of management option (the decision making) is generally not straightforward and depends on a number of factors, including:

- the economic value of the assets that might be protected
- the technical feasibility of different engineering solutions
- the environmental sensitivity
- the cultural and ecological value of the land that might be protected
- pressure from local communities, developers and environmental groups.

Any decision also needs to be informed by the results of objective investigations, such as cost–benefit analysis (CBA) and environmental impact assessments (EIAs).

The decisions made often lead to opposition and generate conflicts of interest among different stakeholders.

When decisions are made these will inevitably create winners and losers.

Coastal planners need to weigh up the options and make a decision that will have the greatest potential impact on protecting the coastline.

The degree to which coastal management is implemented varies around the world depending on a country's wealth. In many developed countries there are greater opportunities to implement coastal management strategies compared to developing countries.

Revision activity

Create two revision cards for the concept of sustainable coastal management.

Exam tip

The implementation of coastal management leads to disagreements among different stakeholders. Make sure you can discuss why these differences exist for specific locations.

Making links

Unfortunately, not all players in coastal environments share the same attitudes. What they seek ranges from outright exploitation of the coast to strict conservation.

Revision activity

In order to be prepared for your exam you should have an understanding of two examples of coastal management. One of these should be in a developing country and the other in the UK.

When creating your notes you should consider including the following:

1 The nature of the coastal risk — recession or flooding
2 Identifying what is at risk — number of people, economic activities, etc.
3 Management actions, if any to date
4 Identifying or anticipating the main winners and losers

Now test yourself

TESTED 

17 What are sediment cells and why is it important that coastal management recognises the existence of such cells?

Answer on p. 222

Exam skills

You should be familiar with the following skills and techniques used in the geographical investigation of coasts:

- interpreting GIS maps, satellite images and aerial photos
- field sketching
- field measurements — erosion, beach characteristics, coastal vegetation
- measures of central tendency
- Student's *t*-test
- index of diversity
- cost–benefit analysis
- environmental impact assessment.

Exam practice

A-level

1 a) Study Figure 1.

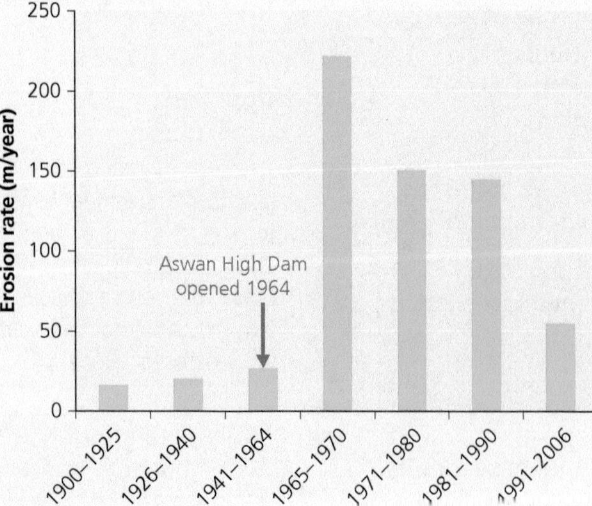

Figure 1 Erosion rates at Rosetta in the Nile Delta, Egypt, 1900–2006

i) Explain the impact of the construction of the Aswan High Dam on the erosion rate at Rosetta in the Nile Delta. (6)

ii) Explain the importance of coastal vegetation. (6)

b) Explain how waves affect beach morphology. (8)

c) Assess the risks and threats posed by the current rise in sea level. (20)

Answers and quick quiz 2B online

Summary

You should now have an understanding of:

- the physical and human importance of the coastal zone
- the physical variety of coastlines and the significance of geology
- the role of coastal vegetation
- the impacts of waves on the coast
- coastal erosion and its landforms
- coastal deposition and its landforms
- sub-aerial processes at work on cliffs
- coastal risks and threats — sea-level changes, storm surges, recession and flooding
- the human consequences of coastal recession and flooding
- hard, soft and sustainable management of the coast
- integrated coastal zone management (ICZM)
- place context studies of sample stretches of coastline in the UK and in the developing world.

3 Globalisation

Globalisation is the process that causes the increased interconnectedness of people and places across the world.

The causes and acceleration of globalisation

The acceleration of globalisation

REVISED 

Key concept

The acceleration of **globalisation** in recent decades can be attributed to four key factors: economic, social, political and cultural (see Table 3.1).

Table 3.1 Four key factors driving the acceleration of globalisation

Economic	Social	Political	Cultural
Driven by the growing influence of **transnational corporations (TNCs)** and advancements in research and ICT.	The rise in the association of people through international migration.	The growing interconnectedness of governments through trade blocs and free trade agreements.	The spread of Western cultures caused by the diffusion of commodities and ideas, for example **glocalisation**.

Transnational corporations (TNCs) Businesses/companies whose operations are spread across the world, and which operate in many nations as both makers and sellers of goods and services.

Glocalisation The changing of the design of products to meet local tastes or laws.

Global connections, flows and interdependence

Modern **globalisation** differs from that which preceded it:

- when the sourcing of goods is further away lengthening the connection between people and places
- when the connection to people and places is deepened creating a greater sense of being connected
- when the connection between people or the distances between places are faster through new technologies.

Globalisation involves building up networks of places and their populations. The connections between places represent different kinds of network flow. These flows are movements of:

- capital: the world's stock markets moving capital (money) on a daily basis around the world
- commodities: the exchange of raw materials through trading between nations
- information: the increase in real-time communication between distant places through the development of the internet
- tourism: places now seem 'closer' through the development of budget airlines creating more opportunities for people to travel to distant places.

Rapid developments in transport and trade

Since the nineteenth century, the world has started to 'shrink' owing to the improvements in transport and accelerated further during the twentieth century from containerisation and developments in air transport (see Figure 3.1).

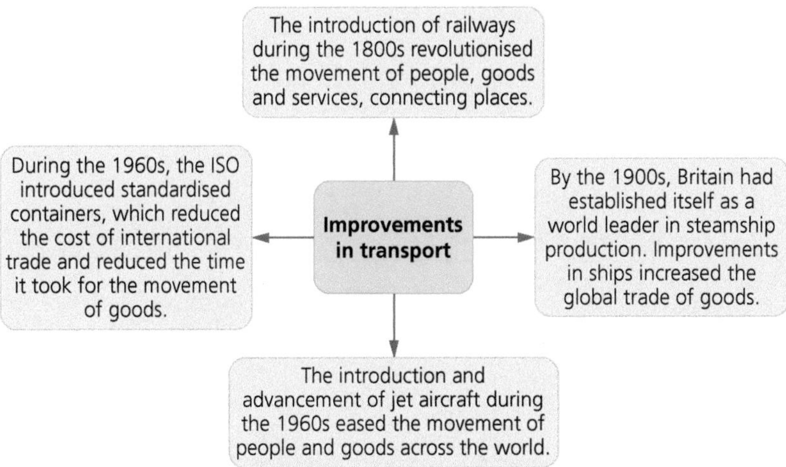

Figure 3.1 Important innovations in transport

ICT and global communications

The advancement and continued development of ICT and global communications is contributing to time-space compression. The series of graphs in Figure 3.2 illustrates the significant rise in flows of information, goods, capital and people.

Flows of information: Individuals using the Internet; Mobile cellular subscriptions; Monthly active Facebook users worldwide

Flows of goods: Merchandise exports; Air transport, freight; Rice imports by the European Union

Flows of capital: Personal remittances, received; Foreign direct investment, net outflows; Net official development assistance received

Flows of people: Air transport, passengers carried; International tourism, arrivals; International migrants, total

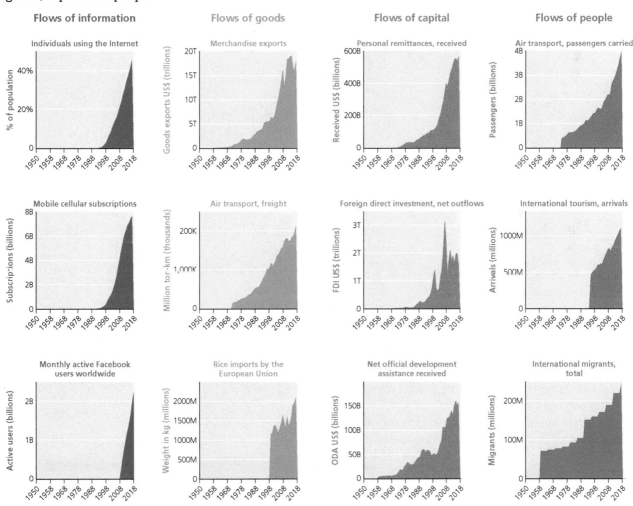

Figure 3.2 The rise in flows of information, goods, capital and people

Exam tip

The role of globalisation in shaping our world is continually changing. Make sure you keep your geographical knowledge up-to-date.

Now test yourself

TESTED

1 Study Figure 3.2. Analyse how flows of information have changed over time.

Answer on p. 222

Decision making: politics and economics

REVISED

The work of international organisations

Making links

Three intergovernmental organisations (IGOs) are major players in the promotion of the global economy (see Figure 3.3):

- the International Monetary Fund (IMF)
- the World Bank
- the World Trade Organization (WTO).

Other important players are TNCs and national governments.

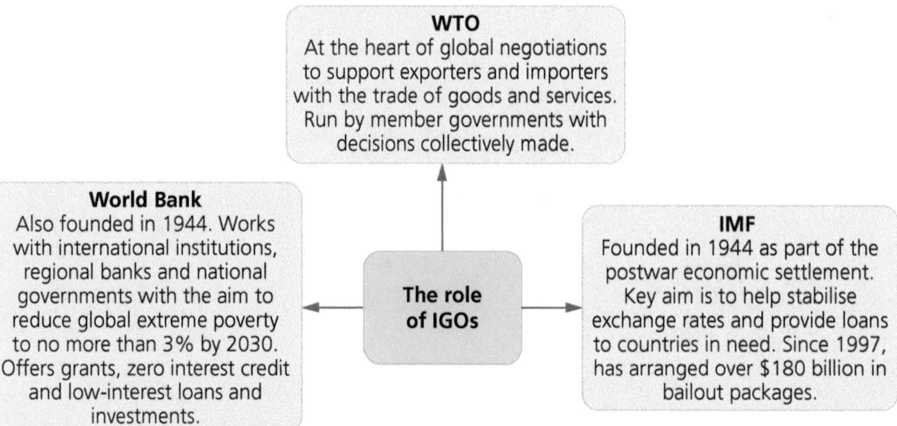

Figure 3.3 The role of IGOs

Now test yourself

TESTED

2 Outline the role of the World Bank.

Answer on p. 222

Exam tip

You need to understand that the growth and spread of TNCs can be encouraged by the actions of governments.

Key players: national governments

National governments are also key in promoting the growth of the global economy through:

- the promotion of free-trade blocs: for example, the European Union (EU) and the Association of Southeast Asian Nations (ASEAN)
- free-market liberalisation: for example, lifting restrictions on the way companies and banks operate
- the promotion of business: encouraging business start-ups and allowing TNCs to grow in size and influence
- privatisation: allowing companies to take over important national services such as railways and energy supply.

Free-trade blocs Voluntary international agreements that encourage the free flow of goods and capital between member countries.

Acceleration of globalisation into new global regions

Special economic zones (SEZs), government subsidies and changing attitudes to foreign direct investment (FDI) have played important roles in the acceleration of globalisation.

China's 1978 'open door policy' accelerated the process of globalisation through the creation of factories dominated by low wages.

This has spearheaded the 'Made in China' movement, with China establishing branch plants and trade relationships with Chinese-owned factories.

China is now recognised as one of the world's most powerful economies, challenging the USA as a future superpower.

Typical mistake

Don't assume that China's economic growth just happened: the Chinese government carefully planned the growth.

Special economic zone (SEZ) An industrial area, often near a coastline, where favourable conditions are created to attract international TNCs.

Foreign direct investment (FDI) A financial injection made by a business into another country's economy, either to build new facilities (factories or shops) or to acquire, or merge with, a firm already based there.

The effects of globalisation

Uneven globalisation

The influence of globalisation on places is uneven as it is driven by different physical and human factors, including the availability of resources and the degree of FDI.

The KOF Index of Globalisation measures the economic, social and political dimensions of globalisation (see Figure 3.4).

Revision activity

Make sure that you understand why globalisation has accelerated in recent decades.

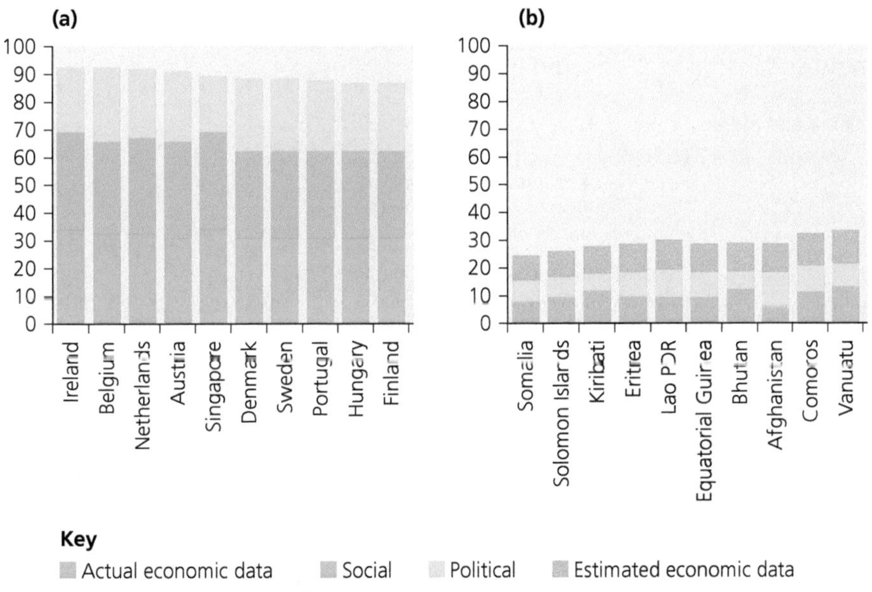

Figure 3.4 KOF Index of Globalisation 2014

TNCs and globalisation

TNCs aim to build their global businesses through the processes of:

- offshoring: relocating factories or offices to other countries to lower overall costs
- outsourcing: contracting alternative companies to produce the required goods and services
- global production networks: establishing chains of connected suppliers of parts and materials for the manufacturing or assembly of the consumer goods.

While the process of FDI brings benefits for places around the world, the degree of this benefit varies. Some places have not benefited to the same degree due to production sites not being suitable and these places not having enough market potential.

TNCs want to maximise their profits and have achieved this by modifying the goods and services they sell to meet local tastes, interests and laws, and by showing respect to religion and culture.

A few of the very poorest nations of the world remain relatively switched off from global networks due to a range of physical, political, economic or environmental reasons. A prime example is North Korea; other examples include countries in the Sahel.

Typical mistake

TNCs might look the same the world over, but on closer investigation many of the products they sell have been adapted for local markets.

Exam tip

Make sure that you can suggest a range of reasons why globalisation has affected some places and organisations more than others and that you can support your ideas with real-world examples.

Now test yourself TESTED

3 List the ways in which TNCs are important players in the process of globalisation.
4 Why are some places 'switched off' from globalisation?

Answers on p. 222-223

The global shift

Winners and losers

REVISED

Globalisation affects countries in different ways, with some benefiting more than others.

Emerging Asia: benefits and costs

You need to understand how the movement of the global economic centre towards the East has created winners and losers in China and India (see Figure 3.5).

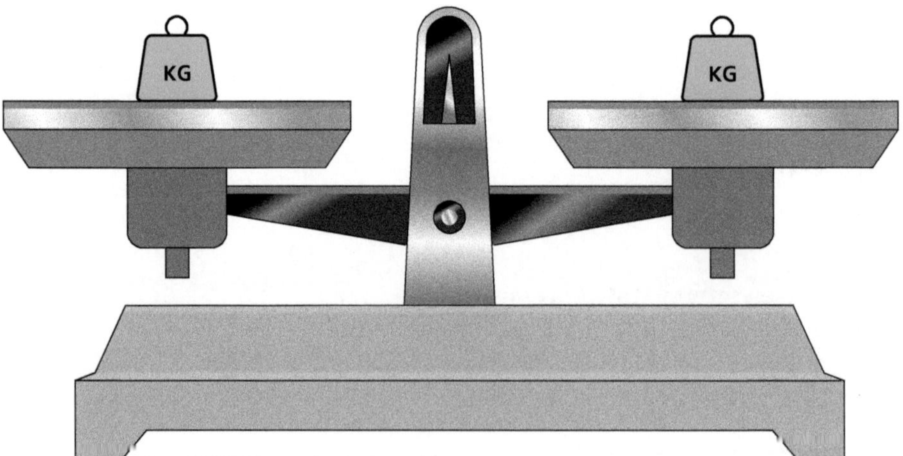

China and India
WINNERS
Created new emerging Chinese markets with businesses looking to expand and grow into all corners of the world.
The standard of living has improved, increasing employment opportunities for its citizens.
Has become a beacon for global tourism, boosting the country's economy.
LOSERS
Rising pollution in the major cities of China and India has led to significant environmental problems.
The standard of working conditions in factories can be low; long hours and poor working conditions are prevalent.
There is rising inequality, with the gap between the poorest and richest widening.
KG
KG

Figure 3.5 Winners and losers in China and India

There is rising inequality with the gap between the poorest and richest widening.

Vulnerable communities

TNCs have contributed towards driving the global shift of manufacturing to vulnerable communities, the key attractions being lower costs and more lenient regulations around health and safety and environmental protection.

Global shift The international relocation of different types of industrial activity, especially manufacturing industries.

Deindustrialisation The decline of regionally important manufacturing industries in terms of either workforce numbers or output and production measures.

Problems of deindustrialisation

The global shift can create social and economic challenges for developed countries.

The deindustrialisation of urban areas has sparked a spiral of decline (see Figure 3.6), which can also be referred to as the negative multiplier effect, resulting in:

+ long-term unemployment rates for the local economy due to the loss of the once booming manufacturing industry
+ an increased feeling of isolation as traditional communities break down due to locals migrating to search for alternative employment
+ a lowering of visual aesthetics due to derelict factories and dilapidation of residential buildings.

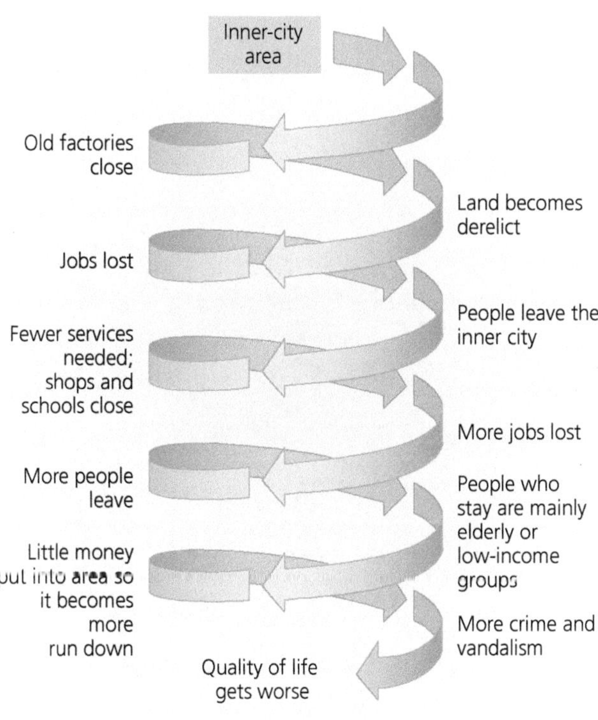

Figure 3.6 The inner-city spiral of decline and deprivation

Revision activity

Create two revision cards that summarise the costs and benefits of globalisation in China and India. Remember, you should know locational detail for these two countries for this part of your A-level course.

Exam tip

You must be able to evaluate a range of costs and benefits caused by globalisation on both people and the environment.

Exam tip

Make sure you understand that global shift has created challenges for some people living in developed countries.

Now test yourself

TESTED

5 How has global shift created winners and losers when it comes to both people and the physical environment?

Answer on p. 223

The scale and pace of economic migration

REVISED

Rural-urban migration and megacity growth

When a city exceeds 10 million inhabitants it is referred to as a **megacity**. These cities achieve this status for two key reasons: rural-urban migration and natural increase. Rural-urban migration is driven by the combination of push and pull factors (see Table 3.2).

Table 3.2 Push and pull factors driving rural-urban migration

Push factors	Pull factors
Fewer opportunities outside of lower-paid employment such as farming	Improved education opportunities
	A wider range of employment opportunities
Pressure on families to survive leading to the younger generation seeking new opportunities	Greater access to doctors and healthcare practitioners
Feeling of isolation	The perception of social cohesion, with more opportunities for entertainment

Figure 3.7 illustrates a selection of current and future megacities.

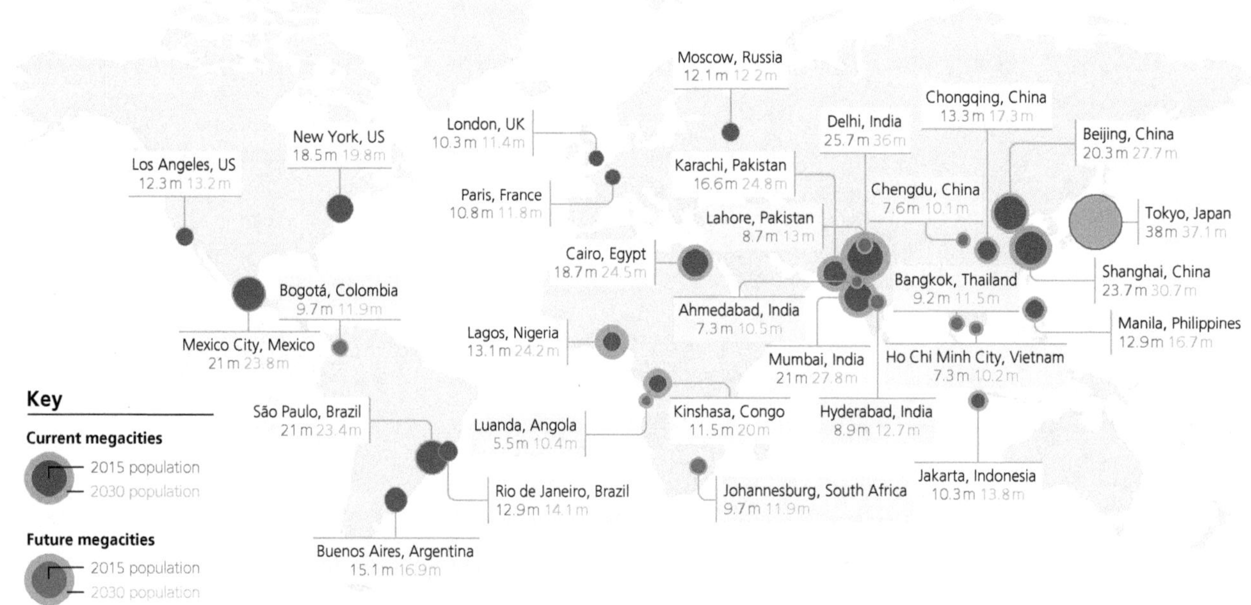

Figure 3.7 Selected current and future megacities, 2015–30

Source: World Urbanization Prospects: The 2014 Revision

Rapid urban growth, particularly in megacities, is creating a range of social and environmental costs, including:

- inadequate provision of housing
- limited access to education and healthcare
- pollution of water and air
- loss of farmland.

Revision activity

Make brief notes about the distribution of megacities shown in Figure 3.7.

Exam tip

Make sure you learn some details about a specific megacity. Mumbai is a good example.

Now test yourself

TESTED

6 Explain how rural-urban migration contributes to the growth of megacities.

Answer on p. 223

International migration into global hubs

Unlike a megacity, a **global hub** is recognised by its influence rather than its population size (see Figure 3.8).

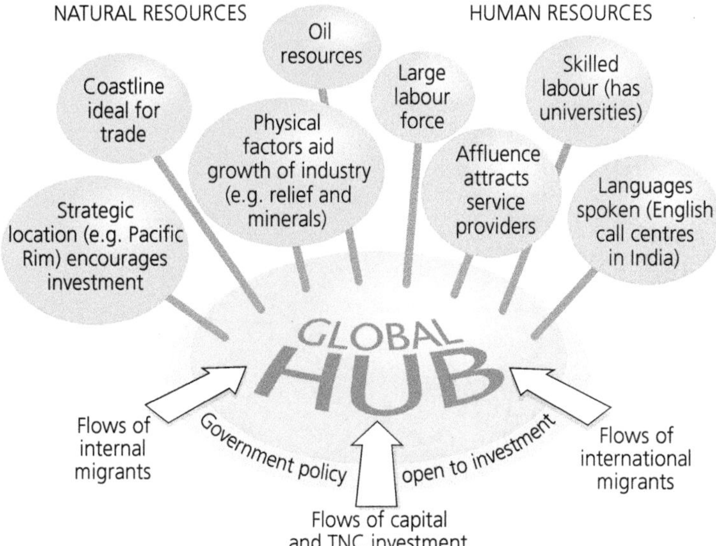

Figure 3.8 Resources involved in the growth of global hubs

Key concept

Global hub A highly connected city, or community, that has drawn global attention and influenced other places through trade, business, governance, research and education.

International migration has contributed towards the growth of **global hubs** in two ways:

- The migration of highly skilled and socially influential people who live as 'global citizens' by having multiple homes in different places around the world, for example the migration of Russian oligarchs to London (see Figure 3.9).
- The migration of both legal and illegal migrants to work in low-waged employment such as on construction sites, as domestic cleaners or factory workers, and in the food industry, for example, the migration of workers from India to the UAE, and from the Philippines to Saudi Arabia.

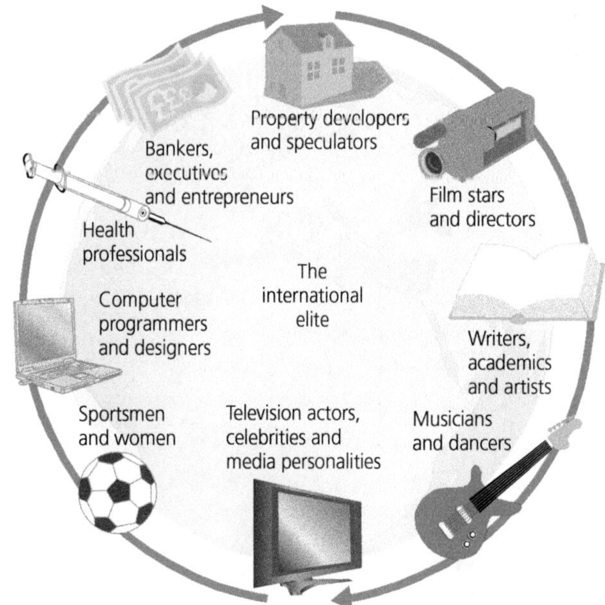

Figure 3.9 Elite migration

Now test yourself TESTED

7 Explain how international migration has increased in global hub cities.

Answer on p. 223

The costs and benefits of migration

International migration increases countries' **interdependence**. It also creates a wide range of costs and benefits of migration, summarised in Figure 3.10.

(a) HOST COUNTRY

BENEFITS

Contributes towards filling the skill gap within the country's labour market.

Migrants contribute towards the host's economy through spending their income on essentials and disposable income on leisure activities.

Migrants can contribute towards providing opportunities for people within the host nation through setting up their own businesses.

COSTS

Pressure on housing as demand rises with the growing number of migrants needing somewhere to live.

Pressure on local health services as migrants may require treatment while living in the host nation.

Tension may develop between migrants and local residents as migrants assimilate to the host nation.

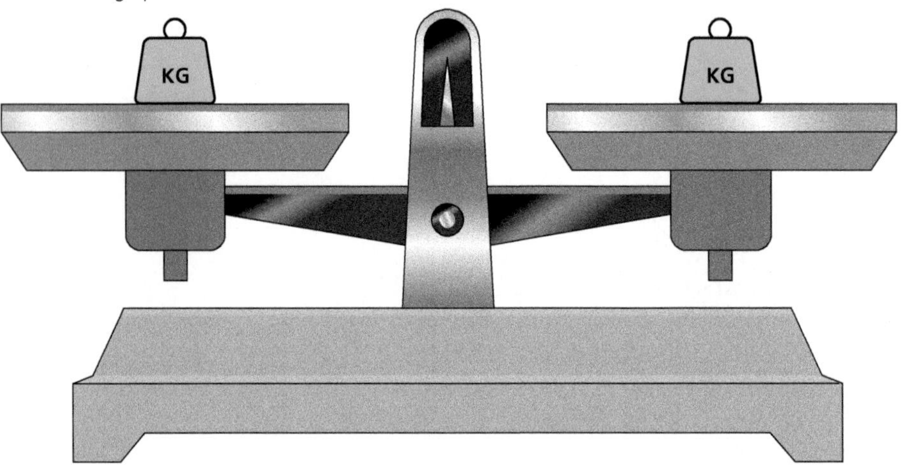

(b) SOURCE COUNTRY

BENEFITS

Migrants send money (remittances) to family, contributing towards the source country economy.

If migrants return to the source country, they bring new skills to fill gaps in the labour market.

Reduces the pressure on key services such as healthcare, with reduced numbers requiring care.

COSTS

Can cause the economy of the source country to shrink with consumer spending reduced.

Potential for key services such as education institutions to close due to a reduction in numbers.

Loss of skilled workers can create a gap in the labour market.

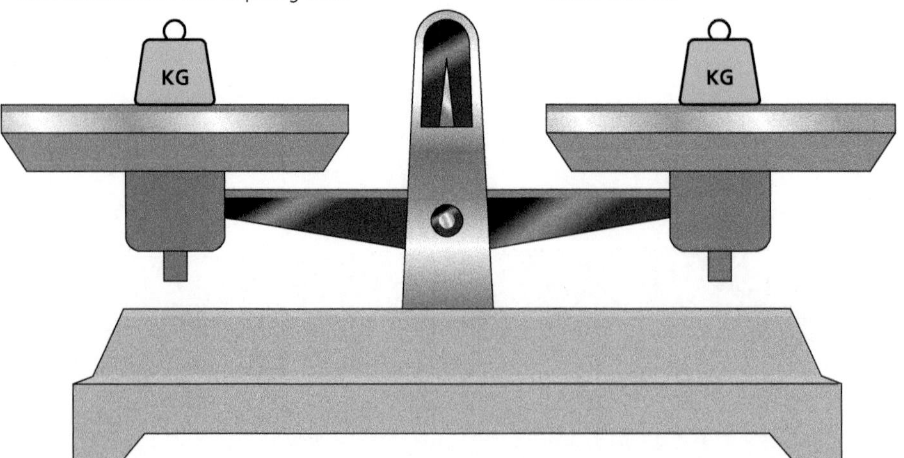

Figure 3.10 Selected costs and benefits of migration for (a) host and (b) source countries

Key concept

Interdependence is very much one of the products of globalisation. It builds up as countries become increasingly reliant on each other in terms of trade as well as financial and political connections. While interdependence offers economic and political security, it does have a downside, namely that a country's continued well-being depends on the economic health of those involved in the same network of interdependence.

Revision activity

Create revision cards for the costs and benefits of migration for both the source and host countries.

Now test yourself

TESTED 

8 Explain some of the advantages and disadvantages for people and the physical environment of the world becoming more interconnected.

Answer on p. 223

The emergence of a global culture

REVISED

Key concept

One of the products of globalisation is the **diffusion of a global culture**. It is being spread by the mass media, the internet and migration. As it spreads, it tends to erode traditional cultures, but occasionally cherry picks, adapts and absorbs select parts of those cultures. Diagnostic features of the emerging global culture are:

- emphasis on consumerism
- a belief in capitalism and personal wealth
- Anglo-Saxon and white
- English (American) the dominant language
- 'American' in terms of tastes and traits, e.g. diet, dress, entertainment, etc.

Making links

The TNCs and global media companies are foremost among the players promoting a global culture. Tourists and migrants can play a part too.

Cultural diffusion The gradual spread of culture from an influential civilisation.

Cultural imperialism Promoting the culture/language of a nation in another nation, which usually occurs with a larger, more powerful nation having influence over a less powerful nation.

Cultural diffusion and its causes

The emergence of a 'Westernised' global culture has led to changes in many traditional cultures (cultural diffusion). Countries like the USA have a significant influence on other places through the use of soft power driven by global media and entertainment (cultural imperialism).

Table 3.3 evaluates the influence of the three different players in the diffusion of a global culture.

Table 3.3 Evaluating the emergence of a global culture

	TNCs	Global media	Migration and tourism
Influence	TNCs play an influential role in shaping a common culture through the dispersal of goods and services. The release of 'global' products, such as the Apple iPhone, are examples of this.	Global media giants such as Disney and Marvel broadcast on a global scale. Marvel's Spiderman, for example, is a globally recognised superhero. Disney's animated films frequently make upwards of $1 billion at the global box office.	Traditional culture change to places is driven by migration and tourism. When people travel for tourism or work they leave behind their cultural footprint on the places they visit.
Evaluation	In many cases, the adoption of products and services by TNCs to meet local market and culture demands can benefit both the company and the place. For example, McDonald's is a distinctive globally recognised brand but sells locally recognisable products which reflect the market and culture its restaurants occupy.	Media provides a glimpse of other cultures through popular TV programmes and period dramas, such as *Downton Abbey*. However, many reality TV programmes are re-filmed to match the national market into which they are broadcast.	Migration and tourism can bring both benefits and costs to host regions and their cultures. The extent to which cultures can change or be influenced varies. Players can perceive the impact to be more or less significant.

The costs of cultural erosion

Different places are experiencing a largely Westernised global culture caused by the erosion of traditional cultures. This is known as **hyperglobalisation**.

Some players, especially those concerned with the loss of traditional cultures and the emerging environmental damage, see this process as a negative change.

In contrast, some players believe that globalisation and cultural erosion has a positive influence, as promoting a global culture prompts freedom of expression, the values of equality and a reduction in discrimination.

Revision activity

Summarise some of the impacts of cultural diffusion and how it has helped create and spread an increasingly 'Westernised' global culture.

Typical mistake

Cultural erosion does not always have to be negative. Some people readily embrace the cultural changes that globalisation can bring.

Revision activity

It is important that you take notes about the impacts of cultural erosion (e.g. loss of language, traditional food, music, clothes, social relations).

Concern about the impacts of globalisation

There is growing opposition to globalisation on a range of geographical scales. Individuals, pressure groups and governments may all have some degree of concern about the cultural impact of globalisation.

Culture can have a significant effect on attitudes towards a variety of issues including resources, heritage and the environment.

Making links

There are strongly polarised views (attitudes) on globalisation, particularly in relation to its environmental and cultural impacts. This is leading to the establishment of both pro- and anti-globalisation protest movements. Strong attitudes against aspects of globalisation exist in various environmental and cultural organisations, such as Greenpeace and UNESCO.

Exam tip

Globalisation is often cited as 'being to blame' for a range of problems — but there are many benefits to be had as well.

Now test yourself TESTED

9 Suggest reasons why some groups of people are concerned about the impacts of globalisation.

Answer on p. 223

The challenges of globalisation

Globalisation, development and the environment

REVISED

- Many still argue that there is a **development gap**, an increasing difference between the rich and poor, which has grown because of globalisation.
- However, the overall global economy has grown and the number of people living in absolute poverty has decreased significantly since 1800.

+ While there is a gap between the world's poorest and richest people, most people now live on a middle income, as outlined by the work conducted by Gapminder (see Figure 3.11).

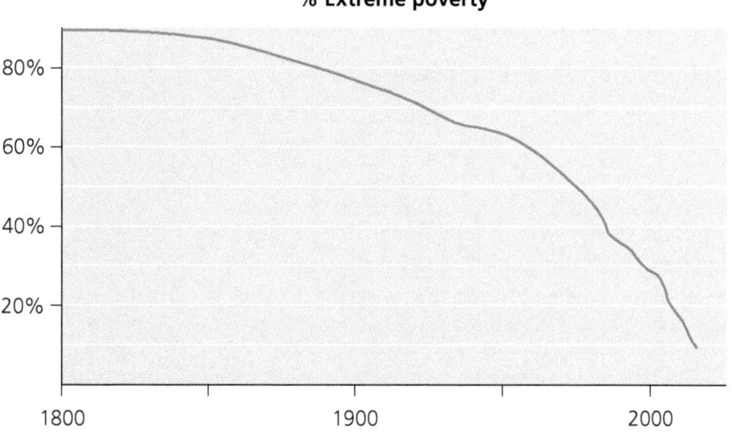

Figure 3.11 The percentage of people in extreme poverty 1800–2010
Source: Gapminder

Now test yourself

10 Study Figure 3.11, which outlines the change in the percentage of people living in extreme poverty between 1800 to 2010. Suggest reasons for the decline in the percentage of people living in extreme poverty.

Answer on p. 223

Economic and social development measures

There are four key economic sectors of employment — primary, secondary, tertiary and quaternary.

As countries develop we see a shifting transition from greater proportions of people working in the primary/secondary sector to greater proportions in the tertiary/quaternary sectors.

We can measure a country's level of development by using both single and composite measures, as shown in Table 3.4.

Table 3.4 Single and composite ways to measure levels of development

Composite measure	Type	What it measures
Single	Gross national income (GNI) per capita	Measurement of a country's income calculated by including all earnings from businesses, residents and international sources. The total is then divided by the population size to provide an average.
	Gross domestic product (GDP) per capita	Calculates all the goods and services produced within a country. The total is then divided by the population to provide a per capita figure.
Composite	Human Development Index (HDI)	Considers several development indicators — life expectancy, GNI and average years in schooling. The formula creates a score from 0–1 with countries closest to 1 indicating a higher HDI.
	Happy Plant Index (HPI)	Combines four elements — well-being, life expectancy, inequality of outcomes and ecological footprint — to show how efficiently residents of different countries use resources to live long, happy lives.
	The Gender Inequality Index (GII)	An example of an inequality index measured by combining three important factors in human development — health, empowerment and the labour market.

Widening inequalities

In recent decades, the widening global economy has contributed towards changing the spatial pattern of global wealth in the following ways:

+ The percentage of people living in extreme poverty has decreased.
+ The wealth in Africa has risen across the continent.
+ There is an increasing wealth divide between countries and within countries.
+ The gap between the world's richest and poorest people is widening.

- Most people now live on a middle income.
- Since the 1970s, many low-income countries (LICs) have developed into medium-income countries (MICs).

Analyses using the **Gini coefficient** show only a rough correlation between a country's level of development and the degree of income equality.

The coefficient can also be used to assess another important dimension, namely inequalities within individual countries.

Key concept

The **Gini coefficient** is a measure of income inequality given as a number between 0 and 100. The higher the value, the greater the degree of income inequality. A value of 0 suggests that everyone has the same income whereas a value of 100 would mean a single individual receives all of a country's income.

Globalisation can also cause serious environmental issues:

- Large-scale agribusiness operations can have long-lasting impacts on biodiversity in some of the world's poorest regions.
- The demand for food due to a rapidly growing global population has resulted in large-scale clearance of forests for agriculture. This has reduced local biodiversity and continues to have a significant impact on climate change.
- The increase in intensive farming practices such as cattle ranching and cash cropping has had detrimental impacts on groundwater supplies and caused the removal of ecologically important mangrove forests.

Differential progress

- Since the 1800s, growth in global GDP has increased, however the rate of growth has fluctuated (see Figure 3.12).
- The USA and Western Europe have experienced long-term growth.
- More recently, Asia and Latin America have witnessed rapid economic growth (see Figure 3.13).
- Developing and emerging countries have borne the greatest brunt of the environmental issues associated with economic growth.
- On the other hand, developed countries have now become increasingly aware of the implication of economic growth and have explored alternative approaches to reduce the environmental risks.

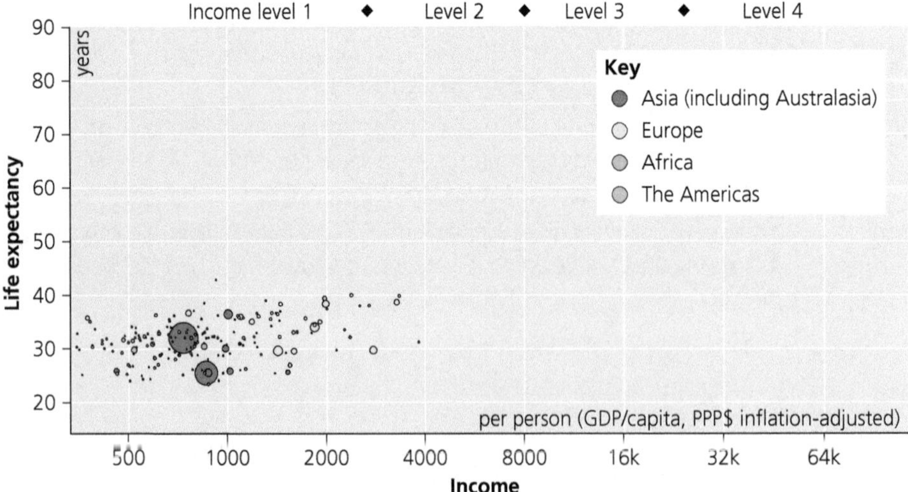

Figure 3.12 Comparing GDP per capita and life expectancy, 1800

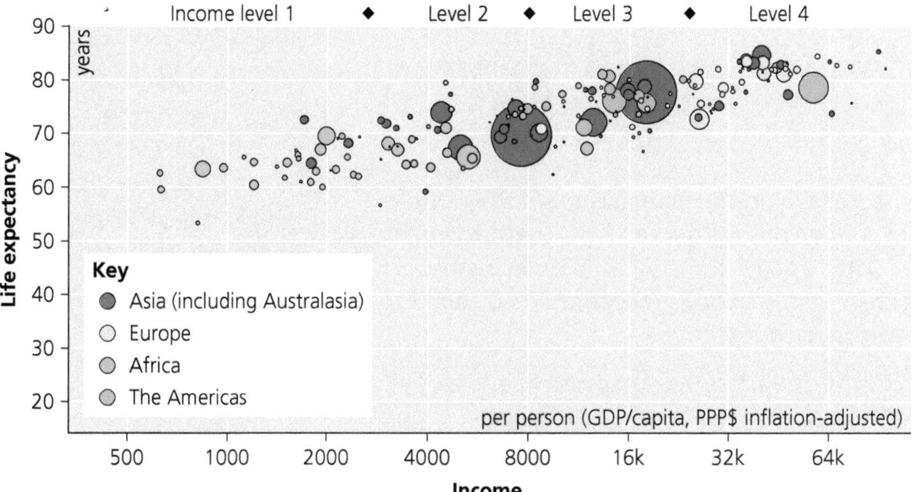

Figure 3.13 Comparing GDP per capita and life expectancy, 2020

Table 3.5 highlights several of the countries with the highest Gini coefficients.

Table 3.5 Several countries with the highest Gini coefficients

Rank	Country	Score
1	Lesotho	0.632
2	South Africa	0.625
3	Haiti	0.608
4	Botswana	0.605
5	Namibia	0.597
6	Zambia	0.575
7	Comoros	0.559
8	Hong Kong	0.539
9	Guatemala	0.530
10	Paraguay	0.517

Source: data from https://worldpopulationreview.com/country-rankings/gini-coefficient-by-country

Now test yourself

11 Study Table 3.5. Calculate the average Gini coefficient for these countries.

Answer on p. 223

TESTED

Social, political and environmental tensions of globalisation

REVISED

The scale and speed at which globalisation has spread has generated a range of tensions — social, political and environmental.

Cultural mixing

Large volumes of international migration are a feature of globalisation, encouraged by open borders, deregulation and FDI. Together with the promotion of a global culture, they have created culturally mixed societies. They have also given rise to thriving diasporas throughout the world.

Diaspora The distribution of people from their original country around the world.

Tensions can easily develop between immigrant groups and host communities, particularly if there are differences in ethnicity:

- Tensions fester over fears that immigrants are taking jobs and homes, and overloading health and social services. This leads to calls to stem the flow of immigration.
- Immigrants feel that they are being discriminated against, which adds to the tension.

In some EU states, nationalist parties have been gaining support because they often oppose immigration; some reject multiculturalism and openly embrace fascism.

Exam tip

Migration can be a rather sensitive issue, as people have different opinions. Be careful to balance different attitudes from a geographical viewpoint.

Controlling the spread of globalisation

While governments are generally positive about the spread of globalisation, there are negative aspects, with some governments exerting greater control over the flow of people, goods and information. This can occur by governments:

- exerting control over citizens' access to online data, although they find it more difficult to control the exchange of information over the 'dark web'
- changing laws to control the quota of economic migrants entering the country, although illegal migration can still occur because it is difficult to keep track of the flow of people across borders.

Internet censorship still exists in China as part of the government's strategy to stifle criticism. The 'Great Firewall of China' means social media giants such as Facebook, Twitter and YouTube are banned in the country, but this hasn't stopped citizens finding ways to access these media sites.

Making links

One of the issues raised by globalisation is immigration. Attitudes towards immigration are becoming increasingly polarised, for example in the UK.

Now test yourself TESTED

13 List the ways in which some countries have attempted to control the spread of globalisation.

Answer on p. 223

Now test yourself

12 Explain how globalisation has contributed to culturally mixed societies and how this can sometimes lead to tensions.

Answer on p. 223

TESTED

Economic migrant A person who moves from one country to another in order to find work and improve their standard of living and quality of life.

Resource nationalism A growing tendency for state governments to take measures ensuring that domestic industries and consumers have priority access to the national resources found within their borders.

Resource nationalism and protecting cultures

Some countries have adopted resource nationalism as a response to globalisation. For example:

- the Venezuelan government took control of ExxonMobil and ConocoPhillips oil operations in their country
- First Quantum, a Canada-based company, was forced to give 65 per cent ownership of a US$550 million copper mining project located in the Democratic Republic of Congo to the government.

Traditional indigenous groups such as those in Canada can be resistant to global forces exploiting their resources, with the fear that important landscapes will be threatened by the extraction of resources.

Typical mistake

Despite some internet restrictions, many people in China enjoy a wide range of freedoms and some citizens argue that control of the internet is good governance.

Revision activity

Draw up a table to summarise some of the social, political and environmental tensions that have resulted from globalisation.

Globalisation, sustainability and localism

REVISED

While the process of globalisation has brought many benefits, it has also led to increased stress on the ability of our planet to keep up with the demand for food, water and energy.

The 'local sourcing' solution

In order to relieve the stress humans place on the planet citizens are actively adopting an ethical consumption approach to meet their needs:

- People are increasingly purchasing food and other key commodities from local butchers, farm shops, farmers' markets, etc., rather than shopping at the larger supermarkets like Asda and Morrisons.
- Alongside the reduction in the purchasing of pre-packed food from bigger brands, people are increasingly growing their own produce, which is contributing towards a reduction in food miles.

Ethical consumption Occurs when the consumer takes into account the costs (economic, social and environmental) of producing food and goods, and of providing services.

Table 3.6 summarises a range of costs and benefits to sourcing food and other commodities locally.

Table 3.6 The costs and benefits of local sourcing

	Costs	Benefits
Consumers	The sourcing of meat and vegetables can be more expensive than visiting the major supermarkets, which can be a struggle for people on lower incomes.	Sourcing from local producers is appealing due to local ethical considerations and a reduction in the use of pesticides.
Producers	The reduction in demand for produce can impact on the economic development of producer countries.	The rising demand for local produce from UK farmers has helped to increase the manufacturing of jams, fruit juices and wine.
Environment	The production of locally grown produce such as tomatoes requires them to be grown in heated greenhouses and polytunnels during the winter, which increases the UK's carbon footprint.	Global campaigns such as 'Think global, act local' have helped to reduce individual carbon footprints.

Now test yourself TESTED

14 How can local sourcing reduce some of the negative environmental impacts associated with globalisation?

Answer on p. 223

Making links

Local pressure groups can do much through a range of actions to promote localism and ethical consumption.

Ethical consumption and fair trade

- Globalisation has led to a shift towards the production and consumption of cheap goods.
- For many citizens, the benefits of consuming cheap goods are overshadowed by the rising ethical issues associated with the exploitation of workers at the expense of saving money.
- This has led to an increased awareness among NGOs and charities of the need to raise these concerns and provide consumers with the option of buying goods that leave an ethical footprint.

Ethical consumption schemes aim to redress the balance (see Table 3.7).

Table 3.7 Evaluating ethical consumption schemes

	Examples of actions	Evaluation
Fair trade	The Fairtrade Foundation's certification scheme offers a guaranteed higher income to farmers and some manufacturers, even if the market price fluctuates. Today, a wide variety of Fairtrade produce is available, including bananas, wine, clothes, tea, coffee and chocolate. The Waitrose Foundation has increased the pay farmers receive during the production process by adopting fairer trading principles.	The Fairtrade label allows consumers to make an informed decision about the products they purchase. However, as the number of schemes grows, it can be difficult to ensure that the money is being distributed fairly. Not all consumers are willing to pay more for goods.
Supply chain monitoring	Large businesses increasingly accept the need to adopt corporate social responsibility models. The largest TNCs have thousands of suppliers. This increases the risk of branded products being linked with worker exploitation. For example, in 2019 it was revealed that factory workers in Bangladesh were being paid 35p an hour producing T-shirts for a Spice Girls campaign linked to raising money for Comic Relief.	Many global brands such as Gap and Nike have put in place new measures and monitoring operations to prevent worker exploitation in their overseas factories and ensure the protection of basic human rights for all workers. However, it can be difficult to monitor every single supplier and control what happens in their workplaces. Many TNCs have large and complicated supply chains, which can make tracking the true life of their products difficult.

	Examples of actions	Evaluation
NGO action	The charity War on Want supported South African fruit pickers. It flew a female worker, Gertruida, to a Tesco shareholder meeting to share her story. She informed the shareholders that the conditions for fruit pickers were awful and there was no toilet for females at the farm where she worked. Tesco agreed to use a different fruit supplier unless conditions improved.	NGOs have limited financial resources. This can inhibit the scale of what they can achieve, or result in slow progress. While NGOs such as Amnesty International work hard to raise awareness of ethical issues, many people remain unaware of, or unconcerned about, worker exploitation, meaning the problems persist.

Recycling and resource consumption

- When manufactured goods reach the end of their life, they are usually sent to landfill. The rise in global consumption of goods has put pressure on the availability of sites to store this waste.
- An alternative is to recycle these goods. This can reduce the production and sourcing of new natural resources. The recycling process does itself, however, require the use of energy and water.
- Recycling is the first step towards the larger goal of a circular economy (see Figure 3.14), as human try to find ways to reduce our ecological footprint.

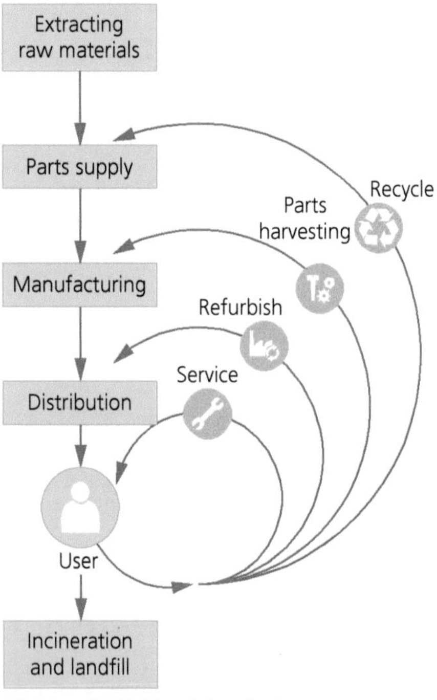

Figure 3.14 The 'circular' economy

Revision activity

Summarise how ethical and environmental concerns about unsustainability have led to increased localism and awareness of the impacts of a consumer society.

Making links

There is uncertainty about the exact environmental consequences of different patterns of consumption.

Now test yourself

TESTED

15 What are the benefits of recycling? Is there a downside?

Answer on p. 223

Exam skills

You should be familiar with the following skills and techniques used in the geographical investigation of globalisation:

- using proportional flow lines to show networks of flows
- ranking and scaling data to create indices
- analysing human and physical features on maps to understand a lack of connectedness
- using population, deprivation and land-use data sets to quantify the impacts of deindustrialisation
- using proportional flow arrows to show global movement of migrants from source to host areas
- analysing global TNC and brand-value data sets to quantify the influence of Western brands
- critically using World Bank and United Nations (UN) data sets to analyse trends in human and economic development, including the use of line graphs, bar charts and trend lines
- plotting Lorenz curves and calculating the Gini coefficient.

Exam practice

A-level

1 Explain how the spread of a global culture impacts on people. (4)

2 Explain one reason why ethical consumption may reduce environmental degradation. (4)

3 Assess the extent to which globalisation is widening the development gap. (12)

4 Assess the extent to which migration has brought more costs than benefits for both host locations. (12)

Answers and quick quiz 3 online

Summary

You should now have an understanding of:

- globalisation, a long-standing process that has accelerated due to rapid developments in transport, communications and businesses
- political and economic decision making, important factors in the acceleration of globalisation
- how globalisation has affected some places and organisations more than others
- the way in which the global shift has created winners and losers for people and the physical environment
- the scale and pace of economic migration, which have increased as the world has become more interconnected, creating consequences for people and the physical environment
- the emergence of a global culture, based on Western ideas, consumption and attitudes towards the physical environment — an outcome of globalisation
- how globalisation has led to dramatic increases in development for some countries, but also widening development gap extremities and disparities in environmental quality
- social, political and environmental tensions, which have resulted from the rapidity of global change caused by globalisation
- ethical and environmental concerns about unsustainability, which have led to increased localism and awareness of the impacts of a consumer society.

4 Shaping places

4A Regenerating places

Local **places** vary both socially and economically, and change can affect places, which can exacerbate inequalities in an area.

Regeneration is aimed at improving a place and hopefully helping to change the perception of an area and improve people's quality of life.

Key concepts

Place is an area which is distinct due to a number of different factors, including its physical and human landscape and economic history. Local inhabitants often feel a connection to a place because of these different characteristics.

Regeneration involves positively transforming the economy of a place that has displayed symptoms of decline, making it viable and sustainable. It frequently goes hand in glove with rebranding (changing people's perceptions of a place in order to promote it) and reimagining (positively changing the standing and reputation of a place through specific improvements). Rebranding is when a place is purposely reinvented for economic reasons, and then marketed with its new identity in order to attract new visitors.

Now test yourself

1 What is regeneration and why is it such an important issue in the UK?

Answer on p. 223

TESTED

How and why places vary

- Place identity is strongly influenced by the nature of the place and what its residents do for a living.
- In particular, places are shaped by internal and external connections.
- The nature of a place also affects the type of work on offer and that work, in turn, affects the type of employee. Compare, for example, the labour needs of a coastal resort, such as Newquay, with those of an industrial town, such as Sheffield.

Connection Any type of physical, social or online link between places. Places may keep some of their characteristics or change them as a result.

Economies vary from place to place

REVISED

Classifying economies and workers

A place is strongly influenced by its local economy and this in turn causes a variation in the socio-economic makeup of an area, for example in health, life expectancy and levels of education.

There are four key employment sectors:

- Primary: producing food crops and raw materials, e.g. farming, mining.
- Secondary: manufacturing finished products.
- Tertiary: providing services. This can be in the private sector, e.g. retail, or public sector, e.g. education and healthcare.
- Quaternary: providing specialist services in finance and law, or industries such as IT.

The emergence of the quaternary sector, also known as the knowledge economy, has been relatively recent (since around 1975).

The UK has seen employment declines in the primary industries and secondary sectors linked to deindustrialisation and the global shift of manufacturing. The tertiary sector now totally dominates the economy. These sectoral shifts are mirrored in the economies of most places.

Deindustrialisation The process of economic and social change due to a reduction in the industrial capacity or activity of a country, region or city. This process is widely experienced in the developed world with the global shift of manufacturing to emerging economies.

Skills activity

Study Figure 4.1.

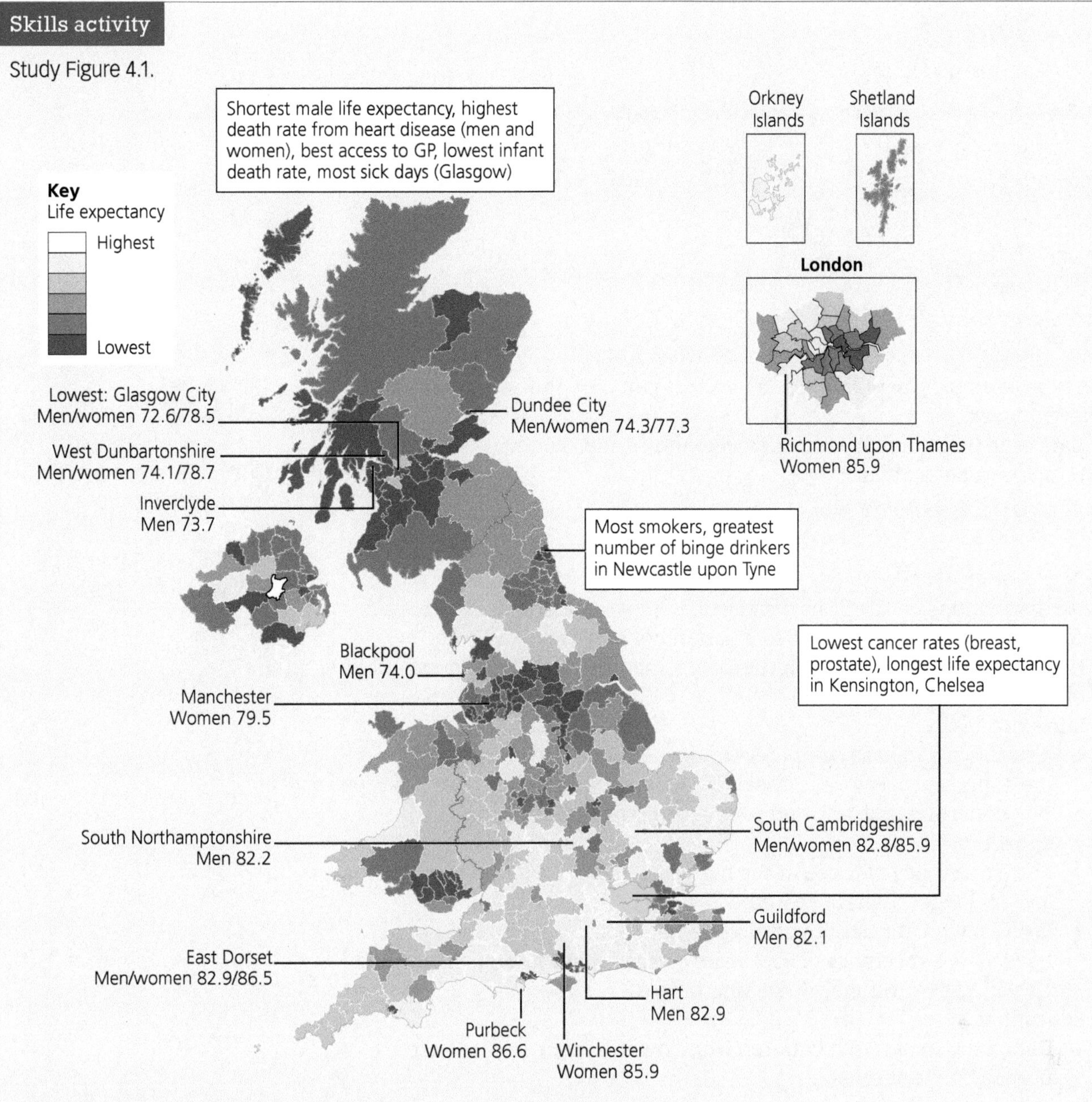

Figure 4.1 Health and life expectancy patterns in the UK, 2016

Describe and suggest reasons for the variations in health and life expectancy patterns between places.

Tips:

+ Where are the highest and lowest levels?
+ Use compass directions when describing areas.
+ Try to come up with two to three geographical reasons as to why this pattern exists.

When a question asks you to 'suggest reasons', you are not expected to know every area in the UK, but you need to use your geographical understanding to help come up with reasons.

Now test yourself

TESTED

2 How has sectoral employment in the UK changed over the past 150 years?

3 How have employment changes affected your local place?

Answers on p. 223

Revision activity

Complete Table 4.1 for the UK.

Table 4.1 Advantages and disadvantages of the decline and growth of UK sectors

	Advantages	Disadvantages
Decline of primary and secondary industry		
Growth of tertiary and quaternary industry		

The type of employment found in all sectors can be classified in several different ways:

- part-time (less than 35 hours per week)/full-time (35 hours per week)
- temporary/permanent
- employed/self-employed.

Exam tip

Make sure you know the different employment sectors and the types of worker, and how the sectoral balance of the UK's economy has changed.

Social impacts

These distinctions on the basis of type of employment and worker have profound impacts on people's lives in a number of ways:

- Health: in general, those with the lowest income have the poorest quality of health.
- Life expectancy:
  - A national database compares the correlations between job type and life expectancy. People in the database are classified by their job type.
  - The data show a clear trend between occupation type and life expectancy.
  - Life expectancy of those in the highest occupation group is on average 5 years longer than those in the lowest occupation group.
  - Women who are classified as higher managerial and professional have a life expectancy of 85.2 years compared with women who are classified as a routine worker and have a life expectancy of 78.5 years.
- Education:
  - Data have shown that children from lower-income families often underachieve at school.
  - This can lead to fewer of them going into further education, which results in them moving into lower-income jobs.
- Lifestyles:
  - Higher salaries lead to more disposable income, but this does not always lead to a happier life.
  - Workers in South East England and London on average have higher salaries, but there is a higher cost of living and housing is much more expensive.

Income inequalities

Key concepts

Inequality is the outcome of uneven distributions. This topic focuses on the uneven economic and social distributions that exist within societies and communities — for example, the uneven distributions of income and wealth, and of quality of life and social opportunities. The factors responsible for such differences are complex and resistant to remedy. One of the main aims behind much regeneration is to reduce such inequalities.

- **Inequalities** in pay levels are linked to differences in the type of employment. Some types of work (in the professions, for example) are more highly paid than others (such as manual work).
- There are huge disparities in incomes and costs of living, both nationally and locally. This has always been the case, but the view is that these inequalities are increasing.
- Quality of life correlates closely with wage and salary levels. This reflects the fact that many of the things that contribute to the overall quality of life are goods (e.g. housing and household equipment) and services (transport and leisure) that have to be paid for.

Quality of life The level of social and economic well-being experienced by individuals and communities, measured by various indicators such as health, longevity, happiness and educational achievement (see Table 4.2).

Table 4.2 Factors affecting quality of life and inequality and ways to measure them

Factors and processes	Possible measures
Economic inequality Employment opportunities, income and type of work	Use of the DataShine Census to understand: + employment/unemployment rates and types + average incomes Purchasing power (shopping basket surveys) The index of multiple deprivation (IMD) Land-use surveys
Social inequality Segregation of people and exclusion or marginalisation of subgroups	Gender, age, health, longevity, disability and educational achievement data from the Office for National Statistics (ONS) and Public Health England www.police.uk and crime data Zoopla's ZED Index Village/community centre activities, such as playgroups, elderly social groups, internet blogs and email lists Placecheck Perception survey of the local area Use of social media
Service inequality Access to public transport, health facilities and food	Functional surveys of services (high/medium/low order) Public transport timetable Use of Ordnance Survey (OS) maps Interviews with local residents/members of the local authority Supermarket and shop location survey
Environmental inequality Derelict land, pollution levels and access to open space, impacting on people's well-being	ONS: central heating provision Building quality surveys Pollution data Environmental quality surveys Use of a decibel metre to test noise pollution levels Building decay survey

Now test yourself TESTED

4 Explain how social characteristics are affected by economic differences.
5 To what extent has economic activity affected your local place?

Answers on p. 224

Typical mistake

It is erroneous to think that high levels of inequality exist only in developing and emerging countries. In fact, strong inequalities prevail in most so-called prosperous societies.

Revision activity

Make sure you know some of the possible measures of inequality shown in Table 4.2.

Rank the measures of inequality from most effective to least and explain your rank order.

Now test yourself

TESTED

6 Explain how and why inequalities in pay can affect quality of life.

Answer on p. 224

Changing functions and characteristics

Functional and demographic changes

Places can change their functions and characteristics over time due to developments in things such as accessibility and connectedness. This can be seen at present with the internet and broadband services changing the way people work and consume.

The function of a place may also change, for example containerisation saw the decline of the West India Docks in East London, an area that was subsequently regenerated as Canary Wharf, London's secondary central business district.

While places change what they do for a living, so too do their populations. In many instances, it is the former conditioning the latter. Typical demographic changes include:

- trends — increasing or decreasing
- rates of change
- increasing ethnicity
- age and gender balances
- socio-economic structure — changes in response to processes such as gentrification, deindustrialisation, deprivation and studentification.

Gentrification The movement of middle-class people back into rundown inner-urban areas, resulting in an improvement of the housing stock and image.

Deprivation A condition in which a person's well-being falls below a level generally regarded as a reasonable minimum. Measuring deprivation usually relies on indicators relating to employment, housing, health and education.

Studentification Social, economic and environmental change brought about by the concentration of students in particular areas and cities, usually located close to universities.

Measuring change

REVISED

Changes within places can be measured in a number of ways, such as:

- land-use conversions
- employment trends
- demographic changes
- levels of deprivation.

The last of these is particularly important and much use is made of the index of multiple deprivation (IMD).

Now test yourself

7 Explain the ways in which geographers can measure the changes in function and characteristics of a place.

8 How has increasing connectedness shaped the economic characteristics of a place you have studied?

Answers on p. 224

Revision activity

Research a regeneration project in your local area and find out the following:

- its purpose
- the impacts it has — social, economic and environmental.

Places and their connections

REVISED

This part of your revision focuses on the two place studies you have completed as part of this topic. Figure 4.2 shows the main components of place, some of which you should have investigated in each of your contrasting places.

The specification recommends that you pay particular attention to:

- the regional and national connections (linkages) of your places
- the international and global connections of your places.

To what extent have these connections brought about change in your places? Have those changes impacted on local people?

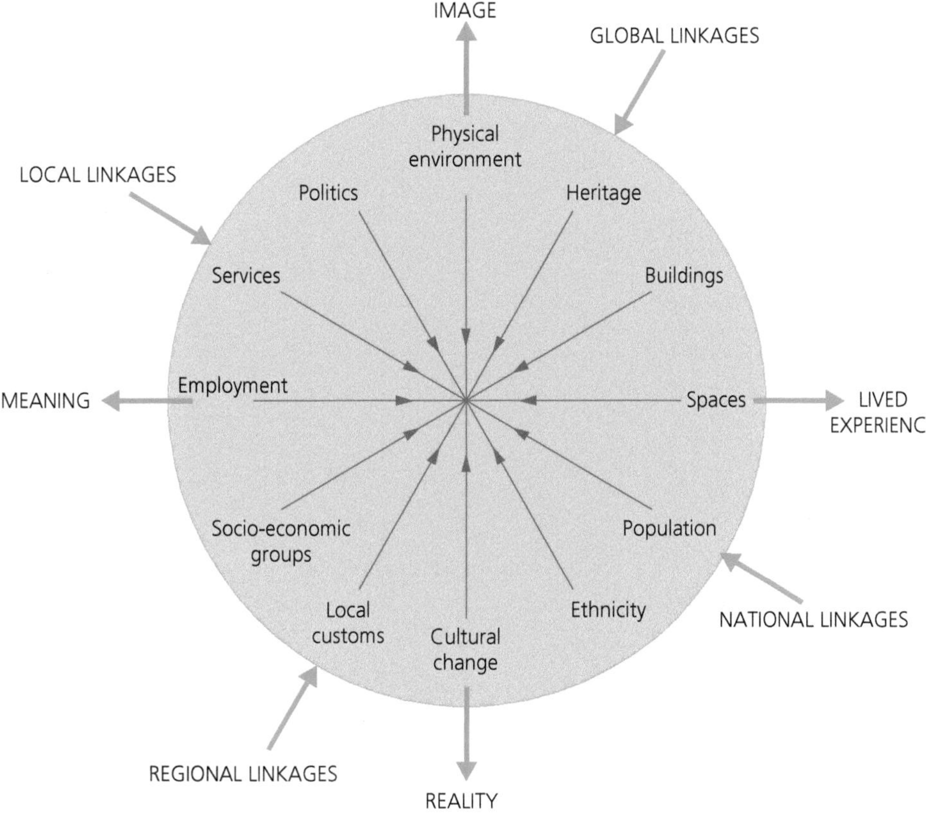

Figure 4.2 Components of place

Your place investigations will have been guided by four groups of questions:

- those that establish the initial identities of your chosen places
- those about how their economic and social characteristics have been shaped by regional and national connections
- those about how their economic and social characteristics have been shaped by international and global connections
- those about how economic and social change have influenced the identities of people living in those places.

Making links

Places in the UK, as elsewhere in the world, are being increasingly influenced by two major players: TNCs and IGOs. This is part and parcel of globalisation.

Revision activity

It is recommended that you summarise your place studies by creating a table of two columns and four rows. There should be one column for each place and one row for each of the four groups of questions: place identity, regional and national influences, international and global influences, and people's identities.

Making links

Attitudes towards place changes can be highly polarised between those that see them as eroding heritage and those that see them as place enriching.

Exam tip

Sense of place and the meaning of place are rather 'slippery' terms. There is very little difference between them. They are so close that they may be taken to mean one and the same thing.

Now test yourself

TESTED

9 Explain how a range of past and present connections has shaped the economic and social characteristics of your two places.

10 Explain how the identity of the place(s) you have studied has been affected by change.

Answers on p. 224

The need for regeneration

First, a reminder about the nature of regeneration. It is first and foremost an exercise in economic improvement. The hope is that the improvement will attract inward investment and create jobs. From this, hopefully there will be a spin-off of social benefits, such as a reduction in poverty and deprivation.

There are two particular challenges associated with regeneration:

- persuading people of the need for it
- agreeing what would be the most appropriate and effective form for that regeneration to take.

Both challenges are made more difficult by the fact that people differ in so many ways, particularly in their **lived experiences**, their **attachment to a place**, their perceptions and who they happen to be.

Key concepts

Lived experience is the accumulated experience of living in a particular place. This can have a profound impact on a person's perceptions, values and identity, as well as on their general development and outlook on the world.

Place attachment is the emotional bond between person and place. It is highly influenced by a person's lived experience and how long they have resided there.

Typical mistake

Planned regeneration does not take place only in towns and cities. Regeneration may be more common in urban areas, but much is now being undertaken in rural areas.

Inequalities and perceptions

REVISED

- The economic and social inequalities that lie rooted in employment and different levels of income affect people's perceptions of places.
- Place has a huge impact on us all, with the lived experience it has to offer and a sense of security that comes with feelings of attachment.
- In addition, our perceptions of places are often coloured by the current fortunes of places.

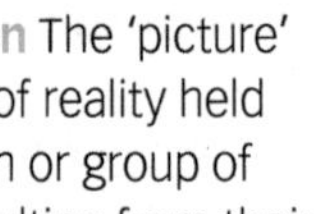

Perception The 'picture' or 'image' of reality held by a person or group of people resulting from their assessment of received information.

Successful places

Places perceived as successful tend to be characterised by:

- high rates of employment
- high rates of in-migration (domestic and international)
- low levels of deprivation.

Such places become self-sustaining as more people and investment are drawn to the opportunities created. However, there may be negative externalities, including:

- increased property prices
- congestion of roads and public transport
- overburdened services, such as education and healthcare.

The perception of residents in such places may differ:

- Younger people in high-earning jobs enjoy the fast pace of life and abundant opportunities.
- Unskilled people, lower earners and the long-term unemployed have more negative views.

- Retirees may view them as too busy and look to other, perhaps less successful places offering a slower pace of life with a more pleasant climate, sheltered accommodation and good access to healthcare.

It has to be emphasised that we may all have different views on what constitutes a successful place. Would everyone agree that London or San Francisco are successful places?

Less successful places

Economic inequality and technological change breed less successful places. For each successful place, there is at least one failing. Examples of such places (which have suffered from the negative multiplier effect) abound and vary in scale from entire regions, such as the Rust Belt in the USA, through to towns (for example, Hartlepool in North East England), to small inner-city areas, for example in Salford.

The symptoms of less successful places are all too obvious and include:

- high levels of poverty and deprivation
- unemployment
- derelict buildings
- graffiti
- crime and vandalism.

There is no doubt that less successful places are generally going to be perceived less favourably. The lived experience may be found wanting and the feelings of place attachment rather weak.

The particular challenge facing less successful places is becoming drawn into the spiral of decline (see Figure 4.3).

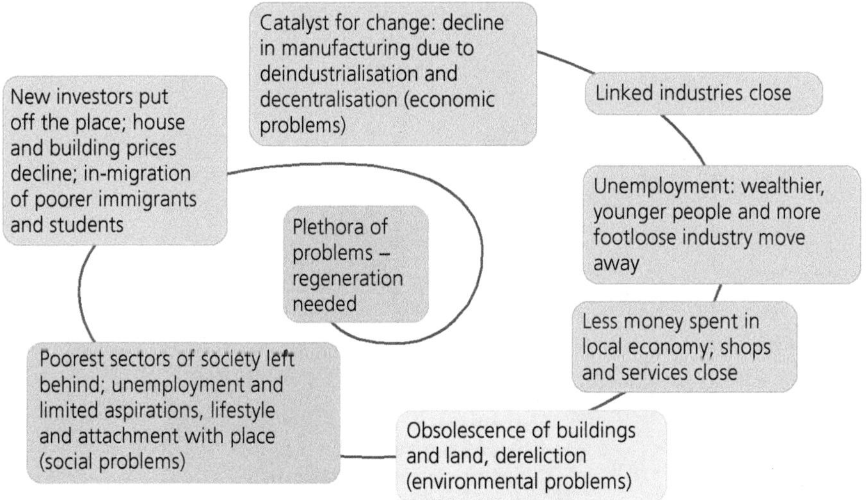

Figure 4.3 The spiral of decline

Revision activity

Research a town or city that has gone into decline, e.g. Detroit, Michigan.

Complete a spider diagram to answer the following:

- Why did the town/city go into decline?
- Where did the population move and why?
- What issues does that town/city now face?

Do your findings link to the spiral of decline shown in Figure 4.3?

Revision activity

Make notes about a successful place. What factors have made it successful? Are there negatives? Classify the successes into economic, social, environmental and political. A good example to use is the San Francisco Bay Area.

Negative multiplier A downward cycle. Change in economic conditions can produce less spending, which results in less investment from businesses and therefore fewer employment opportunities.

Exam tips

You should learn a region that has been negatively affected by economic restructuring, e.g. the Rust Belt region in the USA.

Remember that not all unsuccessful places are urban. Think of the remote rural areas of the UK and those villages that were once prosperous farming communities.

Priorities for regeneration

Economic and social inequalities create a need for regeneration, if only to target those places that are the victims of inequality, such as sink estates, areas left derelict by deindustrialisation and declining villages.

Sink estate Housing estate that experiences deprivation, poverty and crime, such as domestic violence, gang warfare and drugs.

TESTED

11 Produce a mind map showing how economic and social inequalities can affect people's perceptions of different areas.

12 Why are some places economically successful? Use the area you have studied as an example.

Answers on p. 224

The lived experience of place and engagement

REVISED

Level of engagement

One aspect of the lived experience that affects place attachment is the level of engagement. This is the degree to which a person participates in their local community, and feel they belong to a place. It might be seen as a reflection of their place attachment. Indicators of engagement include:

- voting in local and national elections
- membership of, and participation in, local societies
- having a circle of local friends.

The opposite condition to engagement is one of marginalisation and exclusion. The factors shown in Figure 4.4 as affecting levels of engagement can and do work in a negative way. People can feel excluded by ethnicity, gender and deprivation.

Marginalisation The social process of being made to feel apart or excluded from the rest of society. This leads to feelings of belonging to an underclass that is discriminated against.

Levels of engagement

- **Age:** especially if combined with length of residence in a place
- **Ethnicity:** non-white British may differ in their views because of local antipathy or acceptance; older generations may feel just as British as their white counterparts
- **Length of residence:** new migrants and students may have less strong attachments than longstanding locals
- **Gender:** despite modern equality measures, women may still feel less able to go to the pub alone; women or men may also be more active in their local community if home with children.
- **Levels of deprivation:** higher levels may be associated with anti-establishment views; those in temporary accommodation or rented housing may feel less 'at home' than owner occupiers

Figure 4.4 Factors affecting levels of engagement

Lived experience and place attachment

Lived experience and place attachment vary from person to person. The factors of engagement shown in Figure 4.4 also apply to lived experience and place attachment.

Now test yourself

13 Explain what makes people feel marginalised and excluded.

Answer on p. 224

TESTED

Conflicts

Conflicts often occur among contrasting groups in a community, largely because they hold different views about the priorities and strategies for regeneration. These can be caused by:

- a lack of political engagement and representation
- tensions between ethnic groups
- social inequality
- a lack of economic opportunities.

Typical mistake

Marginalisation and exclusion occur only in less successful places — not true!

Now test yourself

TESTED

14 Describe the variations in the lived experience of place and how people's engagement can vary.

15 Explain why regeneration can cause conflict among contrasting groups in the community.

Answers on p. 224

Making links

Players and whole communities may differ in their attitudes to urban regeneration and favour contrasting approaches. Those attitudes are influenced by the degree of attachment.

Revision activity

Complete Table 4.3 to explain how these factors impact levels of engagement to a place.

Table 4.3 Factors that impact levels of engagement to a place

Factor	How attachment to a place varies	Why attachment to a place varies
Age		
Ethnicity		
Gender		
Length of residence		
Level of deprivation		

Evaluating the need for regeneration

REVISED

It is quite possible for the **players** living in the same place to have conflicting opinions, even about whether there is a need for regeneration. To what extent is this the situation in your two chosen places?

Key concept

Players are the people and/or organisations involved in an issue, project or proposal.

Economic: types of employment, unemployment, business rates, property values

Environmental: derelict land, closed shops, boarded-up buildings, graffiti, pollution, lack of green space, transport provision

The need for regeneration

Social: ethnicity, access to education and healthcare, educational achievement, community facilities

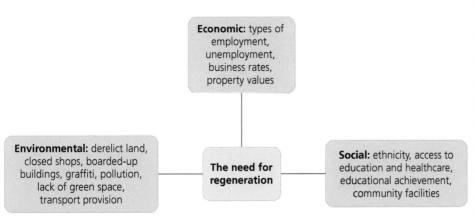

Figure 4.5 Some criteria for assessing the need for regeneration

Figure 4.5 sets out some of the criteria that are widely used when evaluating the need for regeneration.

In investigating the need for regeneration in your two places, you should have:

+ used statistical evidence to determine the need for regeneration
+ used different media to question the need for regeneration
+ identified the factors influencing the perceived need for regeneration.

Exam tips

Remember that one of the most powerful criteria is unemployment. Note that most of the social and environmental criteria are spin-offs from this.

Provided the evidence is relevant, you are recommended to use material from your own place studies to illustrate and support answers to more general questions.

Now test yourself

TESTED

16 Evaluate the need for regeneration in your local place of study.

Answer on p. 224

Revision activity

Make notes summarising the results of your two place investigations, as suggested on page XXX.

How regeneration is managed

The role of national government

REVISED

Investment in infrastructure

In the UK, the government plays a key role in regeneration, largely through investment in the national transport infrastructure. The thinking is that improved accessibility is the key to successful regeneration. Not only is it likely that investment in regeneration is attracted to places with improved accessibility but good accessibility is also crucial to sustaining the regeneration. A current example is the HS2 project, a new high-speed rail link that will hopefully help to regenerate large parts of northern England.

Infrastructure The basic services and systems needed for a place to work effectively, e.g. adequate road networks, telecommunication networks, water and sewerage facilities, affordable housing and access to education.

Factors affecting regeneration policy are shown in Figure 4.6.

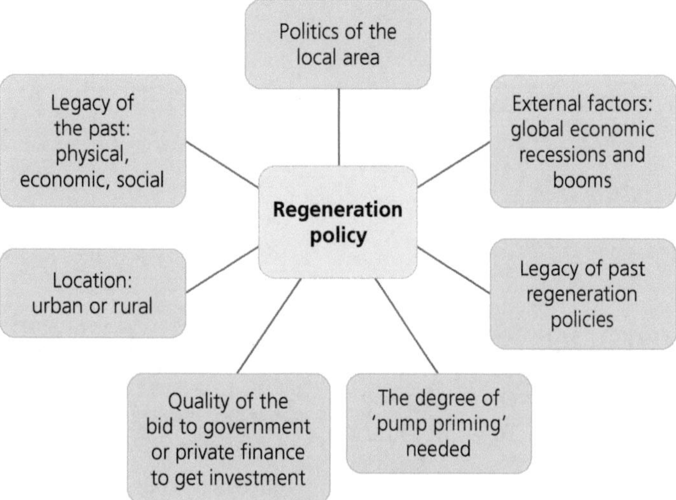

Figure 4.6 Factors affecting regeneration policy

Revision activity

Using Table 4.4, complete a cost–benefit analysis for the proposed HS2 project.

Table 4.4 Costs and benefits for the proposed HS2 project

	Costs	Benefits
Economic		
Social		
Environmental		

From the information you have found, do you support the HS2 project?

Cost–benefit analysis Before a project can go ahead, economic, social and environmental costs are weighed up against the benefits of the project.

Domestic policies

Government domestic policies can also help stimulate regeneration in a variety of ways, for example by:

- relaxing planning laws on, say, developing greenfield sites
- providing incentives to encourage the building of affordable housing
- allowing fracking in the hope that it might play a part in the regeneration of some rural areas.

Revision activity

Research the fracking debate in the UK. Create a mind map with arguments for and against fracking for shale gas.

International policies

The government can also pursue policies at an international level that have a direct bearing on regeneration, for example by:

- deregulating capital markets to encourage overseas and private investment in regeneration schemes
- immigration — there is tension here between the job-generating focus of regeneration and the availability of immigrant workers.

Making links

National governments are influential players in regeneration but they need to encourage and involve investors and developers. Together they have the power to override local needs and opinions.

TESTED

17 How do UK government policy decisions play a key role in regeneration?

18 How does deregulation of capital markets impact on regeneration?

Answers on p. 224

Exam tip

Be clear that you understand that national governments invest in large infrastructure projects in the UK because of the perceived benefits they may have for places.

Typical mistake

Migration is often assumed to be either a good or a bad thing, but the evidence suggests that the issue is much more complicated.

The role of local government

REVISED

Local plans

- The main task facing local authorities is to create a sympathetic business environment that will support regeneration.
- One of the most obvious ways is to produce local plans that clearly designate areas for development or redevelopment, for example as retail, science or industrial parks.
- This is perhaps one of the most crucial decisions a local authority has to make. What should be the lead activity heading up the regeneration — retailing, heritage and tourism, sport or leisure and recreation?
- There is a range of regeneration options from which to choose (see Figure 4.7).
- In some cases, the choice will be conditioned by the history of the proposed regeneration area.

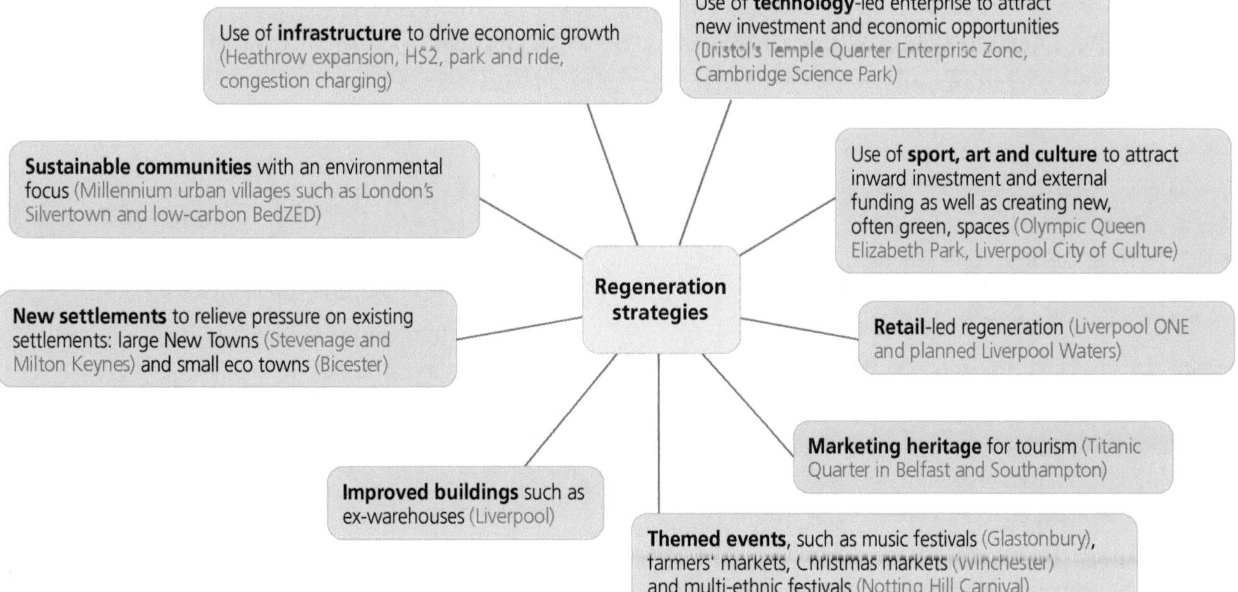

Figure 4.7 Regeneration strategies

Local support

- Local governments hope to receive local support for their regeneration schemes, as for example from the local Chamber of Commerce and the trade unions.
- However, everyone knows that such schemes are not acceptable to everyone.
- Inevitably there will be tensions, as some local interest groups express their support for preservation and conservation over redevelopment, as occurred at the 2012 London Olympics sites.

Revision activity

Using Figure 4.7, note three types of regeneration strategy and come up with an example of each.

Making links

The success or failure of regeneration schemes hinges very much on the actions of local authorities. But their attitudes towards regeneration may well clash with those of other players and lead to conflict.

Now test yourself TESTED

19 Describe the ways in which local government policies can make areas seem attractive for inward investment.

Answer on p. 224

Rebranding

REVISED 

A new look

Using a variety of media, rebranding and re-imaging attempts to change both their outward appearance and public perception, areas can be made to look attractive to potential investors, clients and local people.

Rebranding deindustrialised places

- Rebranding often focuses on the attractiveness of places.
- This is possible even in places bearing the scars of deindustrialisation, either by comprehensive redevelopment and giving a place a totally new identity (for example, London Docklands) or by capitalising on a place's industrial heritage (for example, Telford).
- An added possibility in both instances is to use the regenerated places as venues for prestigious national and international events (for example, Stratford in East London during the 2012 Olympic Games).

Rural rebranding strategies

A wide range of tried and tested strategies has emerged that seeks to rebrand this post-production countryside (where rural areas are less productive than before). Table 4.5 highlights Cornwall as an example of post-production countryside.

Table 4.5 Cornwall: an example of post-production countryside

Reasons for Cornwall's decline	Strategies to rebrand Cornwall
Fishing: Overfishing has caused the decline in many types of fish (e.g. cod) and EU quotas of some Cornish fish stocks were allocated to other European countries. Farming: cheaper imported food caused a decline in farming and a reduction in EU subsidies and government grants. Tin and copper mining: much of the tin and copper has been mined out and tin prices collapsed due to cheaper overseas competition. The last tin mine closed in Cornwall in 1998. Quarrying: large china clay reserves in mid-Cornwall provided 10,000 people with work in the 1960s. By 2015 the labour force had reduced to 800 and most of the operation had been moved overseas.	Farm diversification: many farm shops sell Cornish produce, which helps the local economy. Destination tourism: tourists travel to Cornwall for the food and restaurants, for example those of chefs Rick Stein (Padstow) and Jamie Oliver (Watergate Bay). Other areas nearby then benefit from tourists coming to the area. Cornwall's attractions: Cornwall offers spectacular gardens, for example the Lost Gardens of Heligan and the Eden Project, as well as long coastlines, which are renowned for surfing and other water-based activities, attracting both national and international tourists.

Now test yourself TESTED

20 What evidence do you have of rebranding attempting to represent areas as being more attractive by changing the public's perception of them?

21 How effective has the government been in regenerating rural areas?

Answers on p. 224

Revision activity

Check your notes on rebranding in both urban and rural areas and produce two mind maps showing the strategies used for the places you have studied.

Assessing the success of regeneration

Possible measures

REVISED

The aim of regeneration is to create a **legacy** of increased employment and income, and reduced poverty and deprivation.

Key concept

Legacy refers to the longer-term effects of regeneration schemes. It can be positive or negative. It might be judged on the re-use of landmark buildings, the amount of government support and private investment needed, or whether local people benefit in the long term.

Economic measures

The success of regeneration can be measured in a number of ways.

Possible economic measures are:

- employment: not just the scale of job growth but the mix of job types
- income: several possibilities here — average earnings, business turnovers, number of households on benefits
- poverty: a declining incidence and number of households on benefits.

These and other measures may be used to assess the performance of regeneration in two different ways:

- to compare the same measures both before and after regeneration
- to compare the results from one regeneration project with those of a similar project elsewhere.

Social progress

Social progress relates to how an individual and community improve their relative status in society. It can be measured by assessing social deprivation in an area such as:

- improvement in levels of education
- improvement in health indicators, e.g. risk of premature death
- reduction in levels of crime
- access to affordable housing and local services.

Exam tip

Remember that 'success' can be subjective and that any improvement may not benefit all of the people living in an area.

Quality of the living environment

The living environment can be measured by the quality of the local environment, for example air quality or the quality of housing. Improvements to environmental deprivation can have a knock-on effect on social progress, as health can improve as a result of a cleaner living environment.

One final and important point to be made here is this: as already stated, most regeneration projects are not intended as quick fixes. For this reason, assessments of success or otherwise should not be attempted before the regeneration has had time to bed down and to reveal its strengths and weaknesses. Such a time lapse occurs over years rather than months.

Now test yourself TESTED

22 Check that you understand the four main ways in which the success of regeneration can be measured.

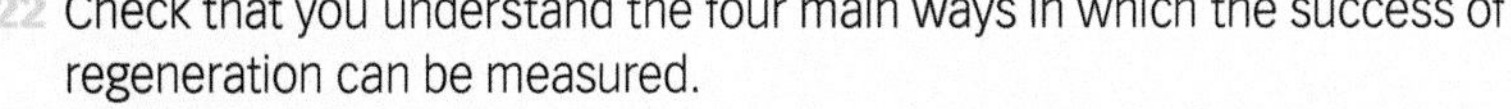

23 Explain how social regeneration has helped to reduce deprivation in an area you have studied.

Answers on p. 225

Skills activity

For an area you have studied, show how regeneration has changed that place.

- Use Google Maps and other online photographs to find photos before and after regeneration has occurred.
- Annotate the photos to explain how regeneration has changed the area.
- Categorise your comments into social, economic and environmental change.

Revision activity

Create a mind map to assess the success of regeneration in an area you have studied. The mind map should have three branches:

- economic success
- social progress
- environmental success.

Are any further improvements needed?

Evaluation by urban stakeholders

REVISED

Stakeholders generally

In the remainder of this and the next section, the specification expects you to take two regeneration schemes (one urban and one rural) and to focus on the following:

- the contested nature of the regeneration proposals — stakeholders for and against, and their particular views
- the impact of national and local policies and strategies in determining the nature of the regeneration
- the evaluation of the regeneration outcomes by different stakeholders.

Stakeholder An individual, group or organisation with a particular interest in the actions and outcomes of a project or issue-solving exercise.

Making judgements about the success of regeneration in any place involves not just those of the actual decision makers but also those of stakeholders — the people, groups and organisations with an interest in the regeneration. In most situations, stakeholders fall into four groups:

- providers: could be landowners, investors, contractors
- users or beneficiaries: those who stand to benefit (or lose out)
- governance: local government officials, enforcers of local bye-laws and national government policy
- influencers: action groups, political parties.

From this, it follows that each stakeholder has their own particular perception of or opinion on what constitutes 'success' and 'failure' (see Figure 4.8). They have their own vested interests and agendas. They have their own criteria for assessing whether a particular scheme has been, or is being, managed successfully or not. Each stakeholder therefore arrives at their own verdict.

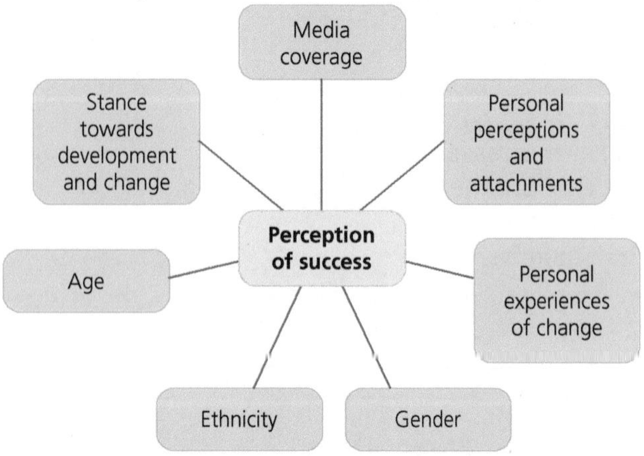

Figure 4.8 Factors influencing perception of success

Table 4.6 identifies the factors likely to influence the viewpoints of different urban stakeholders.

Table 4.6 Viewpoints and roles of different urban stakeholders

	Viewpoints	Roles
National government and planners	Often keen on public service projects where longer-term national goals take priority Can be used to kick-start the economy	Planning permission Often funded by central government to oversee large, nationally important developments, e.g. HS2 and Crossrail
Local councils	Have a duty to tackle inequality in local communities Make local planning decisions (although they do not have control over larger developments) Are supposed to balance out the economic, social and environmental needs of a locality	Local regeneration schemes 'Soft management' helping regeneration, e.g. regeneration continuing after the 2012 London Olympics Permissive arrangements, e.g. roller skating, artists and street performers
Developers	Economic standpoint: support projects that lead to benefits for the company, i.e. profits	Research Funding of schemes
Local businesses	Views may be polarised: those expecting an increased customer base from regeneration will differ from those threatened by it, e.g. local companies on the former industrial site where the London Olympic site was built The local Chamber of Commerce may give majority viewpoints of business leaders	Lobby councils Investment in schemes Compensation to move to allow new development
Local communities	The majority affected by the regeneration project May be represented by a few willing and able to give up their time to be involved in either the local council or a pressure group The legacy of providing affordable housing to those that lived in the area of the London Olympic site has not been honoured	Vote for local and national political parties Form pressure groups Lobby councils

Typical mistake

It is wrong to think that all stakeholders are equal when it comes to influencing the character and management of a regeneration scheme.

Now test yourself

24 Check that you understand the difference between the four types of stakeholder shown in Table 4.6.

25 Explain the roles of different players in urban regeneration.

Answers on p. 225

Stakeholders in an urban regeneration project

A possible urban case study for use in this part of the specification might be Salford Quays, London's Olympic Park or Liverpool Waters.

Making links

Attitudes towards any regeneration proposal are more likely to be positive, but in all places NIMBYism will find a voice and press for 'change, but only when it's somewhere else'.

Revision activity

You will need to prepare notes on your chosen urban regeneration scheme under the three headings: contested nature (i.e. conflicts between stakeholders); impact of national and local policies; evaluation by different stakeholders.

To help you understand which stakeholders agree or disagree with each other, complete the conflict matrix in Table 4.7 for your stakeholders.

Table 4.7 Conflict matrix for your stakeholders

	Stakeholder A				
Stakeholder A		Stakeholder B			
Stakeholder B			Stakeholder C		
Stakeholder C				Stakeholder D	
Stakeholder D					Stakeholder E
Stakeholder E					

Key: ++ Strongly agree + Agree -- Strongly disagree - Disagree

Evaluation by rural stakeholders

REVISED

Stakeholders in a rural regeneration project

A possible rural case study for use in this part of the specification might be Powys regeneration partnership, the North Antrim coast or the Isle of Skye.

Figure 4.9 shows the Egan wheel, which creates an evaluative scoring system that can be used when assessing the outcome of regeneration projects in rural settings.

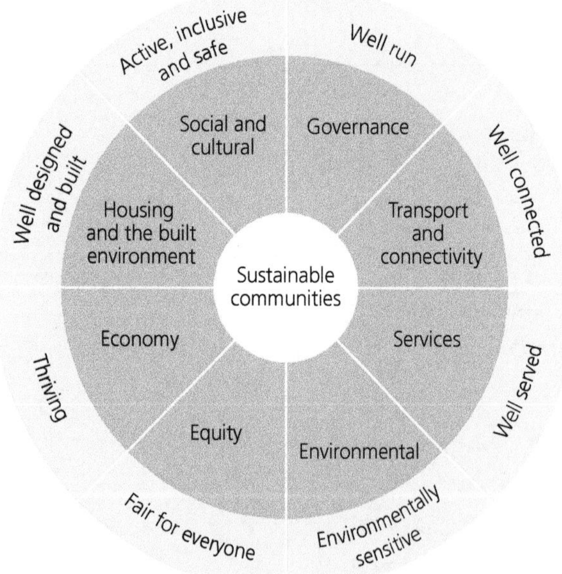

Figure 4.9 The Egan wheel

Now test yourself TESTED

26 Check that you are able to provide examples of stakeholders and their views drawn from your own regeneration investigations.

27 Assess the role of different stakeholders in the success of a rural regeneration project.

Answers on p. 225

Exam tip

Remember that not all stakeholders are equal when it comes to influencing the management of an issue. When studying any issue, try to rank stakeholders according to the strength of their influence.

Revision activity

Create a mind map to assess the success of a rural regeneration project you have studied. Use the following to start you off:

- What are the costs? (social, economic, environmental)
- What are the benefits? (social, economic, environmental)
- How does the regeneration project support local people and the environment?
- How could the regeneration project be improved?

Exam skills

You should be familiar with the skills and techniques used to investigate the regeneration of places:

- use GIS to represent data about place characteristics
- interpret oral accounts relating to lived experience
- use the IMD to investigate spatial variations in deprivation
- use social media to understand how people relate to the places where they live
- test the strength of relationships by scatter graphs and Spearman's rank correlation
- investigate newspaper sources to understand conflicting local views
- evaluate different sources used in conveying the image of a place
- use media sources to explore how a place's identity has been changed by regeneration
- use photographic and map evidence to depict 'before' and 'after' scenarios
- interrogate blogs and other social media to understand different opinions on the performance of a regeneration project.

Exam practice

A-level

1 a) Study Figure 1.

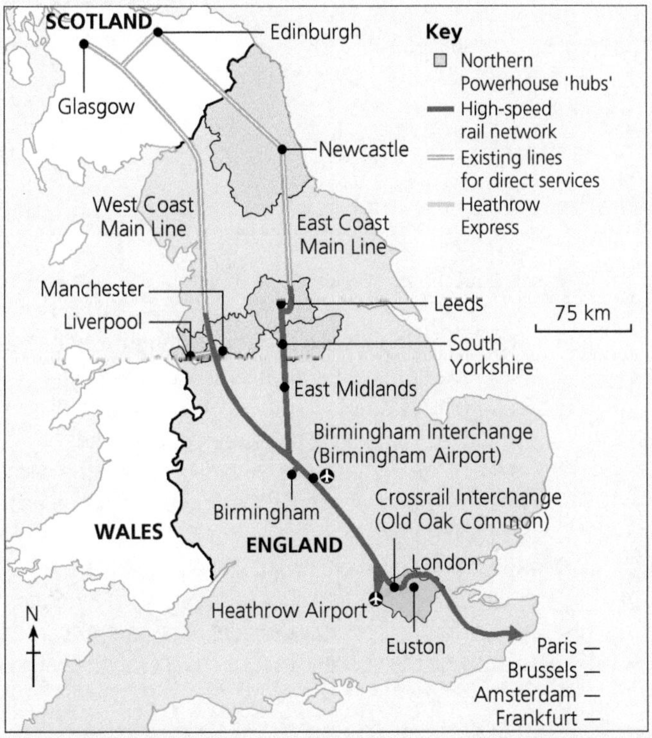

Figure 1 HS2, Heathrow and the Northern Powerhouse

i) Suggest how HS2 will benefit people living in Birmingham. (3)

ii) Suggest reasons as to how infrastructure development regenerates regions in the UK. (6)

b) Explain why different urban stakeholders have different criteria for judging the success of urban regeneration. (6)

c) Evaluate the extent to which the success of regeneration can lead to social progress and an improvement in the living environment. (20)

→

2 a) Study Figure 2.

Figure 2 Detroit, in the USA's Rust Belt

i) Suggest one reason why economic restructuring could have resulted in a spiral of decline in Detroit. (3)

ii) Suggest reasons for the relationship between increasing levels of social deprivation and economic restructuring. (6)

b) Explain how rural rebranding strategies can make these places more attractive to national and international tourists and visitors. (6)

c) Evaluate the importance of the role of the UK government in the success of regeneration projects. (20)

Answers and quick quiz 4A online

Summary

You should now have an understanding of:

- how economies can be classified in different ways and vary from place to place
- the way in which places change their functions and characteristics over time
- past and present connections that shape the economic and social characteristics of places
- economic and social inequalities that affect people's perceptions of an area
- the significant variations in the lived experience within places and levels of engagement with them
- a range of ways to assess the need for regeneration
- UK government policy decisions that play a key role in regeneration
- local government policies that aim to represent areas as being attractive for inward investment
- rebranding attempts to represent areas as being more attractive by changing public perception of them
- how a range of measures is used to assess the success of regeneration: economic, demographic, social and environmental
- the fact that different urban stakeholders have different criteria for judging the success of urban regeneration
- how different rural stakeholders have different criteria for judging the success of rural regeneration.

4B Diverse places

This topic covers four key concepts: **place**, **population structure**, the **rural-urban continuum** and **ethnicity**.

Key concepts

Place is a part of geographical space with a distinctive identity and character that is deeply felt by local inhabitants. The sense of each and every place derives from a unique mix of external connections, natural and human features in the landscape and the people who happen to occupy it.

Population structure is the makeup of a population in terms of different age groups, the balances between those groups and between the sexes within them. Other components include life expectancy, family, size, marital status and ethnicity.

Rural-urban continuum is the scale that moves from a large city or conurbation to rural areas.

Ethnicity is the distinctive culture, religion or language shared by a social group.

Now test yourself TESTED

1 What is ethnicity and why is it such an important issue in many UK places?

Answer on p. 225

Population structure, time and place

Population structure

REVISED

- Population structure refers to the makeup of a population, particularly in terms of age and gender.
- Population structures vary from place to place and over time, as do population distributions. Both variations are particularly noticeable along the rural-urban continuum.

Population density

The place-to-place variations in populations are best shown in terms of population density. In the UK, densities range from zero to more than 5,000 persons per square kilometre (see Figure 4.10). There are many factors affecting population density, the most significant of which include:

- the physical environment, particularly relief and climate
- the economy, agricultural or non-agricultural, with the latter generating higher densities
- population characteristics — youthful structure is likely to raise densities
- planning, by controlling the location and amount of new housing.

Population density The number of people per unit area (usually per square kilometre), i.e. the total population of a given area (a place, city, region or country) divided by its area.

Exam tip

Be sure that you know the difference between density and distribution. Population density is one measure for showing the distribution of population.

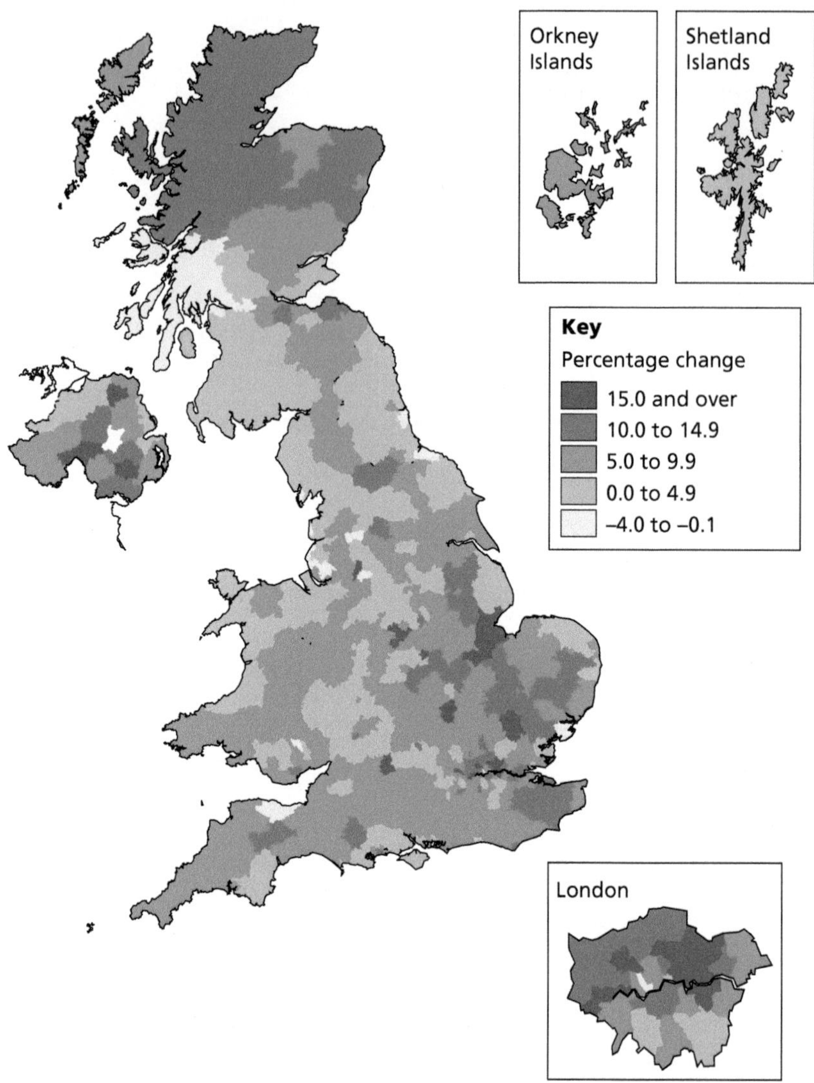

Figure 4.10 The distribution of population in the UK, 2019

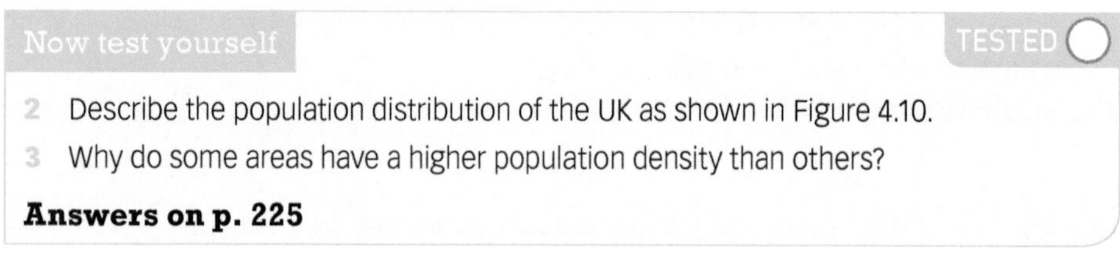

Now test yourself

TESTED

2 Describe the population distribution of the UK as shown in Figure 4.10.

3 Why do some areas have a higher population density than others?

Answers on p. 225

Population dynamics

Over time, populations change, either increasing or decreasing. Change is the outcome of two processes: natural change and net migration (see Figure 4.11).

Natural change The difference between births and deaths over a given period of time. Natural increase happens when the birth rate is higher than the death rate and natural decrease when the death rate is higher than the birth rate.

Net migration The difference between immigration and emigration from an area over a given period. This can be either positive or negative.

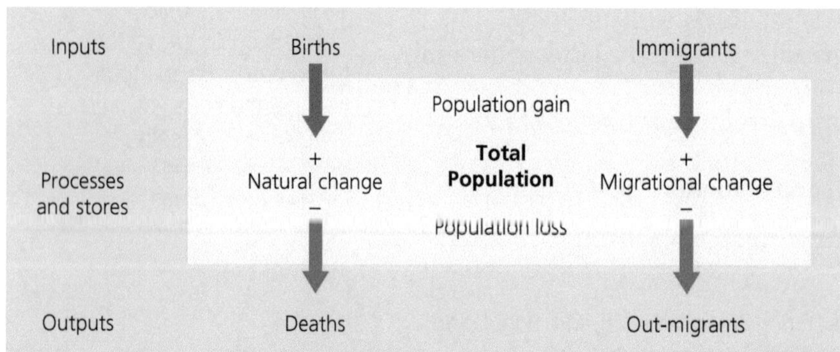

Figure 4.11 Components of population change

Since 1965 the UK has seen a rapidly increasing population. This population growth has not been even — not only are there differences in the population density of the countries that make up the UK, but there are also large regional differences within England.

In England, the fastest population growth can be seen in London and the South East and the slowest population growth in the North East and North West regions of the UK.

Some of the reasons for this are as follows:

- London's expanding knowledge economy attracts highly qualified workers.
- London's booming economy has meant high levels of internal and international migration.
- Traditional industry (coal mining, steel manufacturing, shipbuilding) collapsed in the North East.
- A longer life expectancy in the UK has increased the UK's overall population as people are now living longer.

Now test yourself

4 What are the two components of population change? Which do you think is responsible for more population growth in the UK?

5 Explain the impact of population change on different areas of the UK.

Answers on p. 225

TESTED

Revision activity

It is a good idea to have a glossary of the key terms used for each topic.

Start by finding the definition of the following words:

- Internal migration
- International migration
- Fertility rate
- Birth rate
- Mortality rate

Continue to add to your glossary as you move through your revision.

Population characteristics

REVISED

- In terms of their impact on natural change in a population, age and gender are the two most important components.
- When these two are plotted in the form of a population pyramid, its detailed shape can tell us much about what has happened to the population over the last 75 or more years.
- It also provides some pointers as to how the population is likely to change in the foreseeable future.
- A basic distinction is made between youthful and ageing populations (see Figure 4.12), the former promising population growth and the latter population stagnation or decline. The dependency ratio measures the percentage of dependent people (not of working age) divided by the number of people of working age (economically active).

Population pyramid
A graph to show the distribution of age groups in a population. It is constructed with males on one side and females on the other, with the youngest age group at the base of the pyramid.

Dependency ratio
The ratio of dependants (children 0–15 and over 65s) to the working-age population (16–64).

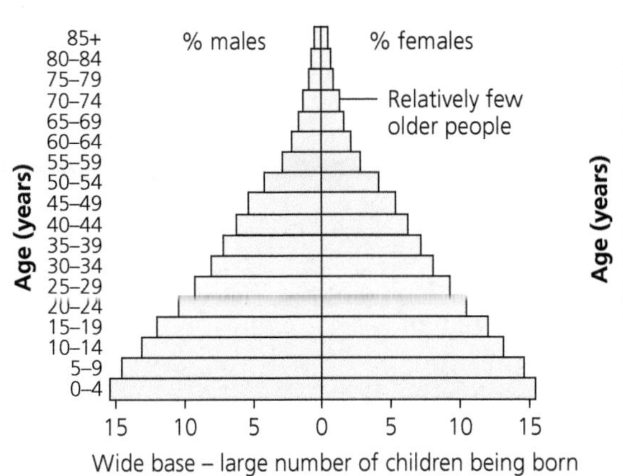

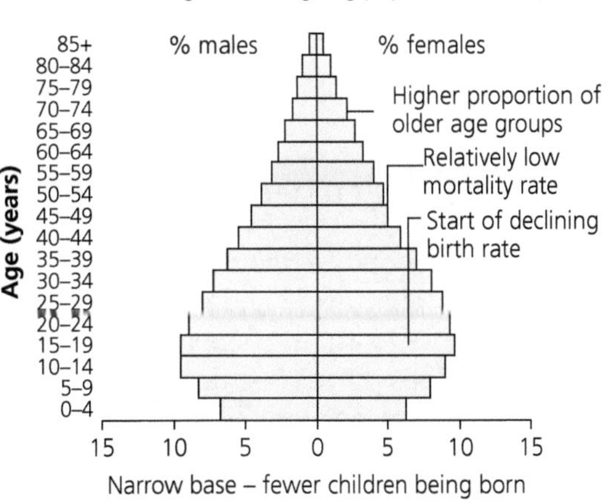

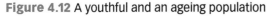

Figure 4.12 A youthful and an ageing population

Now test yourself TESTED

6 How might the recent influx of immigrants into the UK have impacted the country's population pyramid?

Answer on p. 226

Age and gender are not the only components of population structure. Others include:

- family and household size
- marital status
- ethnicity.

These, and other demographic characteristics, such as life expectancy, vary between and within settlements.

- Within-settlement differences are well exemplified by the differences between the populations of places in the inner city and those of places in the suburbs. Compare, for example, the boroughs of Brent and Bromley in Greater London in terms of their age structures and ethnicity.
- Between-settlement differences are well shown along the rural-urban continuum. Compare, for example, the populations of commuter villages with those in remote rural places.

One very obvious demographic and cultural variable is ethnicity. Ethnic diversity is most marked in the inner city and generally declines along the rural-urban continuum.

The ethnicity and cultural diversity of places in the UK are being changed by:

- the impacts of migration
- the social clustering of immigrants
- the pull of major cities
- government planning policies.

Typical mistake

Population structure is not only about age and gender. There are other components (see below) but perhaps of less significance, with the notable exception of ethnicity.

Making links

The actions of governments may foster or suppress cultural diversity. The number of governments falling in the latter category is dwindling fast. It would be true to say that an increasing number of governments are neutral on this issue.

Now test yourself 

7 What is the significance of family size and marital status as components of population structure?

8 How can ethnicity vary within settlements in the UK?

Answers on p. 226

Revision activity

Create a mind map to explain how the UK's cultural diversity has been established.

Use the following as your four main branches:

- social clustering
- accessibility to key cities
- physical factors
- government planning policy.

Place character and connections

REVISED

This part of your revision focuses on the two place studies you have completed as part of this Topic. Figure 4.13 shows the components that you should have investigated in order to analyse the image (real and perceived) of each of your two contrasting places. The specification recommends that you pay particular attention to the external connections and influences that have affected both continuity and change.

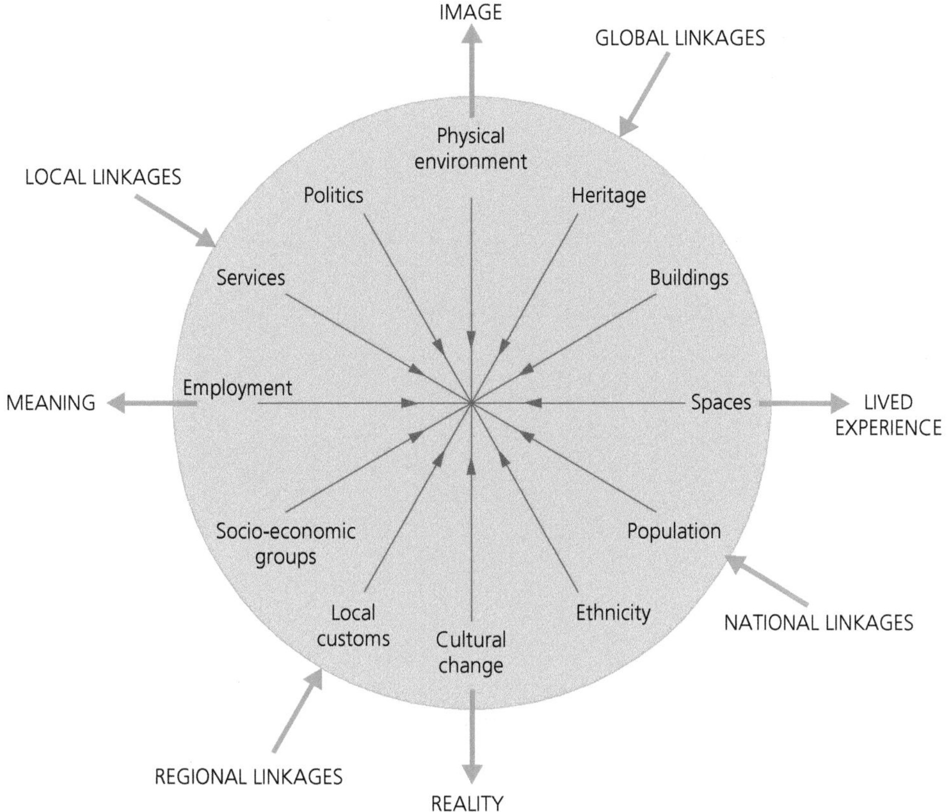

Figure 4.13 Components of place

Revision activity

It is recommended that you summarise your place studies by creating a table of two columns and four rows (see Table 4.8). You should have one column for each place and one row for each of the following aspects of your places: thumbnail portraits and initial impressions; main internal factors; external influences; and recent demographic and cultural change. Fill in each of the eight slots of your table with the appropriate summary notes.

The specification suggests that you pay particular attention to the impact of external influences, such as TNCs and IGOs.

Table 4.8 Place studies summary table

	Place 1	Place 2
Thumbnail portraits and initial impressions		
Main internal factors		
External influences		
Recent demographic and cultural change		

Skills activity

One way we can investigate perceptions of place is to search social media to see how people perceive and relate to your local area.

Start by looking at a range of social media pages and websites relevant to your local area. Ensure you consider the aims of each website, taking into account any bias in the opinions you discover.

Now test yourself

TESTED

9 Of the internal factors shown in Figure 4.13, which do you think have had the most impact on the identity of your two places?

10 Explain how regional and national influences have shaped the characteristics of the places you have studied.

Answers on p. 226

Exam tip

Sense of place and meaning of place are rather 'slippery' terms. There is very little difference between them. They are so close that they may be taken to mean one and the same thing.

Living spaces

The remaining enquiry questions in this topic require a sound grasp of three key concepts: **lived experience, living space** and **perception**. These concepts are essential to understanding this part of the topic, which is about the lived experiences derived from both urban and rural places.

Key concepts

Lived experience is the accumulated experience of living in a particular place. This can have a profound impact on a person's perceptions, values and identity, as well as on their general development and outlook on the world.

Living space refers to all that space given over to the day-to-day needs of a population, from work, shopping and leisure to education, healthcare and entertainment. But most important of all the needs is housing, if only because people spend a large proportion of their lives in residential space.

Perception refers to the 'picture' or 'image' of reality held by a person or group of people resulting from their assessment of received information.

Urban living spaces

REVISED

People's perceptions of cities can be extremely different and have changed over time. During the nineteenth century cities were seen as dangerous places. They were often viewed as dirty, especially the towns and cities that grew out of the Industrial Revolution, which often had poor-quality housing and high levels of pollution.

This image of cities is now very different, and many people are attracted to cities due to the following:

- Higher education opportunities, e.g. Manchester, whose student population accounts for 15 per cent of the city's population.
- Leisure and entertainment opportunities, e.g. live music, shopping facilities, bars and restaurants.
- Job opportunities due to many cities having a growing knowledge economy, which provides higher wage and salary levels.

However, this perception has not changed for all cities in the UK. Many are still seen as undesirable or even threatening by the people who live there or from outside perception. This perception can develop due to the area having:

- high crime rates
- poor environmental quality, e.g. high levels of litter, graffiti, derelict buildings
- high cost of living, especially where house prices have also increased disproportionately in an area (see Figure 4.14)
- racism towards migrants or other minority ethnic groups.

Revision activity

Check through your notes on the reality and popular perceptions of Victorian London or some other UK city at that time. Was the reality really that negative?

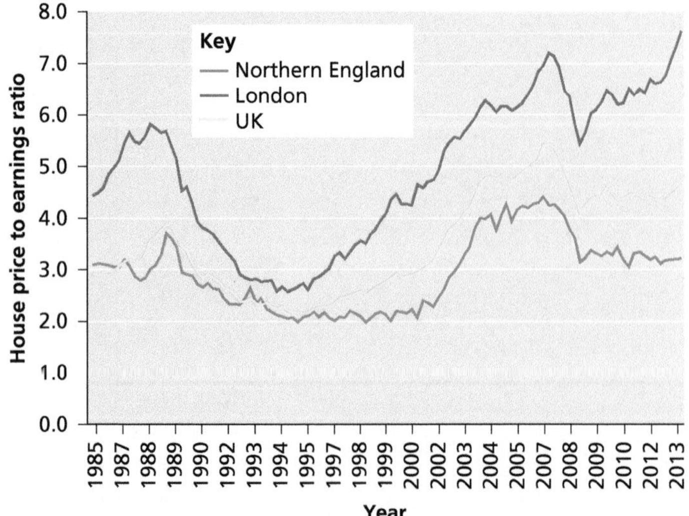

Figure 4.14 House price to earnings ratios, 1985–2013

If you take an urban area, for example Manchester, the suburban and inner-city areas are perceived very differently in terms of whether people desire to live and work there. What are perceived and identified as the pros and cons of the urban lived experience will depend on who you are, your changing values and attitudes, and where you are in the life cycle.

For example, a young family may perceive the suburban life to be more desirable due to:

- more residential space
- better schools and healthcare
- easy access to shops for daily needs
- more green space.

Now test yourself

11 Identify the advantages of living in inner-city areas.

12 Suggest a population group that is likely to find the inner-city living space appealing.

13 Why can cities be seen as 'dangerous places'?

Answers on p. 226

Rural living spaces

REVISED

As seen in the previous chapter, urban areas can be perceived differently and this can be similar for rural areas. Some people are drawn to the tranquillity of rural areas and they are often perceived as idyllic (the rural idyll).

Rural idyll A 'chocolate box' image of quaint villages set in beautiful countryside. A place thought to be free of most of the negative aspects associated with urban living.

For example many people are attracted to Cornwall because of the following:

- It is famous for its 700 km of beautiful coastline.
- The area is known for its local food and drink — its pasties, fresh seafood and award-winning wines.
- The region has a long history and strong heritage.

However, the reality of living in rural areas, such as Cornwall, brings its own challenges:

- Remoteness. Cornwall doesn't have a motorway and has only one main trainline with connections to other parts of the UK.
- Poor public transport can lead to a lack of commuting opportunities, which then gives little access to higher earning opportunities.
- There are limited social opportunities for younger generations.
- The range of services, including health services and access to shops, is limited.

Making links

Attitudes towards suburban and inner-city living not only change over time as part of changing tastes and fashion, they also change as people progress through the life cycle.

Much the same as with urban areas, a range of rural places is spread along the rural-urban continuum.

- Commuter belt: popular with adults of working age with children. Accessibility to urban jobs is the key, plus the perception that here people are able to escape from the downside of urban places. There are fast rates of population growth.
- Accessible rural: popular with retired people and with urban-based day-trippers. Population structure is unlikely to be too disturbed by domestic or international migration.
- Remote rural: victims of depopulation and the spiral of decline in such areas (see Figure 4.15). Recently, some places have experienced a reversal of fortunes, thanks to counter-urbanisation, the communications revolution and tourism.

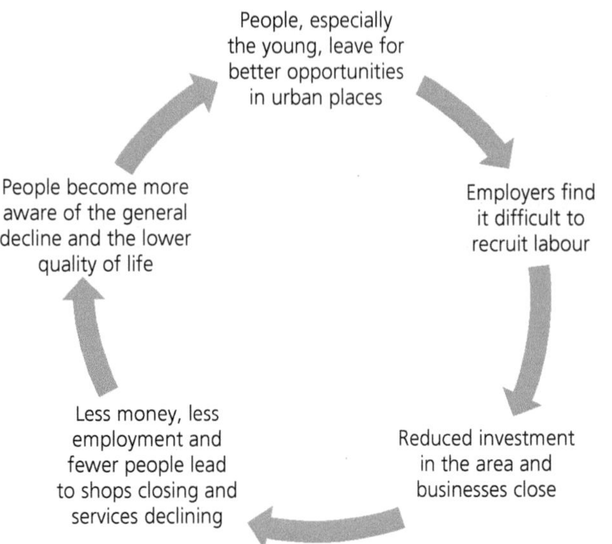

Figure 4.15 The spiral of decline in remote rural places

Revision activity

Refer to your notes of how the nineteenth-century novelist Thomas Hardy portrayed the rural idyll in his fictional county of Wessex. Was it always a 'dream county'?

Now test yourself TESTED

14 Describe the spiral of decline experienced in remote rural areas.

15 Explain how rural places can be described as idyllic.

Answers on p. 226

Typical mistake

Do not assume that everyone sees the rural idyll in exactly the same way. One person's dream can be another's nightmare.

Evaluating perceptions of living spaces

REVISED

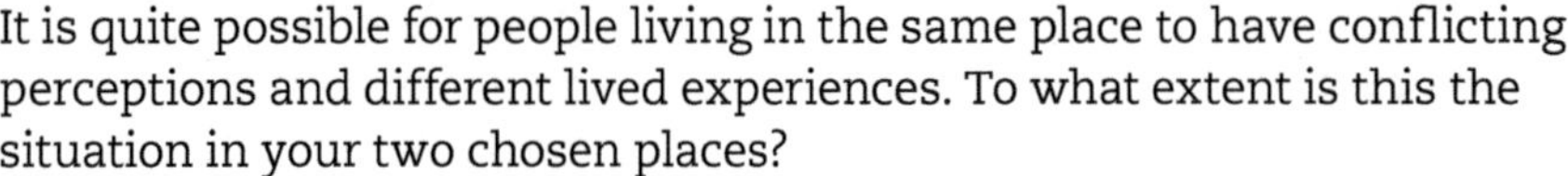

It is quite possible for people living in the same place to have conflicting perceptions and different lived experiences. To what extent is this the situation in your two chosen places?

Investigating this issue is probably best done by means of interviews and questionnaires, researching how the media view your places, and making the link between particular perceptions and lived experiences.

Another investigation required by the specification is that of demographic and cultural change:

- demographic change: relates mainly to total numbers and age structures
- cultural change: relates to shifts in the ethnic mix.

Are either or both of the changes giving rise to tensions?

There is also physical evidence of change, such as the upgrading or deterioration of housing, and the expansion or decline of social and commercial services.

Skills activity

Design an interview to understand people's perception and lived experiences.

- The aim is to create a range of questions that target people from different age groups, backgrounds and/or ethnicities.
- The questions should help you to understand people's perceptions of living and working in inner city and suburban places in an area you know.
- Create a range of structured and semi-structured questions around subjects that help you explore people's perceptions, for example:
  - crime rates
  - job opportunitioo
  - quality of services
  - environmental quality
  - how family-friendly the area is perceived to be.
- Be careful when constructing the questionnaire that your questions allow people to express their own views and that they do not lead the interviewee to think and/or answer in a certain way.

Making links

Urban and rural residents may differ in terms of their attitudes towards places and those attitudes may vary, depending on the type of place.

Now test yourself

TESTED

16 What evidence do you have that local residents hold different views about the place in which you live? Suggest reasons for the differences.

17 How can different media provide contrasting images of the place you have studied?

Answers on p. 226

Revision activity

Be sure you have produced notes summarising the results of your two place investigations, as suggested on page XXX.

Exam tip

Provided the evidence is relevant, you are recommended to use material from your own place studies to illustrate and support answers to more general questions.

Tensions in diverse places

Migration and diversity

Internal

- Migration (internal and international) is an important component of population change.
- In 2013 the UK's net population increased by approximately 400,000:
  - 56 per cent of growth was due to natural increase.
  - 44 per cent of growth was due to migration.
- For much of the twentieth century, a major internal migration within the UK was the so-called 'North–South drift' — the general movement of people from the northern parts of the country to the South East, and to London in particular.
- In 2013 London had the UK's largest natural increase and net international migration.
- However, London also saw the greatest outward internal migration, with a deficit of 55,000.
- This was mainly due to people of middle and retirement age moving from inner-urban areas to the suburbs and to more distant commuter dormitories.
- At the turn of the millennium, the character of decentralising migration began to change with the onset of counter-urbanisation and growing volumes of urban-to-rural retirement moves.

Remember that:

- internal migration within a country is a 'zero sum' phenomenon in that any net gain in one area can occur only if there is a net loss of migrants elsewhere within the country
- migration is two-way traffic
- one outcome of migration is an increase in diversity of the population, as often the people moving into an area can be quite different from those moving out
- this mixing can generate tensions, as for example between newcomers and longstanding residents.

Now test yourself

18 What is counter-urbanisation?

Answer on p. 226

TESTED

Revision activity

- Create a mind map to explain why an area in the UK has had net gain in population and why another area has suffered from net loss.
- Include examples in your mind map.
- Include economic and social causes for each.

Exam tip

Make sure you know an example to illustrate the implications of increased cultural diversity due to migration, for example how culture and society in the UK has changed because of international migration from former colonies (e.g. the Indian sub-continent, the West Indies) and the EU.

International

Since the Second World War, international migration has changed culture and society in the UK. Postwar migration occurred in two major phases.

Post-colonial migration

+ A shortage of workers following the Second World War encouraged workers and their families to migrate to the UK from countries in Africa, the Caribbean and the former Indian Empire (India, Pakistan and Bangladesh) that were or had been British colonies.
+ These immigrant flows converged on the major cities, where job opportunities were most abundant.

Globalisation

+ The UK became attractive due to its links to global brands and global culture.
+ Many migrants became London-based, but substantial numbers migrated to other parts of the UK, influencing local cultures, languages and religions.

The volumes of immigration flows have varied over time according to the booms and slumps in the UK economy.

+ Today immigration into the UK involves significant source areas within the EU, particularly from East European countries like Poland (see Figure 4.16).
+ Jobs, relatively good levels of remuneration and possibly access to social benefits are the main pull factors.
+ The most recent immigrants still largely head for towns and cities.
+ It is interesting to note that even those immigrants finding work in the fields of eastern England actually live in nearby towns, such as Boston (Lincolnshire) and Thetford (Norfolk).

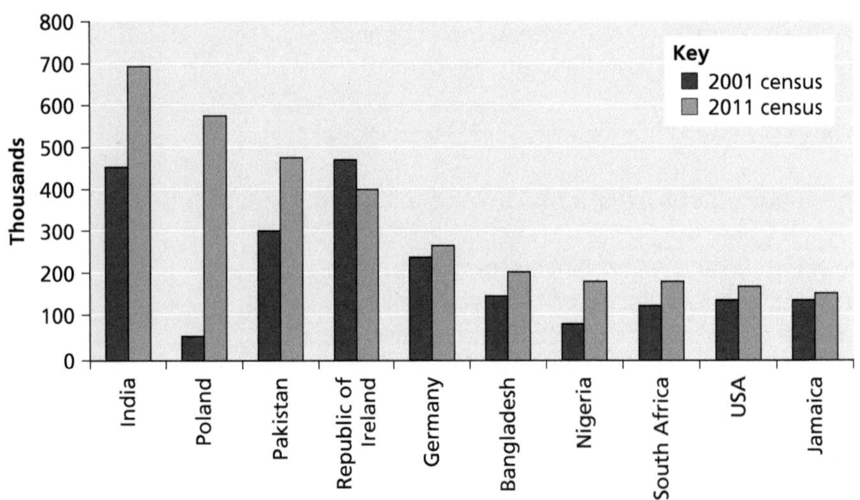

Figure 4.16 Top ten countries of origin by country of birth and nationality, UK 2018

Making links

In many countries, the government is the main player affecting migration flows, for example through its migration policies.

Typical mistake

It is wrong to think that many migrants are choosing to live in rural areas. They may do rural work but they tend to live in towns.

Exam tip

Be clear as to the main factors attracting immigrants to the cities and towns of the UK. Factors include the availability of jobs and cheap housing as well as the snowball effect of choosing to be near to family and friends who have already come from the same source area.

Revision activity

Check your notes about (a) domestic migration within the UK and (b) UK immigration since 1950 (from Commonwealth countries and more recently from Eastern Europe). Highlight the pull factors.

Now test yourself

TESTED

19 How did the main sources of UK immigrants change during the last intercensal period?

20 Explain the significance of international migration on diversity in the UK.

Answers on p. 226

Segregation

REVISED

- The 2011 Census showed that the white British element in the population of England and Wales had fallen to 80 per cent from a figure of 87 per cent in the previous census.
- But the distribution of immigrant ethnic groups is not an even one.
- In London, the white British element is now only 60 per cent of the total population; in all the other regions (with the exception of the South West, North East and Wales) the figure is close to 80 per cent.
- In all regions, the minority ethnic groups are most concentrated in towns and cities, and within those settlements still further segregated into enclaves.
- There are two contradictory views of the possible reasons for this spatial segregation of minority ethnic groups.
  - One argues that it is encouraged by external factors, such as the availability of cheap housing, with the host population 'forcing' ethnic segregation (see Figure 4.17).
  - The other view emphasises internal factors, such as protection and mutual security, that see segregation as being the wish of the segregated groups. For example, London has attracted Russian oligarchs who have purchased expensive London property to escape economic punishments in Russia. This has mostly occurred in segregated enclaves in Central London.

Enclave A group of people surrounded by a group or groups of entirely different people in terms of ethnicity, culture or wealth.

Ethnic segregation Voluntary or enforced separation of people of different cultures or nationalities.

Figure 4.17 Two views of the factors encouraging ethnic segregation

Revision activity

Make notes on at least two examples of ethnic segregation in the UK, paying particular attention to location and causal factors. Suitable London place studies include the ethnic concentration in the borough of Brent, the changing distribution of Jewish communities and the arrival of the Russian oligarchs and their families. The cities of Birmingham and Manchester also offer plenty of place examples.

+ Ethnic segregation has persisted in the UK for some time. This can be seen through markers such as:
  + places of worship (mosques, temples, synagogues, etc.)
  + restaurants and food stores reflecting the culture and food preferences of the local population.
+ It is said that the casual visitor to Southall (London) might be forgiven for thinking they've landed in a part of Mumbai, Dhaka or Karachi, so great is the 'Indianisation' of the urban landscape.

In the long term it appears that there is an increasing degree of assimilation in successive generations and a moving out of the enclave.

Assimilation The process by which different groups within a community intermingle and become more alike. The process particularly applies to the integration of immigrant minority ethnic groups.

Such moves are encouraged by a variety of factors:

+ a greater range of employment opportunities, which leads to improved earnings
+ feeling more safe and secure in the host society
+ a want to become more integrated into the host society.

Even if there is assimilation, the strength of the minority ethnic enclave is not diluted by the outward moves. This is because the people who move out are replaced by newly arrived immigrants belonging to the same minority ethnic group.

Now test yourself TESTED

21 Suggest possible measures of assimilation.

22 Explain why international migrants more commonly live in distinctive places.

Answers on p. 226

Change, tension and conflict

REVISED

+ Change is so often the generator of tension, be it a change in land use, say from residential to commercial, or a change in the people occupying an area of housing.
+ In fewer instances, change can be of such proportions as to lead to conflict.
+ The issue is that in virtually all changing situations there are winners and losers, those who benefit and those who miss out.
+ Any losers inevitably feel aggrieved and even 'cheated'.

Exam tip

For many, tension and conflict are synonymous terms. However, here it is advisable to distinguish between the two, with tension seen as a precursor to conflict.

The shifting pattern of urban land use is the outcome of competition for space. This competition involves two distinct types of living space:

+ competition on sites where there is the potential to build housing or other commodities (services, commerce and industry)
+ competition in residential areas where there is the potential to build more housing.

The fact that most urban areas are expanding is creating an insatiable demand for more space for housing and a range of urban activities.

+ When it comes to bidding on the urban land market, there is a clear pecking order. Retailing and high-order offices are usually the strongest bidders, with housing and recreation at the bottom of the bidding league table.
+ The arrival of immigrants has added to the pressure on an inadequate supply of housing and increased the competition for residential space.
+ The emergence of minority ethnic enclaves in British cities has created a fair amount of ill-feeling on the part of white British residents.
+ So-called 'race riots' (e.g. Notting Hill 1958, Toxteth 1981, Bradford 2001) have been the expressions of mounting tensions.
+ Similarly, the gentrification of residential areas has created, and still is creating, tensions between those incomers investing in the improvement of dwellings and the longstanding residents who feel they are being forced out of their homes.

Gentrification The movement of middle-class people into rundown, inner-urban areas and the associated improvement of the housing stock and area image.

- Much change in urban areas is motivated by market forces. But a significant amount has also been the responsibility of the government (national and local).
- Classic examples include the clearance of slums in the early postwar years and their replacement by high-rise blocks of flats. The lived experience of many occupying those blocks was found to be isolating and alienating.
- Equally, the apparent neglect by the public authorities of areas of acute poverty and deprivation leads to feelings of social exclusion and resentment, particularly among migrant groups (e.g. in parts of Glasgow).

Making links

Planners and developers may make controversial decisions. Their priorities and attitudes may differ from those of local groups.

Revision activity

Check through your notes for examples of the different tension scenarios. Possible place studies include Lewisham (London), Luton and Glasgow.

Create a case study that includes:

- examples of the changes to the place you have chosen
- how it has benefited some groups.
- how the change has provoked hostility from other groups that perceived change as a threat to their culture.

Now test yourself TESTED

23 What evidence do you have of the existence of tensions in your two place studies? What appears to be the cause(s) of those tensions?

24 How have changes to an area created challenges and opportunities for people? Include specific examples in your answer.

Answers on p. 226

Managing cultural and demographic issues

There are many aspects of change in both rural and urban places that need some form of management in order to achieve a satisfactory outcome. What constitutes an issue is important to an understanding of this final part of the topic. Figure 4.18 indicates some demographic and cultural **issues** that would qualify for management.

Management A set of actions that facilitate the transition from one situation to another. More specifically, those actions might be aimed at solving or ameliorating a particular problem or issue.

Key concepts

An **issue** is an important topic or problem for discussion and hopefully solution by means of planning and management. Examples relevant to this topic include immigration, housing, discrimination and segregation, poverty and deprivation.

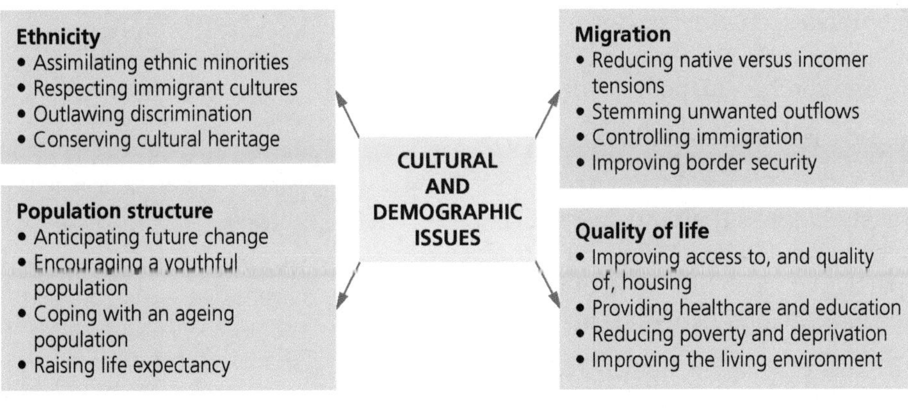

Ethnicity
• Assimilating ethnic minorities
• Respecting immigrant cultures
• Outlawing discrimination
• Conserving cultural heritage
Population structure
• Anticipating future change
• Encouraging a youthful population
• Coping with an ageing population
• Raising life expectancy
CULTURAL AND DEMOGRAPHIC ISSUES
Migration
• Reducing native versus incomer tensions
• Stemming unwanted outflows
• Controlling immigration
• Improving border security
Quality of life
• Improving access to, and quality of, housing
• Providing healthcare and education
• Reducing poverty and deprivation
• Improving the living environment

Figure 4.18 Possible cultural and demographic issues requiring management

Evaluation

REVISED

A crucial stage is reached in the management of an issue when it becomes necessary to take stock of what has been done and to assess whether or not it has achieved or is achieving the original objective. There is a variety of potential measures available here. Which of these to use depends on the issue. Table 4.9 provides three examples.

Table 4.9 Some indicators or measures of issue management

Issue	Possible measures
Assimilation and integration	Number of people moving out from enclaves Voter turnout Incidence of hate crime Involvement in the wider community
Social progress	Trend in index of multiple deprivation (IMD) Unemployment rate Number of households on social benefits Change in life expectancy
Housing provision	Number of new units built Percentage breakdown — for sale and rent Incidence of social housing Council housing waiting list

Now test yourself TESTED

25 Suggest another possible measure for each of the issues in Table 4.9.

26 Explain how ways to manage cultural and demographic inequalities have been successful. Include specific examples in your answer.

Answers on p. 226

Revision activity

Create a mind map to explain how low pay, rising house prices and ethnicity can contribute towards poverty in a city.

Stakeholders

REVISED

A number of groups take an interest in how successfully cultural and demographic issues are managed. These include the decision makers, but also stakeholders such as individuals, groups and organisations with an interest in the issue being managed.

In most situations, stakeholders fall into four groups:

- Providers: could be landowners, investors, contractors.
- Community groups: those who stand to benefit (or lose out).
- Governance: local government officials, enforcers of local bye-laws and national government policy.
- Influencers: action groups, e.g. environmental, political parties.

Each stakeholder, no matter whether the issue relates to an urban or a rural place, has their own vested interests, particular perceptions and agendas.

From this, it follows that each stakeholder has their own particular view or opinion of what constitutes 'success' and 'failure'. They have their own criteria for assessing whether a particular issue has been, or is being, managed successfully or not. Each stakeholder arrives at their own verdict.

In the remaining two sections, the best that can be done is to take two issues (one urban and one rural) and identify the stakeholders as well as causal factors. In both, the emphasis is on the management of change.

Stakeholder An individual, group or organisation with a particular interest in the actions and outcomes of a project or issue-solving exercise.

Typical mistakes

It is wrong to think that everyone agrees on what constitutes a successful outcome in the management of urban and rural issues.

It is wrong to believe that all stakeholders are equal when it comes to influencing the management of an issue.

Now test yourself

TESTED

27 Check that you understand the difference between the four types of stakeholder.

Answer on p. 226

An urban issue

REVISED

In meeting the rising demand for housing in a rapidly growing city surrounded by a green belt and where there is very little land for housing within its boundaries, the issue is where to build the houses once the available land within the city has been developed. Where outside the city should the required housing be built?

The stakeholders fall into two groups: internal and external. Table 4.10 identifies some of them and their particular interests and perceptions.

Table 4.10 A sample of stakeholders

Stakeholders	Particular interest
Internal stakeholders	
City council	Ensuring the shortage of housing does not hinder the growth of the city's economy.
Employers and businesses	Difficulties in recruiting labour as workers are deterred by the shortage and consequent expense of housing.
Social service providers	The rising demand for education and healthcare; recruiting labour to the social services.
Local residents	Concerns related to the impact of more housing on city traffic and quality of living space.
External stakeholders	
National government	Ensuring everything is done to keep the city flourishing.
County council	Anxious to provide the space needed for new housing, but at the same time keen not to alienate local people.
Conservation groups	Protection of designated green belt and concern about the possible impact of housing developments on other rural places and wildlife.
Local residents in areas designated for new housing	Maintaining the status quo and keeping rural areas rural. Concerns related to the impact of new housing on property values.

Making links

What is deemed to be 'successful' depends on the attitudes of different players.

Revision activity

The issue of improving the quality of rural life is only one of a number of rural issues that you might have studied.

For the managed issue you have studied, identify the main stakeholders and set out your notes as in Table 4.10 or Figure 4.19. Other possible issue-oriented place studies include Southall (London) and Oxford.

Now test yourself

TESTED

28 How successful have recent changes been to the diverse urban living space you have studied?

Answer on p. 226

A rural issue

REVISED

How to maintain commercial and social services is a common issue in many rural parts of the UK. Another is how to deal with increasing pockets of deprivation and poverty.

Figure 4.19 illustrates the views of internal and external stakeholders in one such problem area, Breckland in Norfolk. What can be done to improve the quality of life in this local authority area? To what extent do stakeholders' views converge to define a common course of action?

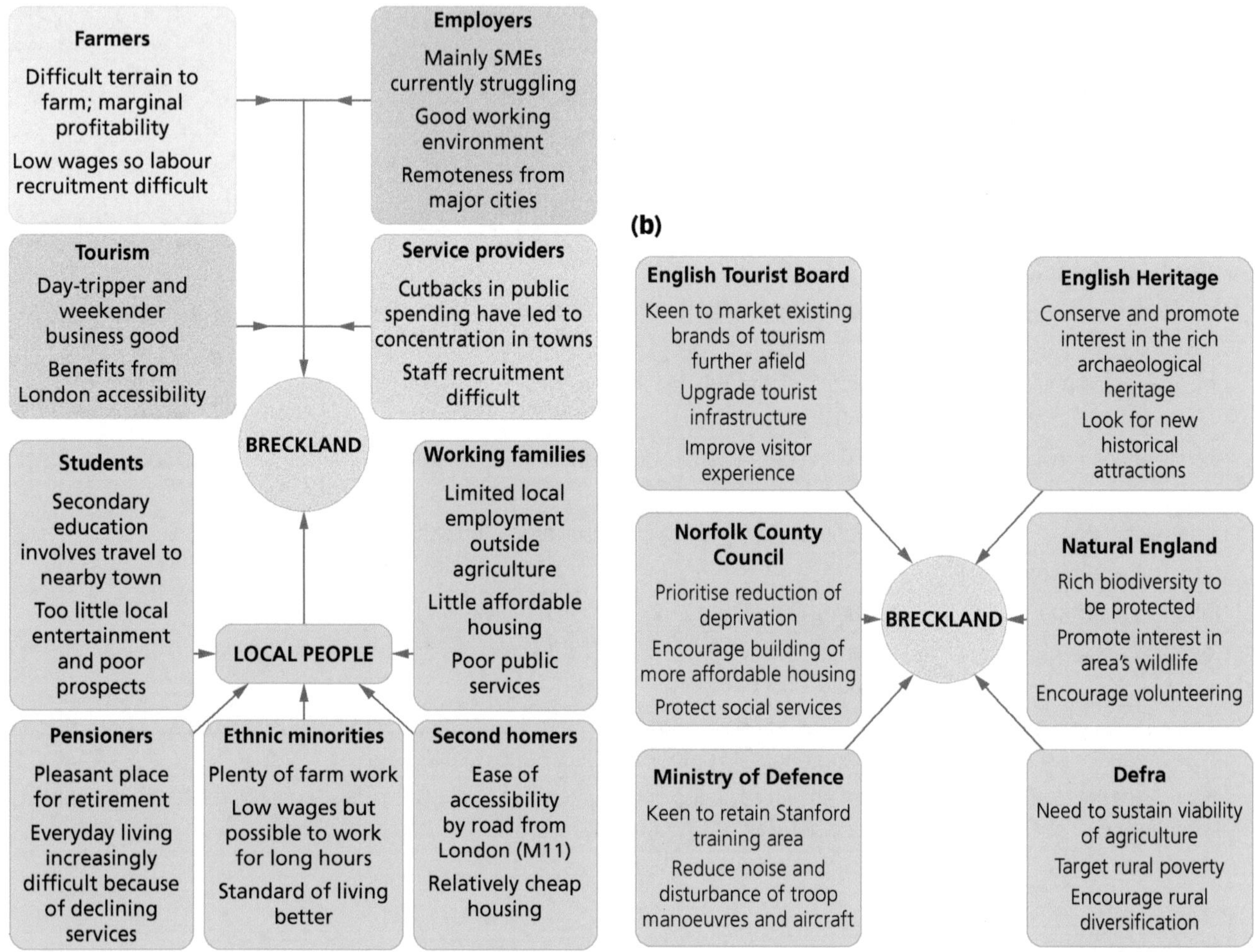

Figure 4.19 Breckland stakeholders: (a) internal and (b) external

Now test yourself TESTED

29 How successful has managing change been for rural communities in the place that you have studied?

30 How have changes at a national and local level impacted on a rural community you have studied?

31 Check that you are able to provide examples of stakeholders drawn from your own place investigations.

Answer on p. 226

Revision activity

Figure 4.19 is only one of a number of rural issues you might have studied. For the managed issue you have studied, identify the main stakeholders and summarise your notes as in either Table 4.10 or Figure 4.19.

Making links

What is deemed to be 'successful' depends on the attitudes of different players. For example, the attitudes of the UK government and environmental stakeholders may be very different when assessing how money should be invested into Cornwall.

Revision activity

For the managed issue you have studied, in a rural and urban area, identify the stakeholders involved in this change and complete the conflict matrix in Table 4.11 to show which stakeholders may agree or disagree with each other.

Table 4.11 Conflict matrix for stakeholders

	Stakeholder A				
Stakeholder A		Stakeholder B			
Stakeholder B			Stakeholder C		
Stakeholder C				Stakeholder D	
Stakeholder D					Stakeholder E
Stakeholder E					

++ Strongly agree + Agree -- Strongly disagree - Disagree

To what extent does your conflict matrix show a success or failure? Explain your answer.

Exam skills summary

You should be familiar with the skills and techniques used to investigate the following aspects of places:

- identity: investigating media images and various forms of qualitative data
- perceptions and lived experiences: interviews, oral accounts and social media
- spatial variations: using GIS; indices of ethnicity and cultural diversity
- change: interpreting old photographs, maps and newspaper reports.

Exam tip

Always remember that not all stakeholders are equal when it comes to influencing the management of an issue. When studying any issue, try to rank stakeholders according to the strength of their influence.

Exam practice

A-level

1 a) Study Figure 1.

Figure 1 A cartoon showing the challenges of rural life

i) Explain why rural locations can be perceived as undesirable. (3)

ii) Suggest reasons why rural places are often perceived as idyllic. (6)

b) Explain the significance of international migration on diversity in the UK. (6)

c) Evaluate the extent to which changes in diverse places can lead to tension and conflict. (20)

2 a) Study Figure 2.

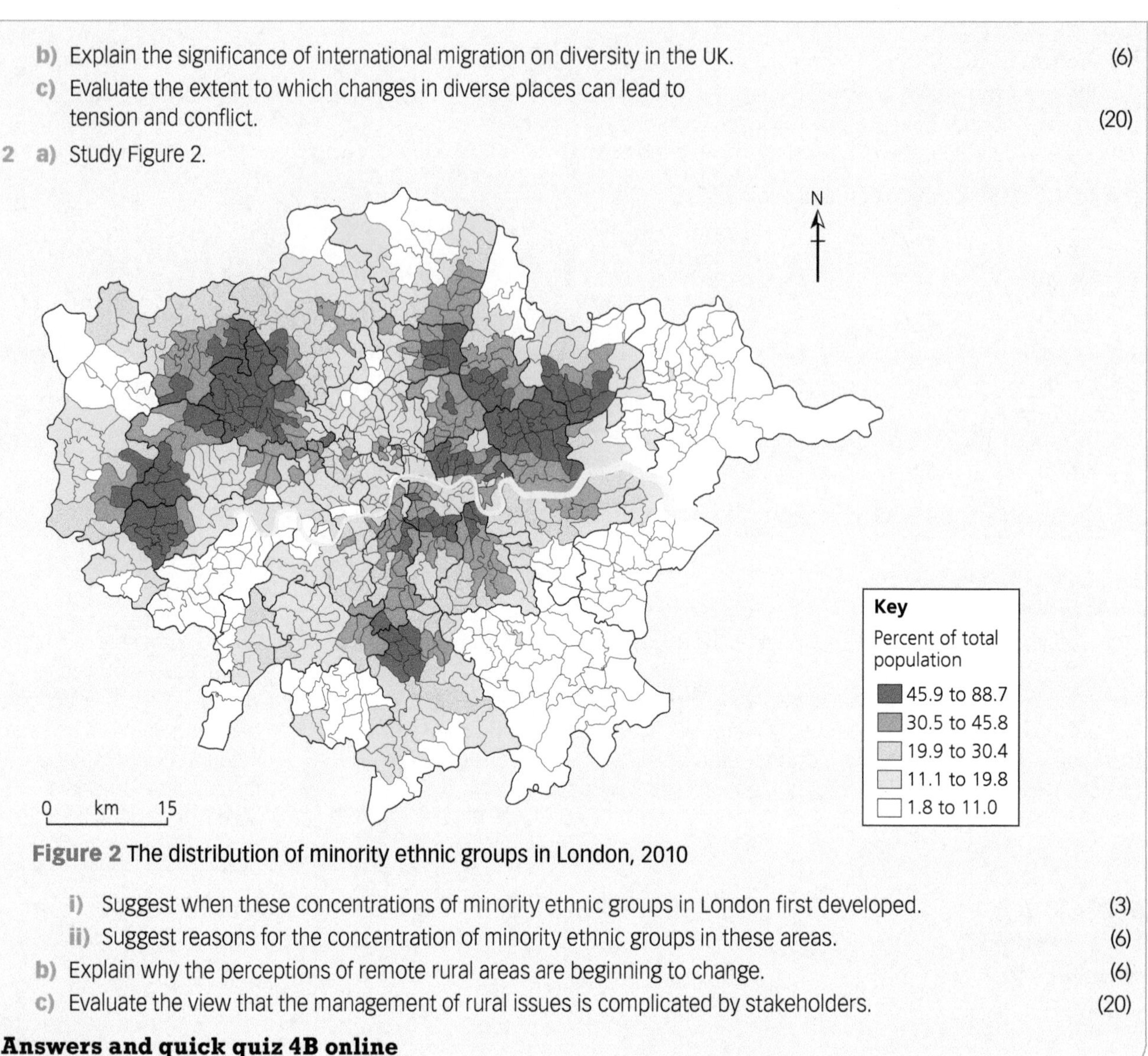

Figure 2 The distribution of minority ethnic groups in London, 2010

i) Suggest when these concentrations of minority ethnic groups in London first developed. (3)

ii) Suggest reasons for the concentration of minority ethnic groups in these areas. (6)

b) Explain why the perceptions of remote rural areas are beginning to change. (6)

c) Evaluate the view that the management of rural issues is complicated by stakeholders. (20)

Answers and quick quiz 4B online

Summary

You should now have an understanding of:

- population numbers, and how densities and structures vary over time and from place to place, particularly along the rural-urban continuum
- the meaning of place and the factors that give places and their inhabitants a sense of identity
- how both urban places and rural places are seen differently by different groups, which is the outcome of variations in lived experiences and perceptions of places
- how society and culture in the UK have increased in diversity largely as a result of immigration, first from Commonwealth countries and more recently from Eastern Europe
- how segregation is a feature of immigrant ethnic groups in the towns and cities of the UK
- how demographic and cultural change can easily lead to tension and even conflict
- demographic and cultural issues requiring management. Any assessment of management depends on the issue as well as on the stakeholders and their criteria.

5 The water cycle and water insecurity

It is important to understand how water circulates at a global level and how the availability of water fluctuates over time and space. Water is vital to human survival and yet **water insecurity** is increasing and so too the likelihood of **water wars**.

Processes operating within the hydrological cycle

The global hydrological cycle

REVISED

Figure 5.1 shows how the **global hydrological cycle** works.

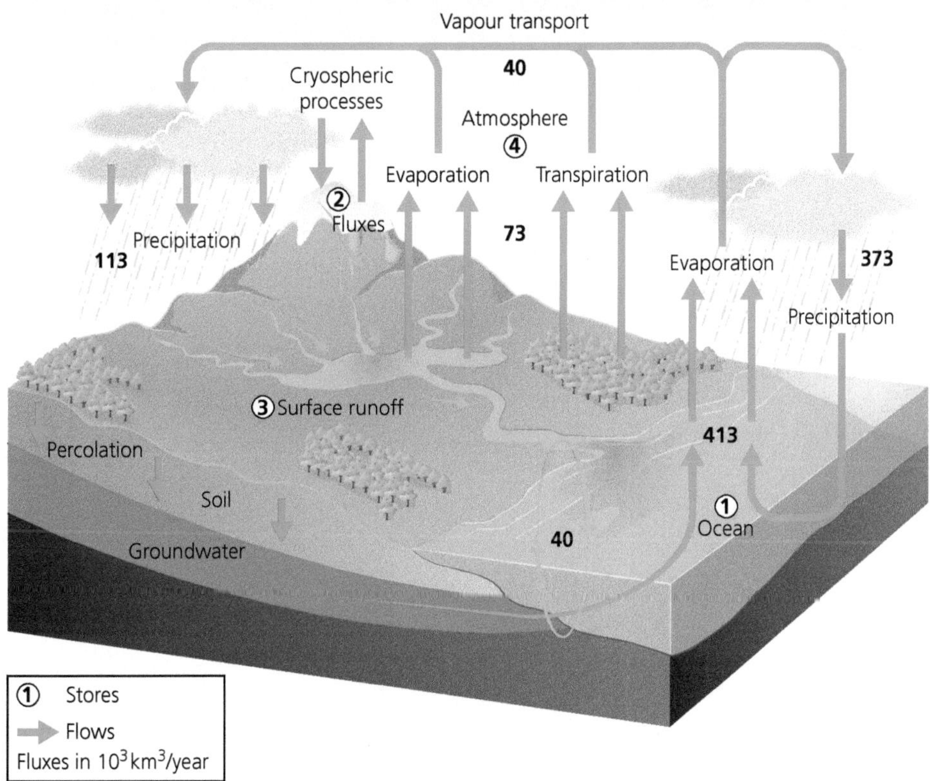

Figure 5.1 The global water cycle

> **Key concept**
>
> The **global hydrological cycle** is a closed system. It does not have any external inputs or outputs. So the volume of water is constant and finite. The system has three components:
>
> - stores: reservoirs where water is held
> - fluxes: the flows which move water between stores
> - processes: the physical mechanisms which drive the fluxes between stores.

> **Typical mistake**
>
> It is wrong to believe that the global hydrological cycle has inputs and outputs. It is a closed system.

The energy motivating the global hydrological cycle comes from two sources: the Sun and gravity.

Stores

The main stores of water and freshwater are shown in the pie charts in Figure 5.2.

With freshwater, a distinction is sometimes made between blue water and green water.

Water stores have different residence times. In general, the larger the store, the longer the residence time. Some stores are non-renewable; for example, fossil water and when the cryosphere melts.

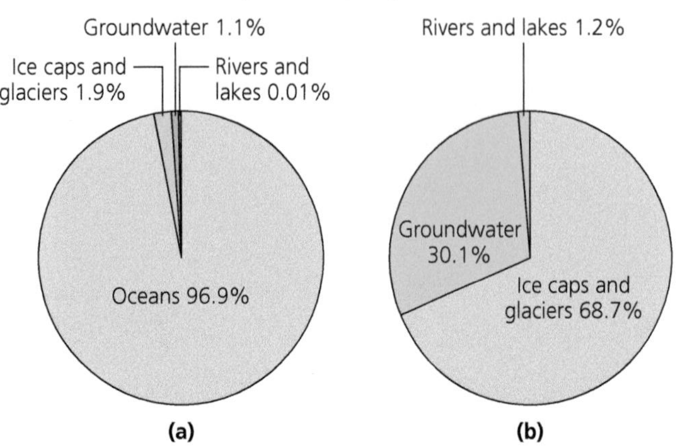

Figure 5.2 The main stores of (a) water and (b) freshwater

Blue water Freshwater stored in rivers, streams and lakes — the visible part of the hydrological cycle.

Green water Freshwater stored in the soil and vegetation — the invisible part of the hydrological cycle.

Residence time The average time a water molecule spends in a store or reservoir.

Fossil water Ancient, deep groundwater from former pluvial (wetter) climatic periods.

Cryosphere Water frozen into ice and snow.

Flows and processes

Flows are the transfers of water from one store to another. They are achieved by processes such as precipitation, evaporation, transpiration, cryospheric exchanges and runoff.

- One of the most important elements of the hydrological cycle for humans is the availability of freshwater.
- The global water budget limits this amount.
- The accessible sources are lakes (natural and artificial), rivers and groundwater.
- Overall only about 1 per cent of global freshwater is easily accessible for human use.

Exam tip

'Fluxes' and 'transfers' are two alternative terms for 'flows'. It is recommended that you stick to the term 'flows'.

Now test yourself TESTED

1 Why is such a small percentage of freshwater accessible for human use?

Answer on p. 227

Exam tip

Remember that:

- climate change impacts on the hydrological cycle, most noticeably on the relative importance of the different stores
- the hydrological cycle is affected in a small way by human interventions, for example by water storage reservoirs and irrigation schemes.

The drainage basin as an open system

REVISED 

- The drainage basin is a subsystem within the global hydrological cycle. It is an open system with inputs and outputs.
- Drainage basins can vary enormously in size. They commonly 'nest' within each other.
- The drainage basin of a major river like the Amazon is made up of the drainage basins of tributary rivers.
- Those tributary basins, in turn, are made up of the even smaller basins of streams draining into those tributaries.

Inputs and outputs

Figure 5.3 illustrates the water inputs and outputs of a drainage basin.

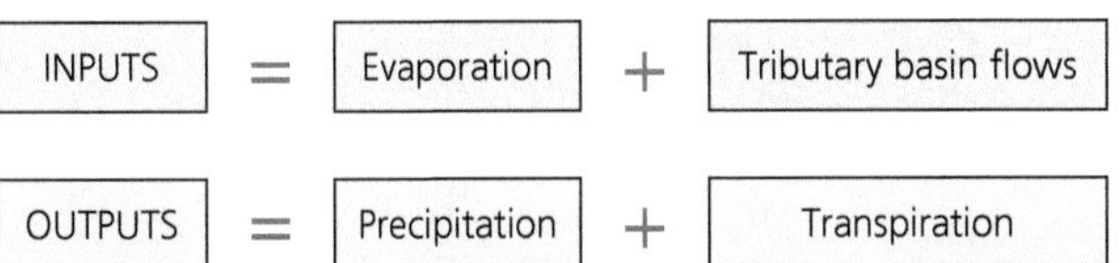

Figure 5.3 The water inputs and outputs of a drainage basin

Flows

Figure 5.4 illustrates the flows of water within a drainage basin.

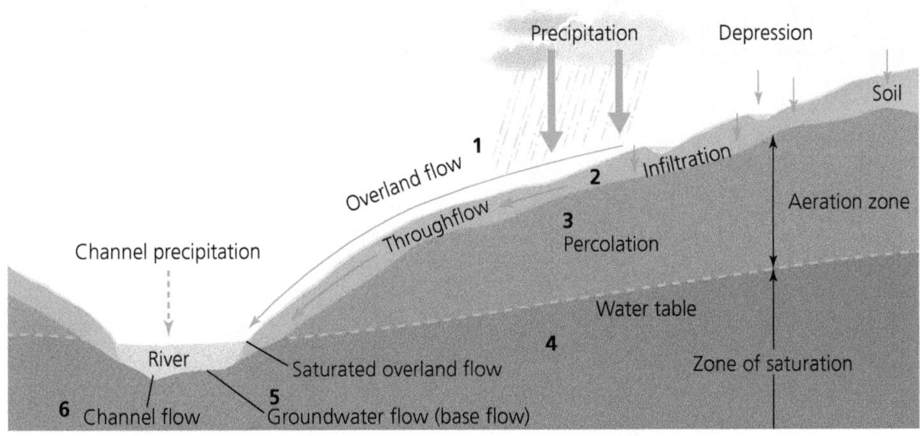

Figure 5.4 How the various flows operate within the drainage basin system

Revision activity

Design a revision card for each of the stores and transfers, adding their definitions to the back of the card.

Physical factors influencing the drainage basin cycle

- Climate is one of the most important factors illustrated in Table 5.1, determining the key input, precipitation, as well as a significant output, evaporation.
- Soils, geology and vegetation affect virtually all transfers, from infiltration and interception to throughflow and overland flow.
- Relief is significant through its impact on precipitation and runoff.

Table 5.1 The impact of physical factors within the drainage basin on inputs, flows and outputs

Physical factor	Impacts
Climate	Has a role in influencing the type and amount of precipitation overall and the amount of evaporation, i.e. the major inputs and outputs, and an impact on the vegetation type
Soils	Determine the amount of infiltration and throughflow and, indirectly, the type of vegetation
Geology	Can impact on subsurface processes such as percolation and groundwater flow (and, therefore, on aquifers) and, indirectly, alters soil formation
Relief	Can impact on the amount of precipitation, and slopes can affect the amount of runoff
Vegetation	Its presence or absence has a major impact on the amount of interception, infiltration and occurrence of overland flow, as well as on transpiration rates

Now test yourself TESTED

2 Summarise how vegetation can play a role in the inputs, flows and outputs of a drainage basin.

Answer on p. 227

Human factors influencing the drainage basin system

Human activity can affect the drainage basin system, particularly the flows. Consider the impacts of deforestation on the aspects of the drainage basin system shown in Figure 5.5.

Evaporation	Runoff	Evapotranspiration	Interception	Infiltration	Groundwater
⬆	⬆	⬇	⬇	⬇	⬇

Figure 5.5 How human activity affects the drainage basin system

When water is abstracted from rivers and lakes for use in irrigation or industry this can upset the natural balance of the drainage basin cycle.

Now test yourself TESTED

3 Explain how interception and evapotranspiration rates are affected by deforestation.

Answer on p. 227

The hydrological cycle at a local scale

REVISED

This section involves looking at three different aspects of a drainage basin's hydrology: water budgets, river regimes and storm hydrographs.

Water budgets (water balances)

Water budgets show the annual balances between inputs (precipitation) and outputs (evapotranspiration and runoff). They are usually expressed by the formula:

$$P = R + E \pm S$$

where P = precipitation, R = runoff or streamflow, E = evapotranspiration, S = changes in storage.

Figure 5.6 is an example of a typical water budget graph.

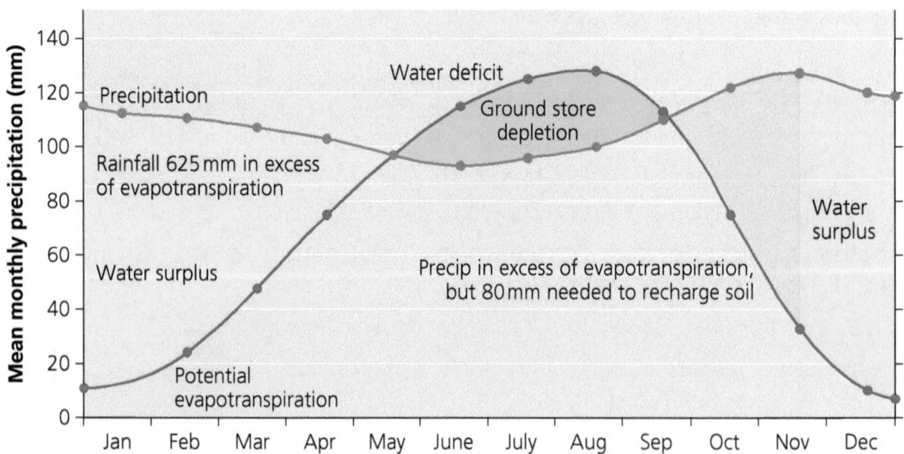

Figure 5.6 A typical water budget graph

Water budgets are useful because:

- They can be used to monitor the amount of water held in stores.
- The base flow provides information on the usually available water.

Water budgets are influenced by the type of climate (not just temperatures and precipitation, but also seasonality).

Runoff is usually divided into surface flow and base flow.

In some places there are severe seasonal differences in surface flow, as for example in monsoonal areas.

Since water budgets impact on the availability of water in the soil, they are therefore of great importance to farmers.

River regimes

- A river regime is the annual variation of the discharge or flow of a river at a particular point along its length.
- It is particularly affected by climate, geology and soils.

Storm hydrographs

Storm hydrographs show how the discharge of a river varies within a short period of time, such as before, during and immediately after the passing of a storm.

- A 'flashy' hydrograph illustrates a river that is likely to flood at a faster rate and has the following characteristics — shorter lag time, high peak discharge, a steep rising limb (see Figure 5.7).
- A 'flat', more gentle hydrograph illustrates a river that is likely to flood at a slower rate with the following characteristics — longer lag time, lower peak discharge, a gently rising limb.

The key factor is the speed with which the rainfall reaches the river. This depends on:

- a range of different physical features of the drainage basin — size, shape, drainage density, rock type, soil, relief and vegetation
- human factors such as land use and urbanisation. Increasing urbanisation changes the relative importance of flows. Most notable is the increasing amount of runoff. This, in turn, increases the risk of flooding.

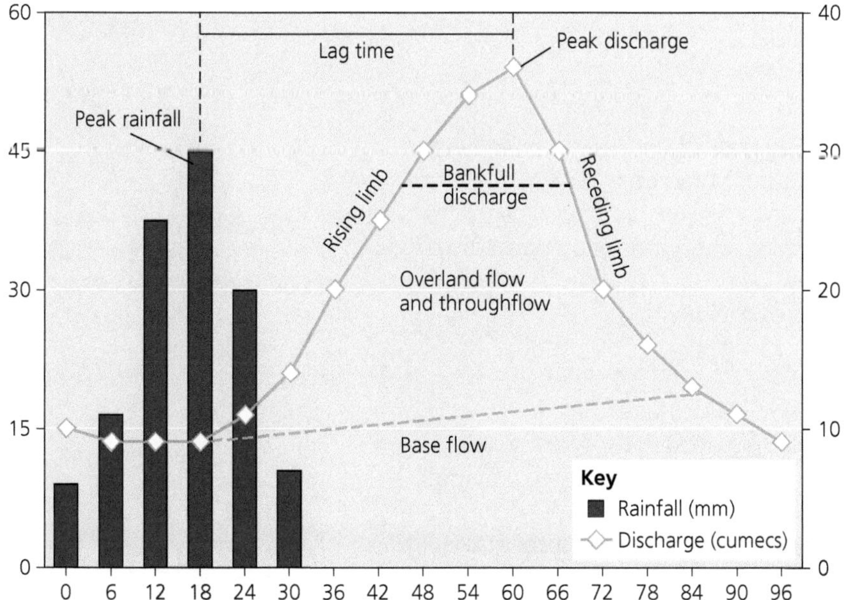

Figure 5.7 An example of a flashy storm hydrograph

Now test yourself

TESTED

4 Explain two factors that influence the shape of a storm hydrograph.

Answer on p. 227

Exam tips

Annotate examples of the water budget characteristics in tropical, temperate and polar locations. You should have a clear understanding of how the characteristics of these water budgets vary in these three locations.

Familiarise yourself with the key features of the two different storm hydrographs and understand the different physical and human factors that cause these differences.

Skills activity

Calculate the lag time for the storm hydrograph shown in Figure 5.7.

Making links

Through their control of land use and land-use changes in a considerable number of countries, planners are major players. The decisions they make, for example about the location and amount of urban growth, can clearly affect the behaviour of drainage basin systems. Their decisions impact most on runoff, water budgets and storm hydrographs.

Exam tip

At this point in the course, you need to consider the role of planners in managing land use, particularly within urban areas. Why should they do it? How might they do it?

Changes in hydrological systems over time

There are three changes to be understood: the short-term but temporary disruptions of both droughts (water deficits) and floods (water surpluses), and the longer-term disruptions caused by climate change.

Droughts or water deficits

REVISED

Causes

- Deficits within the hydrographical cycle often lead to droughts.
- Abnormally prolonged shortfalls in precipitation (meteorological droughts) often correlate with **ENSO** cycles.
- But droughts of increasing frequency and severity may simply be part and parcel of longer-term changes in climate (see below).

Key concept

The **El Niño–Southern Oscillation (ENSO)** occurs in the Pacific Ocean. But it has a global impact on weather patterns, resulting in more intense storms in some places and drought in others. El Niño events are reversals of the normal directions of ocean currents and winds in the Pacific Basin. Such events usually occur every 7 years and last for 18 months.

Now test yourself

5 Summarise the differences between El Niño and La Niña events.

Answer on p. 227

TESTED

Meteorological droughts give rise to two sequentially related droughts:

- Agricultural drought: the shortfall in precipitation leads to a decline in soil moisture and soil water availability. This has a knock-on effect of reducing plant growth and biomass. Crops yields decline, irrigation systems fail and livestock perish.
- Hydrological drought: reduced precipitation and high rates of evaporation lead to reduced stream flows and falling groundwater levels, as well as reduced storage.

Exam tip

Be sure you know the three main human factors contributing to desertification.

Contributory human factors

The desertification of the Sahel (Africa) illustrates how people have helped accelerate the process through a combination of factors:

- overpopulation
- environmental degradation resulting from overgrazing by nomadic tribes and desperate attempts to grow crops
- deforestation as a result of the cutting of fuelwood.

This leads to rural poverty, malnutrition and starvation.

Desertification The degradation of land in arid and semi-arid areas resulting from various factors including climatic variations and human activities.

In Australia, droughts vary considerably:

- While some are intense and short-lived, others last for years. Some are very localised, others impact on huge areas of the country for several years, for example the 'Big Dry' of 2006.
- Australia illustrates the two sequential droughts: agricultural and hydrological.
- More than half the farmland has been negatively affected.
- Water supplies to Australian cities have been threatened as reservoir levels have fallen.
- The pressure on water resources continues to increase due to the growth of population with its high water-consuming lifestyles.

Now test yourself

6 Why are Australians characterised as having a high water-consuming lifestyle?

Answer on p. 227

TESTED

Impacts

- Droughts have many impacts, not just on people but also on various elements of the physical environment.
- Droughts have the potential to cause ecosystem stress and to test ecosystem resilience.

Drought can lead to stress on our forests but over time they can recover. However, wetlands do not recover as easily.

Wetlands cover approximately 10 per cent of the Earth's land surface and are important for a number of reasons:

- they act as temporary stores within the hydrological cycle
- they have very high biological productivity
- they provide a range of valuable goods and services (see Table 5.2)
- they act as giant filters through trapping and recycling nutrients.

Table 5.2 The value of wetlands

Supporting	Provisioning	Regulating	Cultural
Primary production	Fuelwood and peat	Flood control	Aesthetic value
Nutrient cycling	Fisheries	Groundwater recharge/discharge	Recreational opportunities
Food chain	Tourist attractions	Shoreline protection	Cultural heritage
Life support in carbon cycles		Water purification	

Despite these valuable functions, and their resilience to drought, wetlands have been drained, dredged and infilled mainly to create farmland or space for urban development.

However, after more than 50 years of destruction, the ecological value of wetlands is slowly being realised and some are being given official protection (for example, those designated as a result of the Ramsar Convention on Wetlands).

Ecosystem stress Constraints on the development or survival of ecosystems. The constraints can be physical (drought), chemical (pollution) and biological (diseases).

Ecosystem resilience The capacity of an ecosystem to recover from disturbance or to withstand an ongoing pressure, such as drought.

Wetlands Areas where the soil is frequently or permanently waterlogged by fresh, brackish or salt water. The water may be static or flowing. The vegetation may be marsh, fen or peat.

Now test yourself

7 Explain two reasons why wetlands are important.

Answer on p. 227

TESTED

Floods or water surpluses

REVISED

When a drainage basin is unable to cope with the input of vegetation this leads to flooding, which is caused by several interrelating physical factors.

Flood-producing conditions are:

- intense precipitation over a short period (flash flooding)
- sudden snow melt (jökulhlaups)
- unusually heavy and prolonged rainfall (monsoonal rainfall).

Figure 5.8 illustrates additional physical factors involved in flooding.

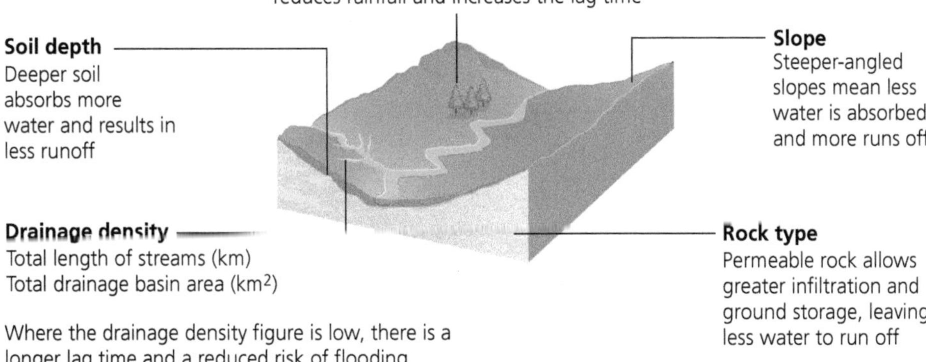
Vegetation
Greater vegetation cover generally produces higher levels of interception, storage and evapotranspiration. This reduces rainfall and increases the lag time
Soil depth
Deeper soil absorbs more water and results in less runoff
Slope
Steeper-angled slopes mean less water is absorbed and more runs off
Drainage density
Total length of streams (km)
Total drainage basin area (km2)
Where the drainage density figure is low, there is a longer lag time and a reduced risk of flooding
Rock type
Permeable rock allows greater infiltration and ground storage, leaving less water to run off

Figure 5.8 Physical factors contributing to flooding

People can exacerbate the flood risk by:

- changing land use and increasing runoff — for example, by deforestation, overgrazing, draining of wetlands and urban development
- mismanagement of rivers — for example, straightening river channels can improve the flow of water at one location but increase the flood risk downstream
- poor maintenance of rivers and ditches.

In recent decades, the UK has been experiencing increasing flood risk from extreme flood events. On 7 November 2019, the UK experienced in one day over half of the average rainfall that would usually fall for the whole of November.

These floods clearly demonstrated the wide range of impacts that this scale of flooding can have (see Table 5.3).

Table 5.3 The impacts of extreme flood events

Living standards	Environmental	Economy	Health	Infrastructure
Damage to dwellings	Degradation of soils and disrupting ecosystems	Significant impacts on the economics of an area	Loss of life and long-term life-changing illnesses	Damage to key infrastructure — road, railways and water supply

Impacts of climate change

REVISED 

Inputs and outputs

- Most scientists agree that climate change (the outcome of global warming and ENSOs) is enhancing and accelerating the global hydrological cycle.
- In the future, it is expected that climate change will have direct consequences for water availability, with changes in temperature, precipitation and evaporation varying around the world.

Now test yourself

8 Explain the link between more water circulating around the global hydrological cycle and more energy in the atmosphere.

Answer on p. 227

TESTED

Stores and flows

Climate change is already affecting stores by:

- decreasing the total amount of water held in the form of snow and ice
- deepening the active layer of the permafrost
- lowering water levels in lakes and reservoirs
- reducing wetland storage
- reducing soil moisture.

As for flows, it looks as if more climate extremes will be reflected in an increase in hydrological extremes.

- There will be more high flows (floods) in some parts of the world and more low flows (droughts) in others.
- More intense rainfall in some locations will increase runoff rates and reduce infiltration.

Typical mistake

It is not the case that climate change only causes more droughts. This is true for some climatic regions, but there are some regions where the prognosis is for more floods.

Future uncertainty

Climate change is the outcome of (a) short-term oscillations (such as ENSO cycles) and (b) long-term shifts that are part of global warming.

There is uncertainty about how exactly climate and hydrological cycles will change in specific locations, particularly in the long term.

This is raising concerns about water management generally. We need to find the answers to the following key questions:

- How reliable are our projections about drought and flood risks?
- Can we accurately factor in the possibility of more extreme weather events?
- How accurately can we forecast human pressure on water resources?

Figure 5.9 illustrates the range of impacts of short-term climate change on water supply.

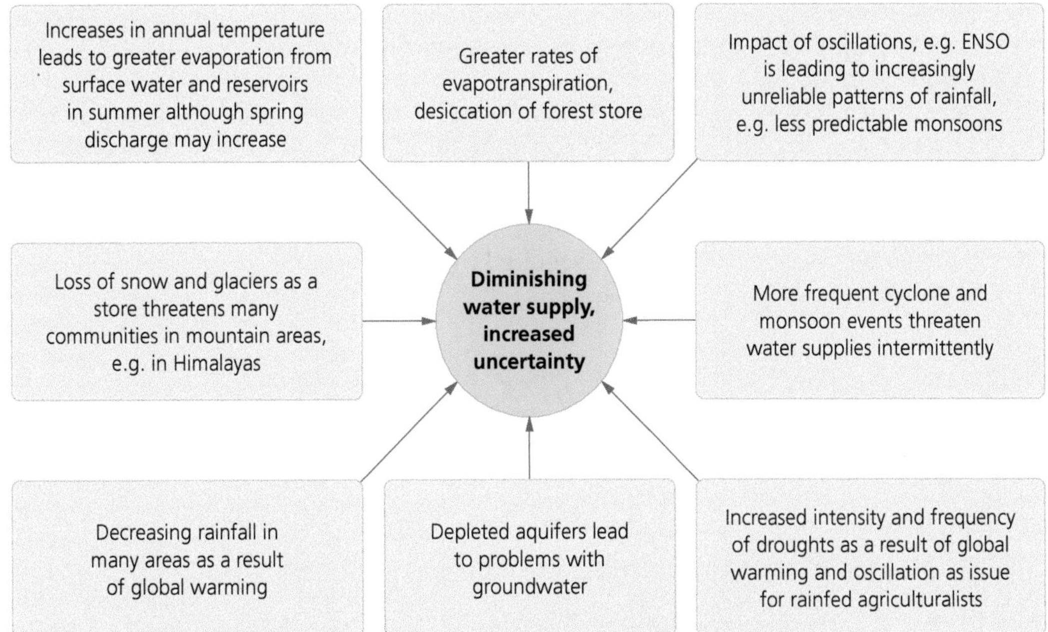

Figure 5.9 The impacts of short-term climate change on water supply

Making links

The threat that most frightens decision makers and players in the water supply industry is the impact of short-term climate change on water supply (see Figure 5.9). Might the possible outcome be a downward path into water insecurity? Is it possible to predict with any accuracy the risks of droughts and floods?

Now test yourself TESTED

9 Make a list of those factors that create uncertainty about future water supplies.

Answer on p. 227

Water insecurity

The causes of water insecurity

REVISED

A growing mismatch between water supply and demand

It is this growing mismatch between water supply and demand that is creating an increasing amount of **water insecurity** in the world.

Key concept

Water insecurity begins to exist when available water is less than 1,700 cubic metres per person per day. This marks the start of water stress. Below 1,000 cubic metres per person per day, water stress gives way to water scarcity.

Remember that the amount of freshwater in the global hydrological cycle is finite.

It is estimated that currently 60 per cent of the world's accessible freshwater is being used.

In theory, this leaves plenty more for future use. But the situation is not that simple:

- There is a mismatch between where water is available and where the demand is — 66 per cent of the world's population lives in areas receiving only 25 per cent of the world's annual rainfall.
- There is a widening water availability gap as a result of rising water demand and dwindling water supplies. The reason for the diminishing water supplies is mainly the over-exploitation of groundwater stores for irrigation.

These are the main drivers of water insecurity. They have resulted in a number of pressure points (see Figure 5.10).

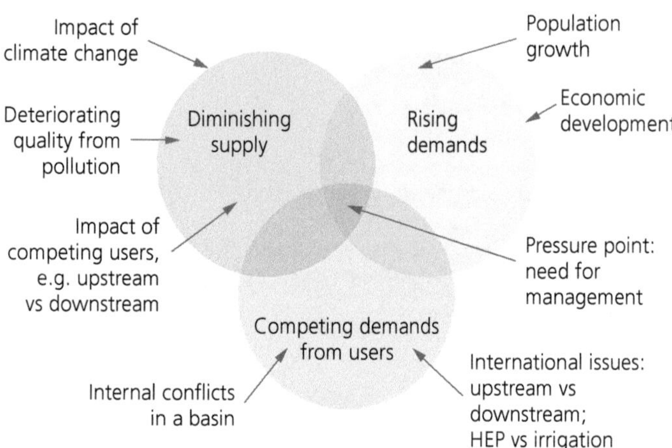

Figure 5.10 Water pressure points

The widening water availability gap means that the world is becoming divided between the 'have-nots' (largely developing countries, particularly in sub-Saharan Africa) and the 'haves' (largely developed countries in temperate latitudes).

Now test yourself

10 What are the main drivers of water insecurity?

Answer on p. 227

TESTED

The physical causes of water insecurity

In 2020, estimates suggested that one-third of people are now affected by water insecurity where the supply is affected by several physical factors:

- Climate: this determines the global distribution of water supply via precipitation. But precipitation varies not only from place to place but also seasonally. Added to this is the impact of global warming which, through the medium of more frequent droughts, is making rainfall receipts less reliable.
- Salt-water encroachment: rising sea levels and coastal erosion are leading to the contamination (salinisation) of freshwater by seawater in coastal areas.
- Over-abstraction of water from rivers, lakes and groundwater: the last of these is contributing to salt-water encroachment in coastal areas.

The human causes of water insecurity

In the case of the demand situation, the causes are largely human, as shown in Table 5.4.

Table 5.4 Human causes of water demand

Population growth	Improved living standards	Industrialisation	Commercial agriculture
The increase in the number of inhabitants on the planet is putting stress on the limited water resources.	The rise in living standards is increasing the average domestic water consumption.	Demand for manufacturing to meet rising global demands is putting pressure on water use.	As food demands rise there is greater use of water for intensive farming practices.

- One of the key factors affecting water security is access to safe water.
- If the climate change scenario becomes reality, by 2030, it is estimated that water scarcity will cause the displacement of between 24 million to 700 million people.
- In 2016, it was predicted that approximately 4 billion people, representing almost two-thirds of the world's population, were experiencing water scarcity during at least 1 month of the year.

Water security Exists when a population has sustainable access to adequate quantities of water of acceptable quality. Adequate, that is, in sustaining human well-being and socio-economic development and for ensuring protection against water-borne pollution and disease.

Safe water Water that is sufficiently clean to be fit for human consumption and use.

Making links

A critical task facing water managers is to predict water scarcity. It is necessary to do this at three spatial scales: globally, nationally and regionally. Key players here, apart from the commissioning governments, are meteorologists (forecasting climate change), hydrologists (assessing water stores and stocks), and demographers and economists (forecasting demand).

The consequences and risks of water insecurity

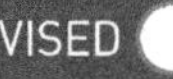
REVISED

Water scarcity

An important distinction between physical and economic water scarcity comes into focus here:

- Physical scarcity: the imbalance between water supply and demand, which results in an increasing percentage of available water being consumed. The threshold of physical scarcity is the 75 per cent mark, i.e. more than three-quarters of blue-water supplies are being used.
- Economic scarcity: here the shortfall in available water is related to shortfalls in human resources such as capital, technology and sound governance. The assumption is that the water potential is there, but it waits to be exploited.

Figure 5.11 shows the distribution of both types of water scarcity.

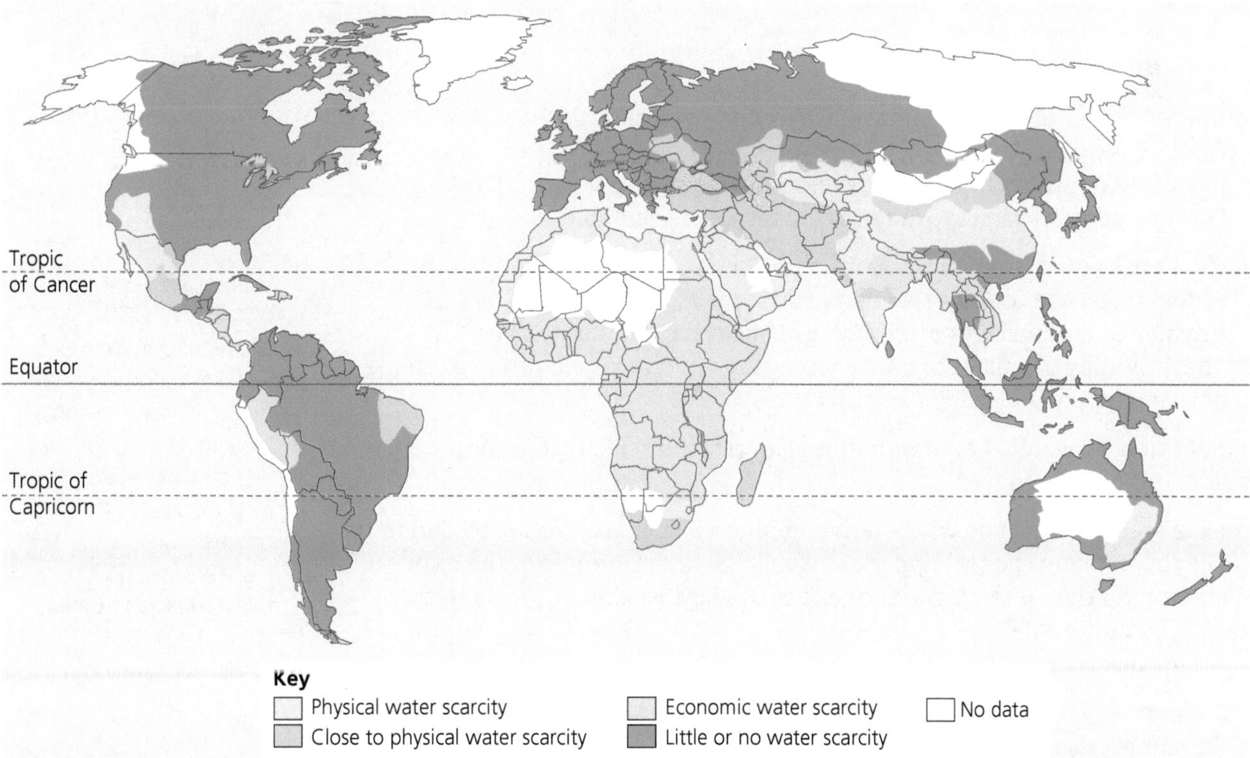

Figure 5.11 The global distribution of water scarcity

It is easy to think that water supply is somehow free — a gift from nature. Yet in developed and some emerging countries, water is big business. Considerable investments are made in such activities as abstraction, storage, processing and the delivery of water to points of consumption. In short, water comes at a price, and that price is likely to rise with increasing water scarcity. That price also varies spatially.

Revision activity

Make summary notes based on Figure 5.11 about the occurrence of the two types of water scarcity.

The importance of water

The consequences and risks of water insecurity all stem from the basic importance of water to people and their progress. The key uses are:

- human well-being: safe water for consumption, food preparation, sanitation and hygiene
- energy generation
- irrigation
- industry and mining: processing, discharging effluent.

Typical mistake

Do not think that water insecurity is only a problem in drought-threatened developing countries.

Now test yourself

11 Why does the price of water vary spatially?

Answer on p. 227

TESTED

In Figure 5.12, note the steep rise in the consumption of water by agriculture during the second half of the twentieth century. By comparison, the rise in the use of water by industry has been modest.

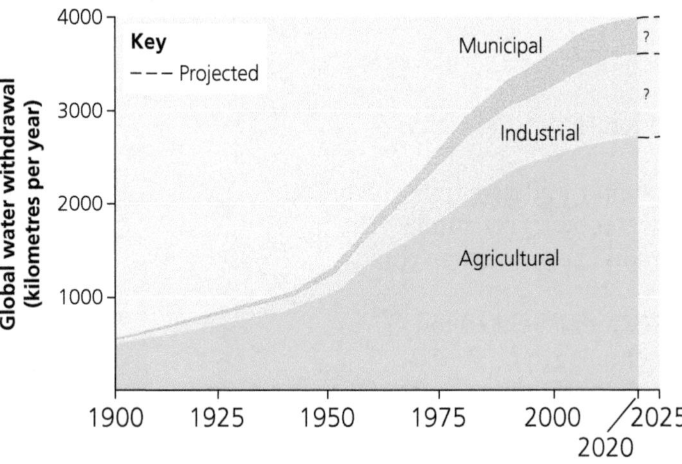

Figure 5.12 Trends in sectoral water use

Competition and conflict

Competition for water increases as the demand overtakes the available supply.

- Within a country, competition develops between users.
- But the users (players) are not equal in terms of their ability.

Conflict arises more from the way water is used.

- For example, use of a river for domestic supply is very likely to be threatened if industry is discharging effluent into the same river.
- The drowning of valleys to create storage reservoirs is a common example of water supply becoming a conflict issue.

Revision activity

Be sure you have examples of the following:

- a controversial regional/local reservoir scheme
- a shared international river (upstream vs downstream)
- a shared international river (common frontier).

Competition for water at an international level is more likely to lead to conflict, perhaps even war.

Possible scenarios include where a river is shared, such as the Nile by Sudan and Egypt. But here the real upper hand belongs to the owners of the upper reaches of the river (Ethiopia in the case of the Blue Nile and Uganda in the case of the White Nile).

In theory Ethiopia and Uganda could abstract as much water as they want and so deprive Sudan of an adequate amount of river water. Sudan could then do the same to Egypt.

There is also the potential for conflict where a river forms the border between two countries, as with the part of the Rhine between France and Germany or the Rio Grande between Mexico and the USA.

As yet, there have been no water wars. But it might not be too long before military force is used to protect a country's access to water.

Now test yourself

12 Why might it be only a short time before there is an outbreak of water wars?

Answer on p. 227

TESTED

Making links

The main players involved in water management are:

- political: IGOs, national governments, lobbyists and pressure groups
- economic: large water-using businesses (TNCs), funders of mega water projects (World Bank, IMF)
- social: concern about access to safe water (WaterAid, Practical Action)
- environmental: conservationists keen to protect water resources and wetlands (WWF, UNESCO).

Different approaches to managing water supply

REVISED

Four broadly different but not discrete approaches to managing water supply are indicated in the specification: hard engineering, sustainable schemes of restoration and conservation, integrated water drainage basin management, and water sharing treaties.

Hard engineering

- The classic examples of hard engineering are the many water transfer schemes that divert water from one drainage basin to another.
- These usually require constructing huge dams for collecting the water and a large canal to carry that water from an area of surplus to an area of deficit.
- The most spectacular scheme to date is the South–North Transfer Project in China. Such schemes have been found to have their environmental and social costs (see Figure 5.13).

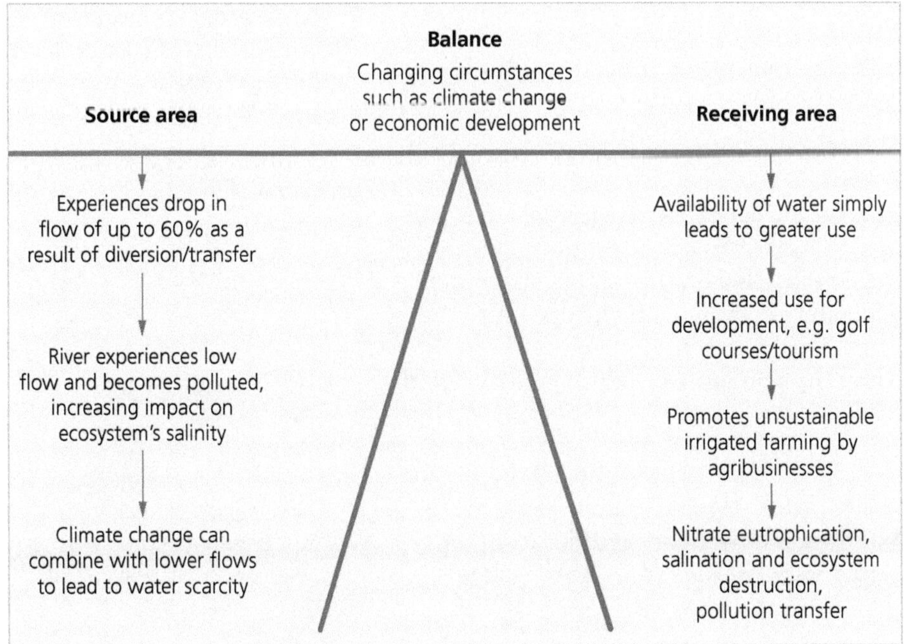

Figure 5.13 Water transfer issues

Desalinisation is a rather different hard-engineering approach, which has the advantage of drawing on the seas rather than freshwater supplies. Recent breakthroughs in technology (for example, the reverse osmosis process) have made desalinisation more cost effective, less energy intensive and easier to implement on a large scale.

Revision activity

Name four of the top ten countries with the largest desalinisation capacities.

Desalinisation The conversion of salt water into freshwater through the partial or complete extraction of dissolved solids.

Sustainable schemes

- This approach is very much focused on **water sustainability** and water conservation, as well as on the efficiency of water use and the recycling of water.
- Faced with an acute shortage of water, Singapore has led the way in filtration technology that makes possible the recycling of dirty water.

Key concept

Water sustainability involves ensuring there are adequate supplies of available water for the benefit of future generations. Water sustainability has three different aspects:

- environmental: freedom from pollution, so available as safe water
- economic: ensuring a secure water supply to all users at an affordable price; maximising efficiency of water usage and minimising wastage
- socio-cultural: ensuring equitable distribution of water (a) to poor and disadvantaged groups and (b) within and between countries.

Now test yourself

13 What is the difference between environmental and economic water sustainability?

Answer on p. 227

TESTED

Making links

Attitudes to water supply vary with climate. Where there is abundant precipitation, there is little concern about levels of water consumption and sustainability. However, where water is scarce and demand pressures supply, very different attitudes prevail: sustainability and efficiency of use become priorities.

Integrated drainage basin management (IDBM)

With this approach, river basins are treated holistically. The aims are to:

- encourage co-operation between basin users and players
- protect the environmental quality of the river catchment
- ensure water is used with maximum efficiency
- distribute water equitably between its users.

Exam tip

Be sure you have case study details of:

(i) a catchment shared by a number of countries (e.g. the Nile)

(ii) a river forming a national frontier (e.g. the Rio Grande).

Water sharing treaties

The challenge here is to resolve or neutralise the potential for hostilities over shared water.

Notable advances in achieving the required international co-operation include:

- the Helsinki Rules, which introduced the concepts of 'equitable use' and 'equitable shares'
- the UNECE Water Convention, which focused on the joint management and conservation of shared freshwater ecosystems in Europe and neighbouring regions
- the EU Water Framework and Directives on issues such as pollution and hydropower.

Revision activity

Make a note of what you think are the strengths and weaknesses of each of the four different approaches to water supply management.

Making links

When dealing with controversial issues, the players indicated here tend to fall into distinct camps: social vs political players, and economic vs environmental players.

The world is slowly beginning to appreciate:

- the real value of water as a resource
- the increasing scarcity of water in the face of increasing demand
- the potential of water to become a source of international conflict
- the critical need for international co-operation where there are shared waters.

Exam skills

You should be familiar with the skills and techniques used in the following investigations of the water cycle and water insecurity:

+ analysing flows within hydrological and drainage basin systems
+ comparing river regime discharges
+ constructing and analysing water budget graphs
+ comparing storm hydrographs
+ interrogating large databases in terms of trends in floods and droughts
+ interpreting synoptic charts relating to drought and flood conditions
+ analysing global distributions of water stress and water scarcity
+ interpreting indices of water poverty
+ identifying seasonal variations in the impact of dams on river discharges.

Exam practice

A-level

* Note that Topics 5 and 6 are examined together in one composite question.

1 a) Study Figure 1.

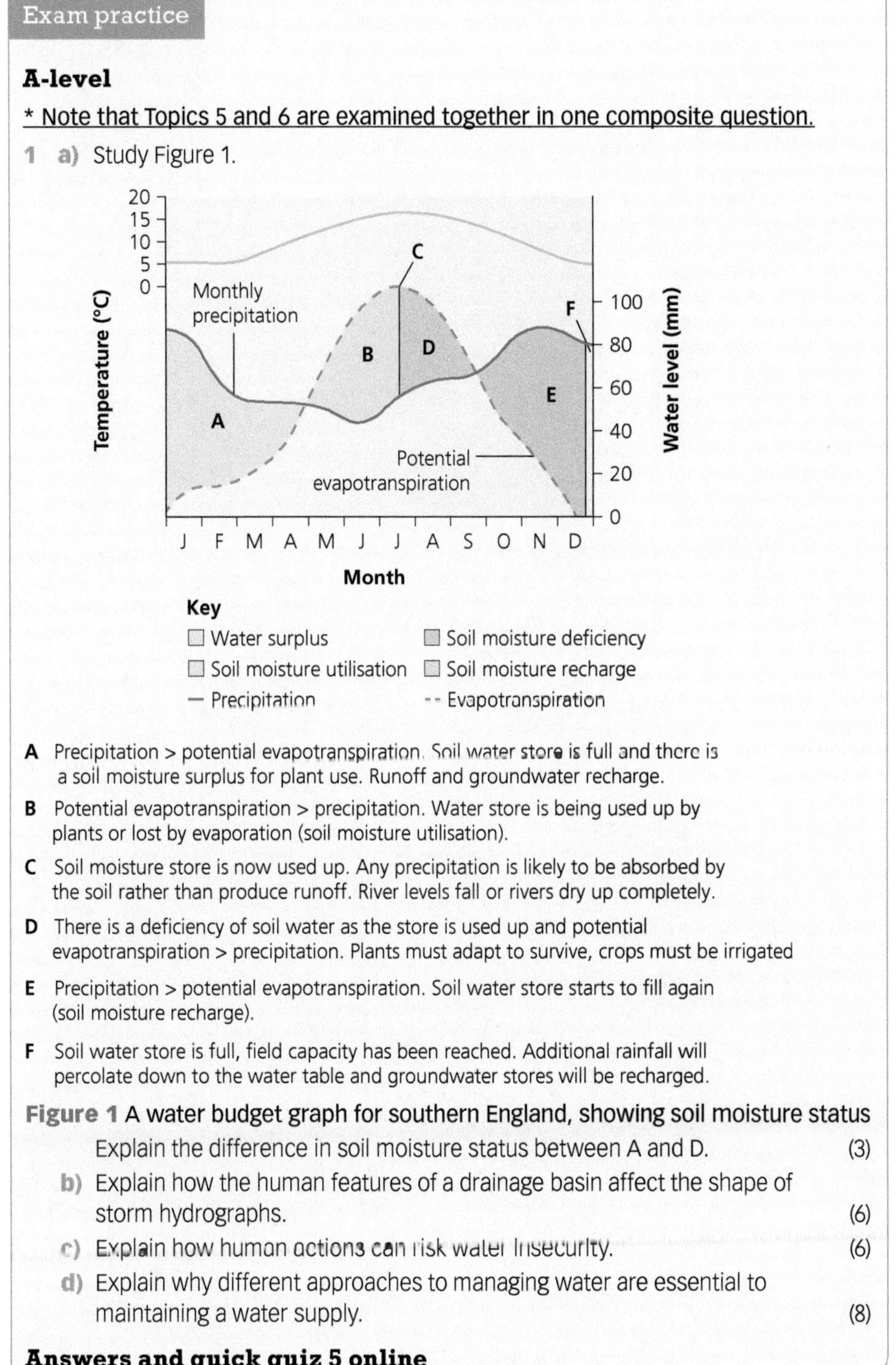

A Precipitation > potential evapotranspiration. Soil water store is full and there is a soil moisture surplus for plant use. Runoff and groundwater recharge.

B Potential evapotranspiration > precipitation. Water store is being used up by plants or lost by evaporation (soil moisture utilisation).

C Soil moisture store is now used up. Any precipitation is likely to be absorbed by the soil rather than produce runoff. River levels fall or rivers dry up completely.

D There is a deficiency of soil water as the store is used up and potential evapotranspiration > precipitation. Plants must adapt to survive, crops must be irrigated

E Precipitation > potential evapotranspiration. Soil water store starts to fill again (soil moisture recharge).

F Soil water store is full, field capacity has been reached. Additional rainfall will percolate down to the water table and groundwater stores will be recharged.

Figure 1 A water budget graph for southern England, showing soil moisture status

Explain the difference in soil moisture status between A and D. (3)

b) Explain how the human features of a drainage basin affect the shape of storm hydrographs. (6)

c) Explain how human actions can risk water insecurity. (6)

d) Explain why different approaches to managing water are essential to maintaining a water supply. (8)

Answers and quick quiz 5 online

Summary

You should now have an understanding of:

- the nature of human development
- the global hydrological cycle — its stores, fluxes (transfers) and processes
- water in the hydrological cycle available for human use
- physical and human factors affecting drainage basin systems
- water budgets and their impacts
- river regimes and their impacts
- factors affecting storm hydrographs
- the causes and impacts of droughts
- the causes and impacts of floods
- global warming, short-term oscillations and climate change
- the impacts of climate change on trends in precipitation and evaporation
- the impacts of climate change on hydrological stores and flows
- the growing mismatch between water supply and demand
- the causes of water insecurity
- the consequences and risks associated with water insecurity
- the importance of water supply
- the challenges of shared waters
- different approaches to the management of water supply.

6 The carbon cycle and energy security

- A balanced carbon cycle is vital in maintaining the health and balance of our planet.
- Operating at both different spatial and time scales, carbon moves between different stores driven by physical processes including photosynthesis and diffusion.
- The burning of fossil fuels and mass global deforestation has driven critical threshold changes to important carbon stores, upsetting the natural balance of the carbon cycle.

The carbon cycle and terrestrial health

Terrestrial carbon stores

The carbon cycle

Key concepts

The **carbon cycle** (see Figure 6.1) is a closed system: it does not have any external inputs or outputs. So the total amount of carbon is constant and finite. The system has three components:

- stores: reservoirs where carbon is held
- fluxes: the flows that move carbon between stores, from one sphere to another
- processes: the physical mechanisms that drive the fluxes between stores.

Carbon stores function as sources (adding carbon to the atmosphere) and sinks (removing carbon from the atmosphere).

The **carbon stores** are:

- the atmosphere: gases such as carbon dioxide and methane
- the hydrosphere (oceans, lakes, etc.): dissolved carbon dioxide
- the lithosphere: carbonates in limestone and fossil fuels
- the biosphere: living and dead organisms.

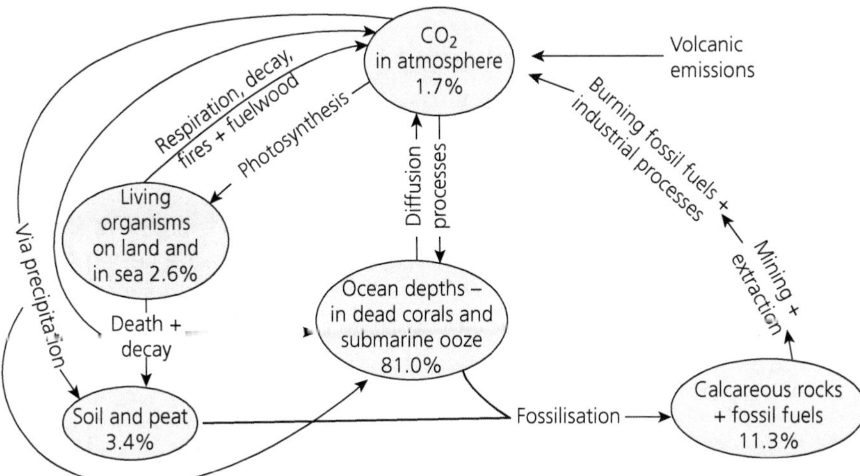

Figure 6.1 The carbon cycle

Revision activity

It is important you understand the processes, stores and fluxes involved in the carbon cycle. After you have learnt about the three components that make up the carbon cycle, try drawing it out from memory on a blank piece of paper. When you look back at the cycle diagram, add in what you have missed in a different colour.

Burning fossil fuels, fuelwood and respiration from plants and animals contributes towards the amount of carbon stored in the atmosphere.

The amount is reduced by precipitation and photosynthesis (see Figure 6.1).

Since fluxes are flows, they are measured in terms of their rates. There are significant differences in these rates (see Table 6.1).

Table 6.1 A sample of carbon fluxes

Flux	Rate (PgC per year)
Photosynthesis	103
Respiration	50
Volcanic eruption gases	50
Diffusion from ocean	9.3
Vegetation to soil decomposition	5
Weathering and erosion	0.9
Sedimentation/fossilisation	0.2

Geological carbon

Figure 6.2 shows the process of geological carbon.

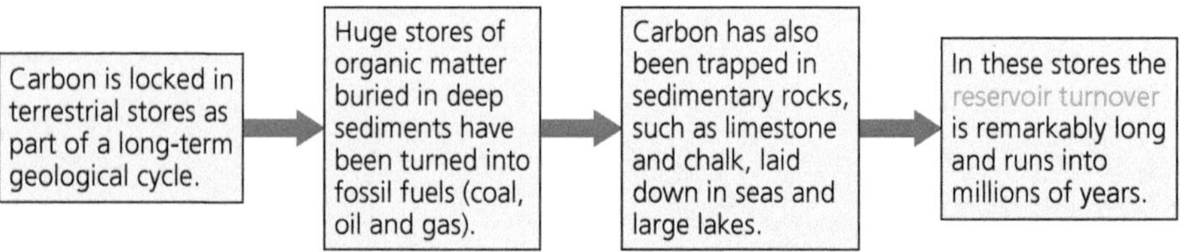

Figure 6.2 Geological carbon

Reservoir turnover The rate at which carbon enters and leaves a store. It is measured by the mass of carbon in any store divided by the exchange fluxes.

Geological processes release carbon into the atmosphere by volcanic activity at tectonic plate boundaries and through the chemical weathering of rocks.

Now test yourself TESTED

1 Where are the main carbon stores and in what form is the carbon held in each?

Answer on p. 227

Typical mistake

Limestone and chalk are not the only rocks to contain carbon.

Biological processes sequestering carbon

REVISED

- When carbon moves into carbon stores this is referred to as sequestering and contributes towards reducing the amount of carbon dioxide in the atmosphere.
- Photosynthesis is the key process responsible for sequestering, but this process is not only confined to land-based plants. Phytoplankton also sequester atmospheric carbon during photosynthesis in surface ocean waters (see Figure 6.3).

Sequestering The long-term storage of carbon dioxide and other forms of carbon.

Photosynthesis The process by which plants capture carbon dioxide from the atmosphere and then store (sequester) it as carbon in their stems and roots. Some is also stored in the soil.

Now test yourself TESTED

2 Define sequestering.

Answer on p. 227

Oceanic sequestering

Crustaceans (shellfish) depend on carbon extracted from the sea and lakes. But much of the carbon is returned to the sea when they die.

The movement of carbon within the oceans is controlled:

- vertically by carbon cycle pumps — there are three of them (see Figure 6.3) and between them they deliver carbon dioxide to the sea floor and to the ocean surface for release into the atmosphere
- horizontally by thermohaline circulation — this is a global mechanism involving surface and deep ocean currents, which are driven by differences in temperature and salinity. So far as the carbon cycle is concerned, warm surface waters are depleted of carbon dioxide by evaporation. But they are enriched again as the conveyor belt circulation drags them along as deep or bottom layers.

Carbon cycle pumps The processes operating in oceans that circulate and store carbon.

Thermohaline circulation The global system of surface and deep-water currents within the oceans driven by differences in temperature and salinity.

Now test yourself

3 Explain how carbon moves within oceans.

Answer on p. 227

TESTED

Exam tip

Oceanic sequestering of carbon is important but more difficult to grasp. Be sure you have a general understanding.

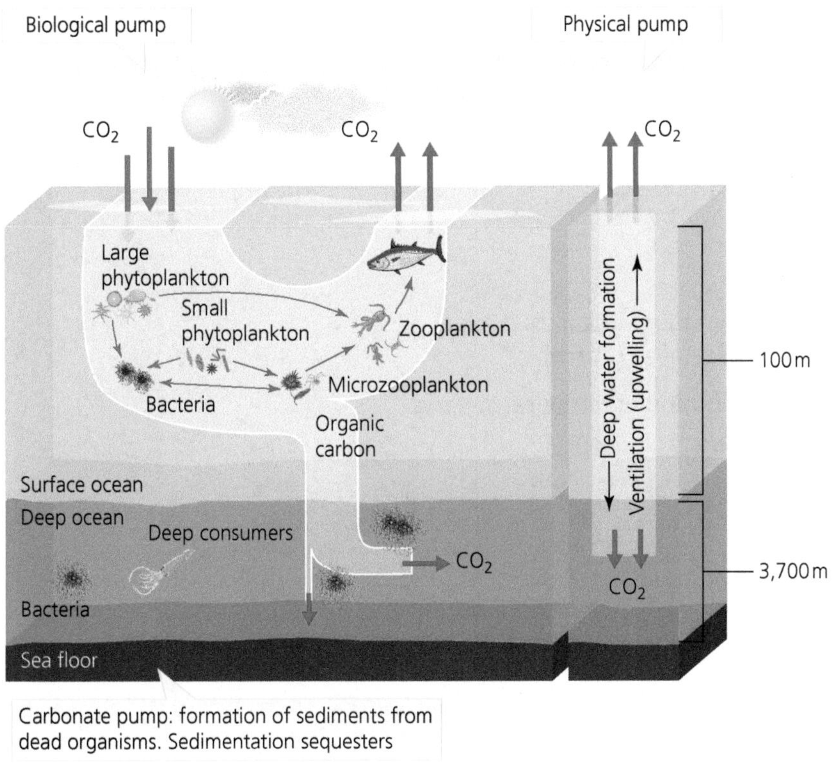

Figure 6.3 Oceanic carbon pumps

Terrestrial sequestering

- Terrestrial primary producers sequester carbon during photosynthesis. Some of this carbon is returned to the atmosphere by respiration from consumer organisms.
- Sequestering is a process common to all biomes and ecosystems, but they all differ in terms of their carbon productivity and their carbon storage capacities.
- Storage is mainly in plants, particularly trees, and soils. The longevity of some trees means that they can serve as stores for tens or hundreds of years.
- Globally, the most productive biomes are the tropical forests. But carbon fluxes with ecosystems also vary with time: diurnally (most active during the day) and seasonally (most active during the spring and summer).

Biological carbon

- Approximately 20–30 per cent of carbon is stored as dead organic matter in soils.
- The ability of soils to store organic carbon is dependent upon how the soils are used, the type of soil and the localised climate.
- Carbon that is not stored in soils is returned to the atmosphere through biological weathering.
- Litter-fall leads to a transfer of carbon from the plant to the soil.

Now test yourself

4 Which biome has the highest carbon productivity and the greatest carbon storage capacity?

5 Suggest a definition of biological carbon.

Answers on p. 227

TESTED

Carbon balance and human activities

- A balanced carbon cycle is important in sustaining other systems of the Earth.
- Through its control of the amount of CO_2 in the atmosphere, the carbon cycle plays a key role in regulating global temperatures and therefore climate.
- These changes, in turn, affect the hydrological cycle. Any increase in the concentration of atmospheric carbon affects the natural **greenhouse effect**.

Key concept

In the **greenhouse effect**, the greenhouse gases (carbon dioxide, methane, nitrous oxide) form a layer that is crucial to controlling the temperature of the Earth. As the amount of carbon dioxide in the atmosphere has increased over the last 250 years, so the blanketing effect of greenhouse gases has increased.

Natural greenhouse effect

The concentration of atmospheric carbon (carbon dioxide and methane) strongly influences the natural greenhouse effect (see Figure 6.4).

This concentration determines the global distribution of temperature and precipitation.

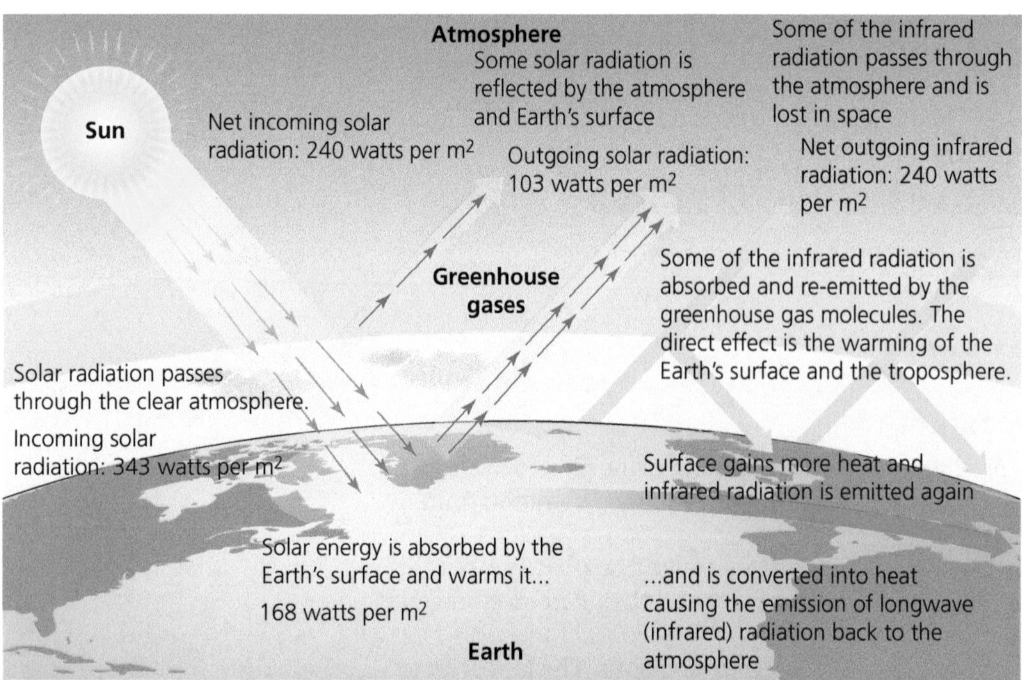
Atmosphere
Sun
Net incoming solar radiation: 240 watts per m2
Some solar radiation is reflected by the atmosphere and Earth's surface
Outgoing solar radiation: 103 watts per m2
Some of the infrared radiation passes through the atmosphere and is lost in space
Net outgoing infrared radiation: 240 watts per m2
Greenhouse gases
Some of the infrared radiation is absorbed and re-emitted by the greenhouse gas molecules. The direct effect is the warming of the Earth's surface and the troposphere.
Solar radiation passes through the clear atmosphere.
Incoming solar radiation: 343 watts per m2
Surface gains more heat and infrared radiation is emitted again
Solar energy is absorbed by the Earth's surface and warms it...
168 watts per m2
...and is converted into heat causing the emission of longwave (infrared) radiation back to the atmosphere
Earth

Figure 6.4 The greenhouse effect

- The Earth's climate is driven by incoming short-wave radiation.
- Approximately 31 per cent is reflected back into space by the atmosphere.
- Nearly half of the remaining 69 per cent is absorbed at the Earth's surface, especially by the oceans, while the other half is re-radiated into space as long-wave radiation.
- A large proportion of this long wave radiation is deflected back to the Earth's surface by clouds and greenhouse gases, however.
- It is this trapping of long-wave radiation that creates the natural greenhouse effect.
- Clearly, if the amount of carbon dioxide and methane in the atmosphere increases (for example, as a result of the burning of fossil fuels), then the trapping of more long-wave radiation will raise global temperatures.

Exam tip

It is vital that you have a sound grasp of the greenhouse effect and how it has become the driver of climate change.

Significant regulators of atmospheric composition

- Oceanic and terrestrial photosynthesis play an important role in regulating the composition of the atmosphere.
- On land, a key factor is soil health, which depends on the amount of carbon stored in it.
- Carbon helps give the soil its water-retention ability. A healthy soil will, of course, enhance ecosystem productivity.
- That, in turn, will mean more carbon stored in biomass and more carbon being sequestered from the atmosphere.
- Of course, partially offsetting this will be respiration of carbon dioxide into the atmosphere.

Revision activity

Draw an annotated diagram showing carbon cycling in soils.

The burning of fossil fuels

Figure 6.5 illustrates the combustion of fossil fuels.

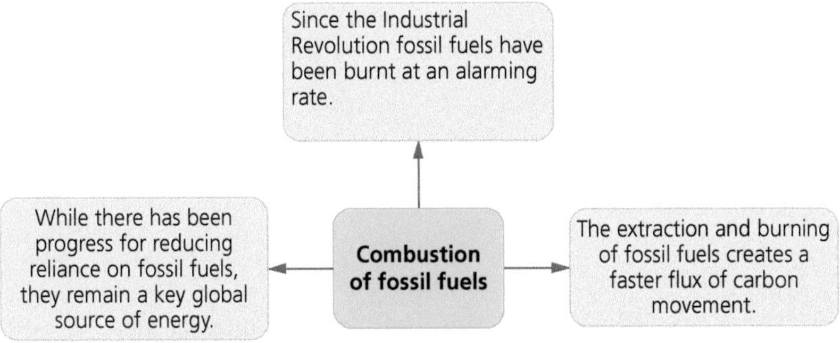

Figure 6.5 The combustion of fossil fuels

Fossil fuel implications

When fossil fuels are burnt this has a significant impact at both global and regional scales, upsetting the natural balance of the carbon cycle through altering carbon pathways and carbon stores.

Carbon pathways The routes taken by flows of carbon between stores.

Table 6.2 summarises the implications of burning fossil fuels on climate, ecosystems and the hydrological cycle.

Table 6.2 Burning fossil fuels

Climate	Ecosystems	The hydrological cycle
On a regional scale, some areas will become warmer and others cooler, affecting liveability. Rising mean sea levels increase the risk of flooding.	Alterations in the biodiversity affect the goods and services provided by global ecosystems. Pressure on marine ecosystems alters oxygen and acidification levels.	A rise in evaporation rates increases moisture in the atmosphere. Water locked in glaciers is reduced. El Niño-Southern Oscillation (ENSO) increases in intensity.

Now test yourself TESTED

6 Summarise the implications of fossil fuel consumption.

Answer on p. 227

Consequences of the increasing demand for energy

It is the continued burning of fossil fuels to meet the increasing demand for energy that is disturbing the carbon cycle.

Energy security

REVISED

Energy security (see Figure 6.6) is a key goal for almost all countries, particularly in this era of growing concern about the use of fossil fuels and the link to climate change.

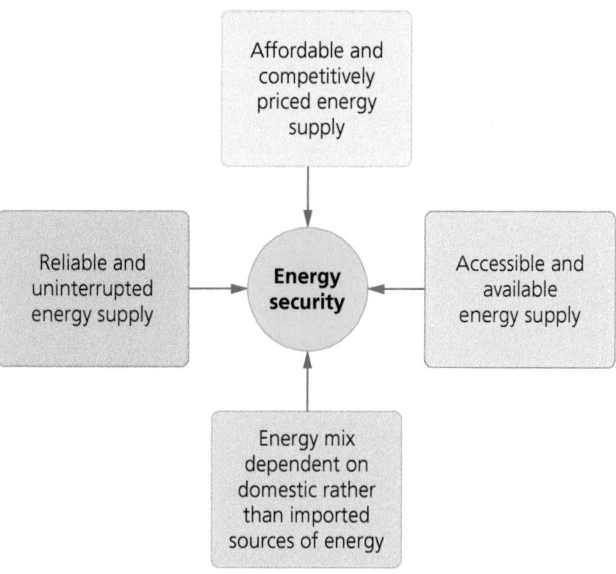

Figure 6.6 Energy security

Key concept

Energy security refers to the uninterrupted availability of energy resources at an affordable price. Long-term energy security comes from prudent investments in energy supply and in line with predictions of future demand. In the short term, energy security is about ensuring an uninterrupted energy supply, and the ability of the energy system to react promptly to any sudden change in the balance between energy demand and supply.

Rising consumption and energy mix

The rising consumption of energy reflects three factors:

- the growth in the global population
- the rising standards of living being created by economic development
- the essential nature of energy to everyday life.

Remember that consumption is a function of demand: the greater the demand, the higher the consumption. But there are other factors at work.

Figure 6.7 shows the rise in global energy consumption since 1994 and how the energy mix has changed.

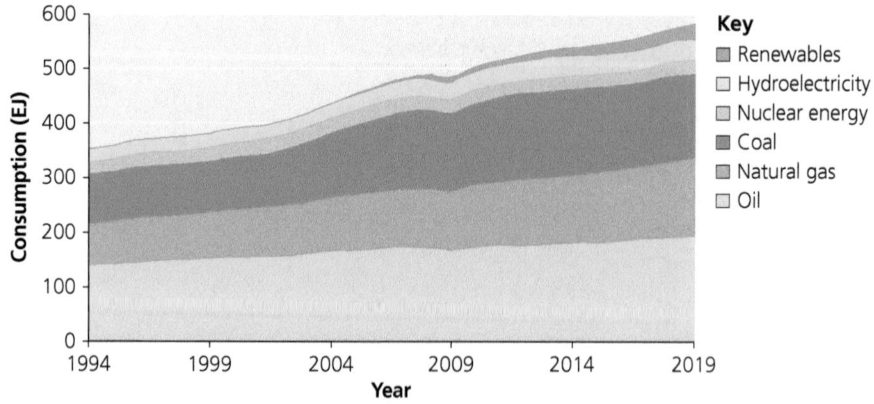

Figure 6.7 World energy consumption, 1994–2019

Now test yourself

7 Why do rising living standards lead to higher energy consumption?

8 Classify the fuels in Figure 6.7 according to the threefold scheme: fossil, recyclable and renewable.

Answers on p. 227

TESTED

Energy mix The combination of different available energy sources used to meet a country's total energy demand

The energy mix has three important dimensions and potential tensions:

- Domestic vs international sources: clearly the more the ratio favours the domestic input, the better it is in terms of energy security. However, it might be the case that overseas sources are cheaper.
- Primary energy vs secondary energy: there might be cost advantages in the reliance on primary energy, but modern economies and societies are highly dependent on electricity.
- Renewable vs non-renewable: given the increasing concern about the burning of fossil fuels and global warming, there is clearly considerable pressure to increase renewable input. However, changing this mix is easier said than done. For example, it may involve covering large areas of both land and sea with wind and solar farms.

Primary energy The forms of energy found in nature have not been subjected to any conversion or transformation process, for example water (watermills) and wind power (windmills) or the burning of gas in heating the home and by motor vehicles.

Secondary energy Derived from the conversion of primary energy. The most important secondary energy is electricity, which is derived from many primary sources.

Access to energy resources

Access to and the consumption of energy depend on a range of factors (see Figure 6.8).

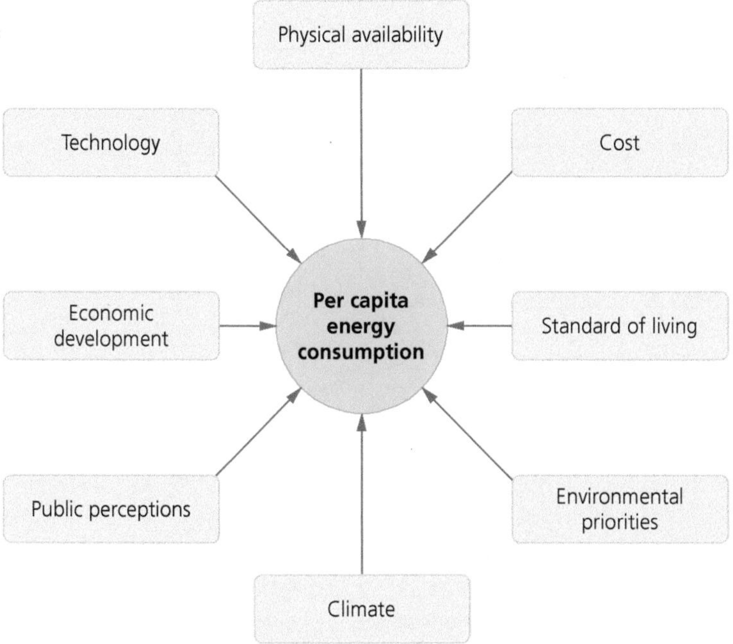

Figure 6.8 Some factors affecting per capita energy consumption

Now test yourself

9 Classify each of the following sources of energy as either primary or secondary: geothermal, solar, biofuels and nuclear.

10 Name three TNCs involved along the complex length of the energy pathway.

Answers on p. 227

TESTED

Energy players

Meeting the demand for energy involves creating **energy pathways** running from the energy producers to the energy consumers. At both ends of such pathways there are influential players (governments, organisations, companies and individuals) with particular involvements in the energy business.

Making links

Major players and the energy pathways are shown in Figure 6.9.

ENERGY PATHWAY

ENERGY SUPPLY → ENERGY DEMAND

Players	Players	Players
TNCs	TNCs	TNCs
OPEC	Shipping companies	Energy companies
Governments	Pipeline controllers	Governments
		Consumers

Figure 6.9 Players and the energy pathway

Revision activity

What is OPEC and who belongs to it?

Reliance on fossil fuels

REVISED

Most countries today still derive the greater part of their energy supply from fossil fuels. China and the USA are the major fossil fuel consumers. In 2019, China consumed over 4.5 metric tonnes compared to the USA's consumption of 2.1 metric tonnes.

Mismatch between fossil fuel supply and demand

The three fossil fuels need to be considered individually in assessing any mismatch:

- Coal: small mismatch, as main producers of coal tend to be the main consumers.
- Oil: a considerable mismatch — the main suppliers are members of OPEC and the consumers are located in Europe. All four BRICs are producers.
- Gas: supply is dominated by the USA and Russia. Major importers are Western European countries and Japan.

Typical mistake

It is wrong to think that the move to oil and gas was caused by the exhaustion of coal reserves.

Energy pathways

Energy pathways have been defined above. Again, the three main fossil fuels should be considered separately:

- Coal: there is still a significant global trade. Three of the main coal producers (China, India and the USA) also import coal. Australia and Indonesia are major exporters to Japan, South Korea and Taiwan.
- Oil: the Middle East is the major hub of oil exports.
- Gas: the pathways are not that different from those of oil but there is a major pathway from Russia to Europe.

Revision activity

Find out why in 2016 the UK began to import shale gas.

Making links

Players in the harnessing of unconventional fossil fuels are:

- exploration companies
- environmental groups
- affected communities
- governments.

Unconventional sources of carbon energy

There are four main 'unconventional' sources of fossil fuel:

- Tar sands: Canada is a major exploiter.
- Oil shale: little exploitation as yet.
- Shale gas: exploitation requires controversial **fracking**; the USA is the leading producer and exporter. There is widespread potential.
- Deepwater oil: technological advances in drilling are beginning to make this a viable source. Brazil is leading the way at the moment.

The negative with all these sources is that their use still threatens the carbon cycle. Exploitation methods — pumping, mining and fracking — also pose risks for fragile environments. Yet it is possible that production costs might be lower than those of conventional sources of oil and gas. This is beginning to look likely for shale gas. But first there needs to be clear evidence that fracking is safe.

Now test yourself

11 What might a government's attitude be to the exploitation of unconventional fossil fuels?

Answer on p. 227

TESTED

Alternatives to fossil fuels

REVISED 

Renewables and recyclables

In order to drive the global reduction of carbon emissions and reduce the reliance on the use of fossil fuels as the primary energy source, we must increase our use of alternative sources of 'clean' energy. There are two key types:

- renewable sources: hydro, wind, solar, geothermal and tidal energy
- recyclable sources: nuclear power and biofuels.

Figure 6.10 confirms how well the renewables and recyclables compare with the four fossil fuels. But note that they are not entirely 'clean' in terms of greenhouse gas emissions.

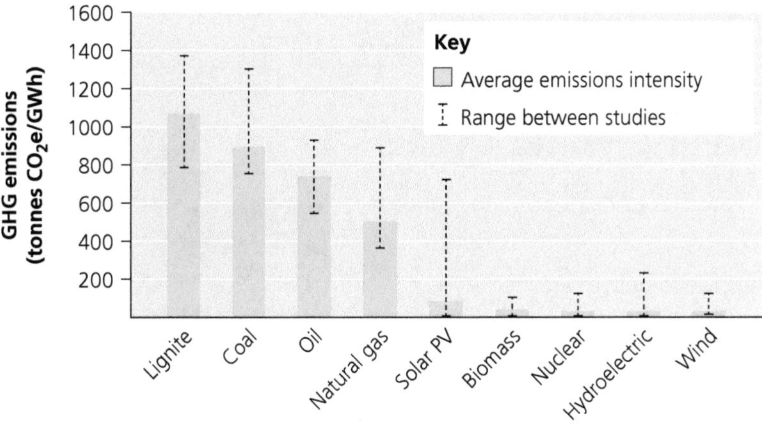

Figure 6.10 Greenhouse gas emissions of the nine electricity-generating fuels

Typical mistake

It is not the case that renewable sources of energy will completely replace the energy currently derived from fossil fuels. The best hope is to reduce the use of fossil fuels and so lower carbon emissions to 'safe' levels.

Now test yourself

12 How is it that the renewables and recyclables are all shown in Figure 6.10 as making emissions?

Answer on p. 227

TESTED

As regards greater use of renewables, there are some negatives:

- Not all countries have physical geographies that contain all or most of these natural resources.
- Increasing the exploitation of these resources is likely to have significant impacts on the environment and on communities.
- People often go off the idea of renewables in the face of a proposal to build a wind or solar farm close to home.
- The construction costs involved in harnessing renewables are high, as are the maintenance costs. So renewable energy is not as cheap as many people like to think.

There are two sources of recyclable energy: nuclear power and biofuels. Several countries have limited choice but to develop the use of these energy sources, despite the associated risks.

Exam tip

Be sure you are familiar with the risks and advantages of using nuclear energy to generate electricity.

Biofuels

The burning of fuelwood has the longest history. However, some biofuel crops are beginning now to make their mark on the energy scene:

- Primary biofuels include fuelwood, wood chips and pellets that are used in an unprocessed state for heating, cooking or electricity generation.
- Secondary biofuels are derived from the processing of crops, such as sugar cane, soybeans, maize and cashew. Two types of fuel are extracted (bio-alcohol and biodiesel), both of which are being used as a vehicle fuel and to generate electricity. Brazil is the world leader here.

Primary biofuels have the disadvantage of encouraging either deforestation or the use of arable land to grow trees.

The latter problem also exists with the growing of biofuel crops. Is this a sensible option in an increasingly hungry world? Added to this, there is some uncertainty as to how 'carbon neutral' they are.

Revision activities

Find out why some people believe that the growing of biofuel crops is not carbon neutral.

Create an annotated diagram to explain how carbon capture works.

Radical technologies

Radical technologies are another route to providing a more sustainable energy future. There are two technologies here that would reduce carbon emissions:

- Carbon capture and storage: its potential effectiveness is hindered by the complex technology involved and doubts about the security of long-term storage.
- Hydrogen fuel cells: this promising technology makes use of an abundant element and is most likely to be used as a source of heat and electricity for buildings and as a power source for electric vehicles.

Links to the global climate system

Human threats to the carbon and water cycles

REVISED

The natural balance of the carbon and water cycles is under increasing threat from human activity.

The biosphere's ability to sequester carbon dioxide is being reduced by land conversion.

Exam tip

Be sure you understand how land conversion affects global warming. Have some specific examples ready.

Changes in land use

- Land conversion is taking place at an increasing rate in order to meet the growing demand for food, energy and other resources.
- Any land conversion is almost bound to affect both cycles, specifically reducing both carbon and water stores, as well as the health of soils.
- Of the range of land conversions, the most serious is undoubtedly deforestation (see below).
- Also serious is the conversion of grasslands into cropland, the building of dams and reservoirs, and opencast mining.

Land conversion Clearing a natural ecosystem and using the space it occupied for a different purpose.

Ocean acidification The decrease in the pH of the oceans caused by the uptake of carbon dioxide from the atmosphere.

Acidification of the oceans

- Our oceans play a key role in carbon sinks but their ability to conduct this vital function for our planet is diminishing due to changes in their pH.
- Over time, this is resulting in ocean acidification, which affects marine ecosystems.
- Warming temperatures, more intense tropical storms and increased levels of pollution are causing greater degradation to these ecosystems.
- The rate of ocean acidification (see Figure 6.11) is crucial to the potential for oceanic organisms to adapt to and survive the changing conditions.

Revision activity

Make brief notes on ocean acidification based on Figure 6.11.

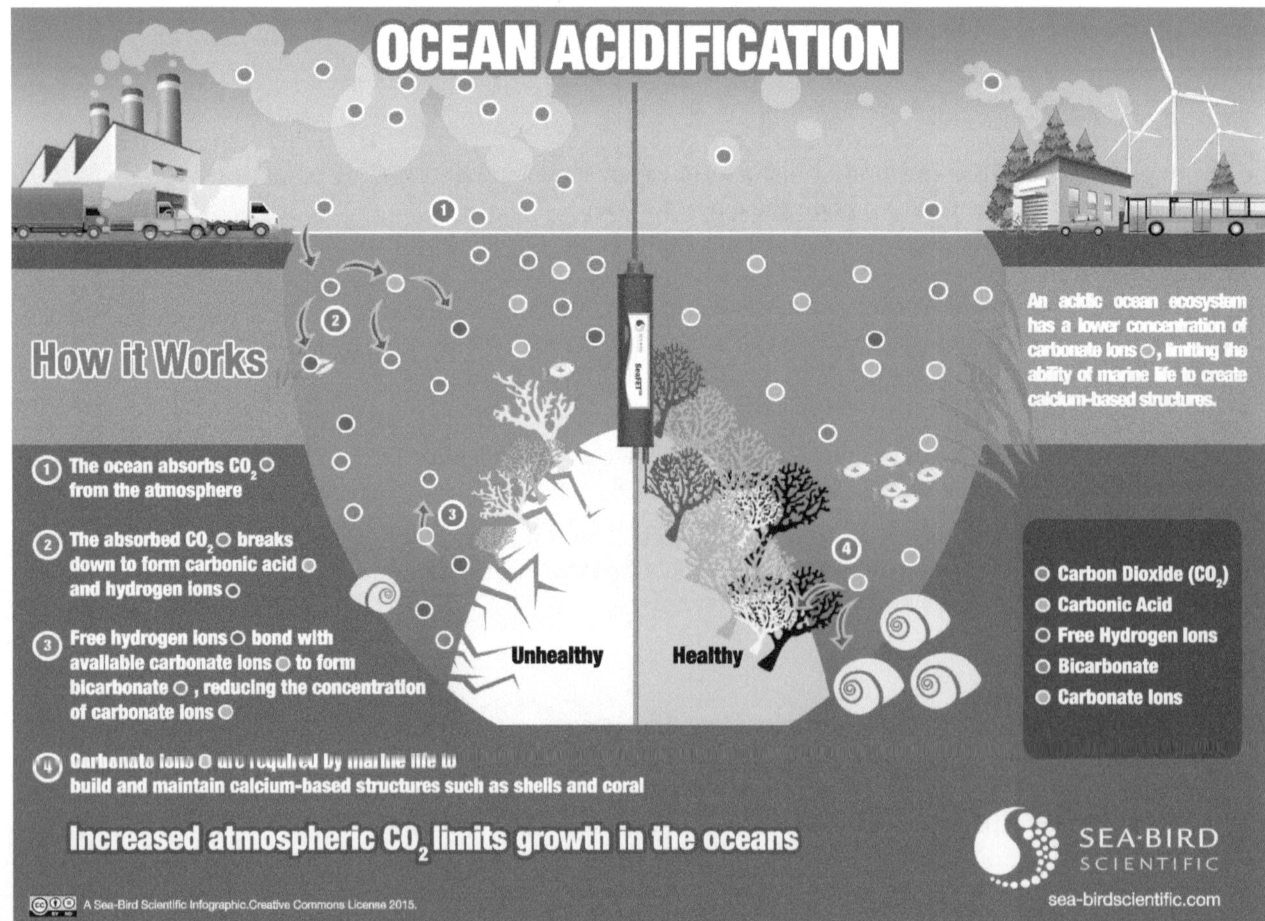

Figure 6.11 Ocean acidification

Climate change

- There is now unequivocal evidence that humans and their activities are enhancing the greenhouse effect and that this is leading to climate change.
- Scientists have indicated that a 2°C rise in global temperatures might lead to a change of climate zone for 5 per cent of the Earth's land area.
- Evidence already suggests that subtropical deserts are expanding and there is greater frequency of stormy weather in the mid-latitudes.

Degraded water and carbon cycles and human well-being

REVISED

Deforestation

- Of all the possible land conversions none has a greater impact on the carbon and water cycles than deforestation (see Figure 6.12).
- Through their ecosystem services, forests are particularly important to human well-being.
- In 2020, the Food and Agriculture Organization of the United Nations (FAO) State of the World's Forest Report indicated global forest area had decreased from 32.5 per cent to 30.8 per cent between 1990 and 2020.
- This amounted to a loss of 178 million hectares of forest, which is equivalent to roughly the size of Libya (see Figure 6.13).

Impact on water cycle

- Reduced intercepted rainfall storage by plants; infiltration to soil and groundwater changes.
- Increased raindrop erosion and surface runoff, with more sediment eroded and transported into rivers.
- Increased local 'downwind' aridity from loss of ecosystem input into water cycle through evapotranspiration.

Impact on carbon cycle

- Reduction in storage in soil and biomass, especially above ground.
- Reduction of CO_2 intake through photosynthesis flux.
- Increased carbon influx to atmosphere by burning and decomposing vegetation.

Figure 6.12 The impacts of deforestation on the carbon and water cycles

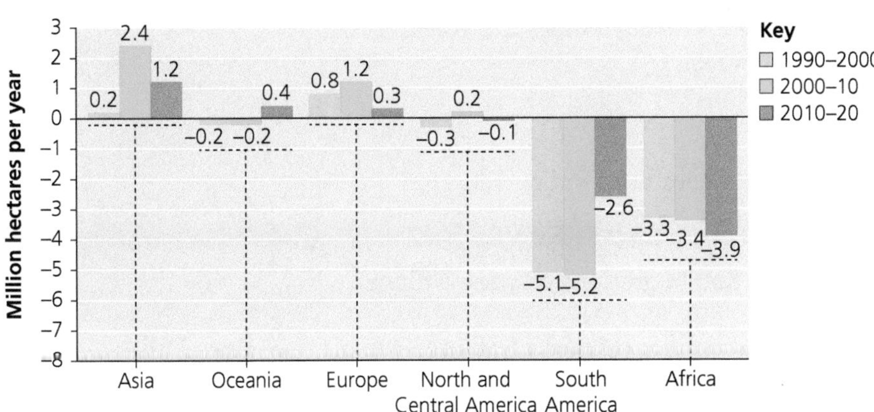

Figure 6.13 Annual forest area net change, by decade and region, 1990–2020 (source: FAO)

Exam tips

You should be sure that you understand the three bulleted impacts on both the water and carbon cycles shown in Figure 6.12.

Make sure you are well informed about the ecosystem services of forests, particularly of the tropical rainforests.

For many, the exploitation of environmental resources is a pathway to environmental destruction. However, optimists cling to the **environmental Kuznets curve**, which offers the hope that exploitation will eventually lead to protection.

Key concept

The **environmental Kuznets curve** is a hypothetical relationship between environmental quality and economic development (see Figure 6.14). Environmental degradation worsens with economic growth until a point is reached when attitudes towards the environment change towards conservation. The critical threshold or tipping point is thought to coincide with a point on the rising curve of income.

Now test yourself

13 Why might it be said that the environmental Kuznets curve takes a rather optimistic view of the global situation?

Answer on p. 228

TESTED

Making links

Global consumers undoubtedly have different attitudes to the fate of the environment. It is likely that the developing and emerging countries favour making best use of resources (why should they be prevented from doing so?), while developed countries are beginning to realise the need to protect and conserve resources.

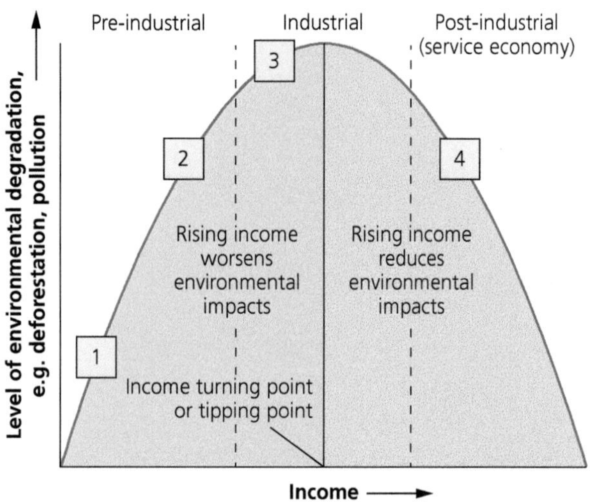

Key

1. UK pre-Industrial Revolution, remote Amazonia today, Indonesia pre-1970s
2. Indonesia today, China in the twentieth century
3. China today
4. UK today

Figure 6.14 The environmental Kuznets curve and tipping points

Higher temperatures and evaporation

Increased temperatures will clearly raise evaporation rates and the quantity of water vapour in the atmosphere. This, in turn, will have a range of impacts on the water cycle, changing:

- precipitation patterns
- river regimes
- water stores (cryosphere and drainage basin).

These sorts of change will have impacts on agriculture, hydropower and the cryosphere. The melting of the Arctic Sea ice is one obvious sign of the last, and one that could have profound repercussions.

Ocean health

- Global warming is also affecting ocean temperatures and currents, according to the World Wide Fund for Nature (WWF).
- The changes in the health of our oceans have implications for the distribution, abundance, breeding cycles and migrations of marine plants and animals.
- The well-being of the millions of people who rely directly and indirectly on marine products for food and income is being adversely affected.
- Hardest hit countries are likely to include Japan and Iceland and many small developing island states, such as the Maldives and St Lucia. It is likely, too, that tourism will suffer.

The risks of further global warming

REVISED

Future uncertainties

There are uncertainties about future global warming and possible contributory factors:

- Uncertainties about natural factors, such as the role of carbon sinks and their capacity to cope with change; possible feedback mechanisms, such as carbon releases from peatlands and permafrost, as well as tipping points relating to forest dieback and the reversal of thermohaline circulation.
- Uncertainties about human factors, such as the future rates of global economic and population growth, the planned reduction in global emissions and the exploitation of renewable energy sources.

Making links

There is uncertainty over global projections about carbon emissions (futures) and their impact on global systems through climate change.

Now test yourself TESTED

14 Since 1959, the mean annual atmospheric carbon dioxide levels have changed. In 1959, the mean annual atmospheric CO_2 was 315.97 ppm and by 2019 this had increased to 411.43 ppm.

Calculate the overall percentage change.

Answer on p. 228

Adaptation strategies

Table 6.3 identifies possible adaptation strategies that humans can adopt to potentially deal with climate change.

Adaptation In the present context, any action that reacts and adjusts to changing climate conditions.

Table 6.3 Possible adaptation strategies to deal with climate change

Water conservation and management	Increased resilience of agriculture systems	Land-use planning	Flood-risk management	Solar radiation management
Reducing and using water more efficiently	Devising crops that are drought resistant	Pre-planning land development to reduce building on flood-prone locations	Reviewing and expanding the use of engineering methods to strengthen risks against flooding	Use of orbiting satellites to reflect inward radiation back into space

Rebalancing the carbon cycle

Can this rebalancing be achieved through mitigation? Possible methods include:

- carbon taxation
- switching to renewable energy sources
- increasing energy efficiency
- afforestation
- carbon capture and storage.

But the success of any of these forms of mitigation requires international agreements at a global scale as well as concerted actions at a national level.

Experience so far has shown the extreme difficulty of persuading governments to sign up to global-scale agreements (e.g. the Kyoto and Paris agreements), while the individual methods listed above have been found to be problematic even when applied at a national level.

Mitigation In the present context, any actions that either reduce or eliminate the long-term risk and hazards of climate change.

Making links

Governments, TNCs and individuals have contrasting attitudes to the challenges presented by the release of stored carbon and the consequent global warming.

Now test yourself TESTED

15 What is the difference between adaptation and mitigation? Give some examples of each type of action.

Answer on p. 228

Exam tip

Be prepared to give examples of how and why people might hold conflicting views on global warming.

Exam skills

You should be familiar with the skills and techniques used in the following investigations of the carbon cycle and energy security:

- proportional flow diagrams to show carbon fluxes and energy pathways
- interpreting maps showing the global distributions of temperature and precipitation
- analysing the energy mixes of different countries by means of graphs
- interpreting maps showing global energy and flows
- comparing carbon emissions from different energy sources
- using GIS to map land-use changes and deforestation
- analysing maps to identify areas at most risk from climate change
- plotting graphs of carbon levels and calculating means and rates of change.

Exam practice

A-level

* Note that Topics 5 and 6 are examined together in one composite question.

1 a) Study Figure 1.

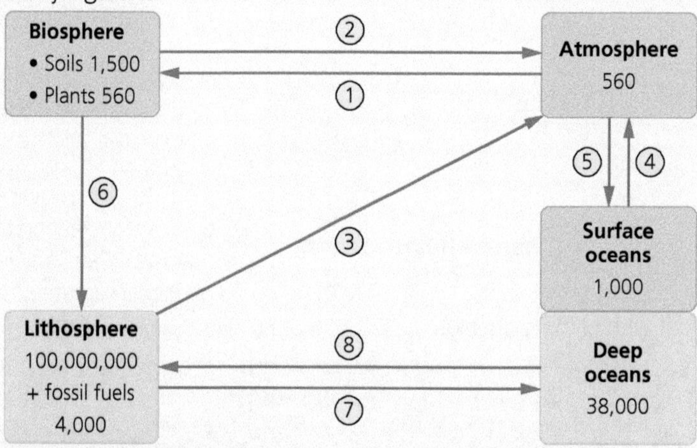

Figure 1 The carbon cycle

Assess the importance of the oceans in the carbon cycle. (12)

b) Evaluate the merits of mitigation as a strategy to rebalance the carbon cycle. (20)

c) Explain how degradation of the carbon cycle affects human well-being. (8)

d) Explain why energy security still relies on the use of fossil fuels. (8)

e) Study Figure 2.

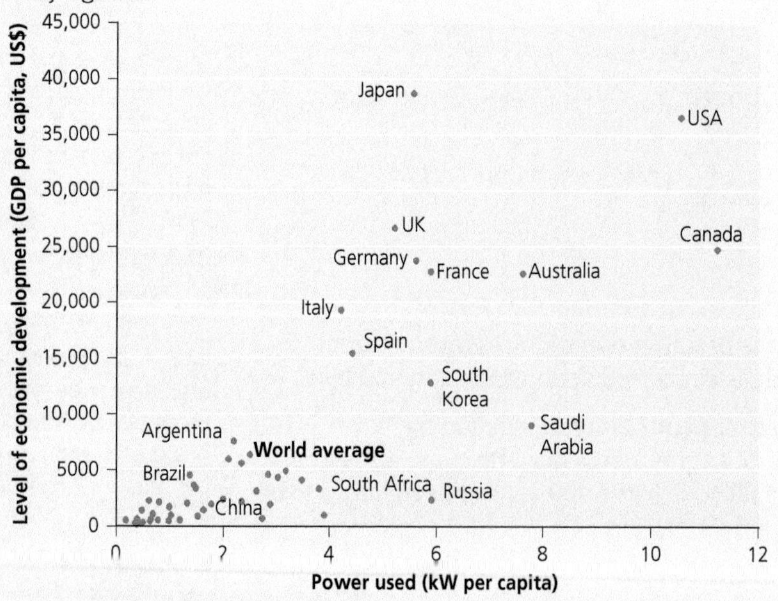

Figure 2 Energy consumption and economic development in a sample of countries

Explain one reason for the relationship between energy consumption and economic development. (4)

Answers and quick quiz 6 online

Summary

You should now have an understanding of:

- the global carbon cycle — its stores and fluxes
- geological carbon and its release into the atmosphere
- biological processes sequestering carbon on land and in the oceans
- the importance of a balanced carbon cycle to the health of other Earth systems
- the importance of the natural greenhouse effect
- the impact of burning fossil fuels on climate, ecosystems and the hydrological cycle
- the rising demand for energy
- energy security and the energy mix
- energy players and their roles in securing and supplying energy
- the continuing reliance of economic development on fossil fuels
- the security of energy pathways
- the development of unconventional fossil fuel energy
- exploiting renewable and recyclable energy sources
- the technological challenge of reducing carbon emissions
- human degradation of the carbon and water cycles and the impacts on human well-being
- the uncertainties surrounding future carbon emissions
- responses to global warming — adaptation, mitigation and global agreements.

7 Superpowers

Changing superpowers

Key concept

Superpowers may be distinguished by a number of characteristics, for example by their military strength, economic power and territorial extent. Their pattern of dominance has changed over time. Today the established superpowers are beginning to be challenged by some of the emerging countries. The spheres of influence of superpowers are frequently contested, resulting in geopolitical tensions and even conflicts.

Defining superpowers

REVISED

A **superpower** is a dominant nation or state with the ability to project and impose its influence outside its borders.

The term 'superpower' dates from the Second World War when it was used to describe the global influence held by the USA, the USSR and the British Empire.

Hyperpower A state that is dominant in all aspects of power (economic, political, cultural, military) — for example, the USA from 1990 to 2010.

Four orders of superpower may be recognised. Three are shown in Figure 7.1:

+ hyperpower
+ global superpower
+ emerging superpower
+ regional superpower.

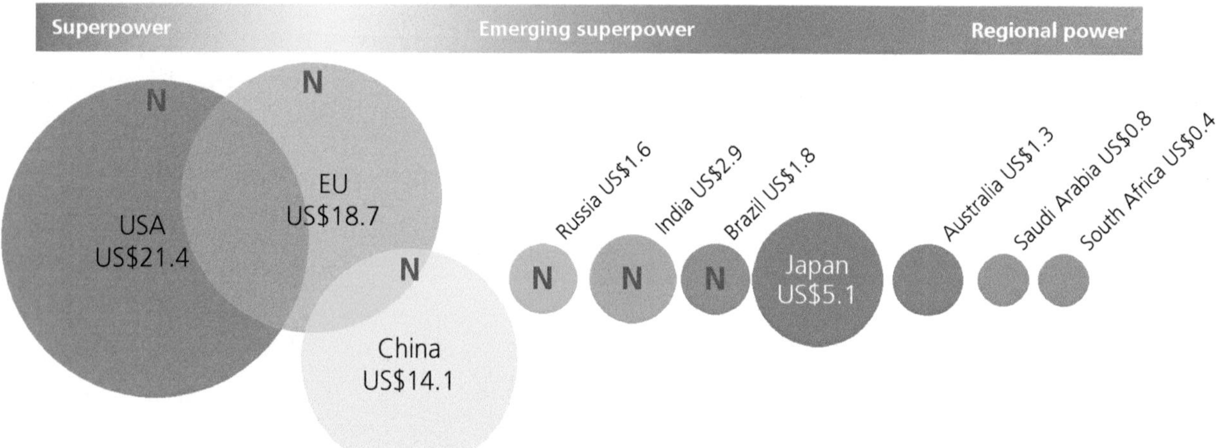

Figure 7.1 The power spectrum of countries by GDP (trillions of dollars), 2019

Note: N = nuclear weapons

Superpower status can depend on five pillars of power. Some nations have all of these pillars, whereas other nations are strong in some types of power but weaker in others:

+ Economic power: a large and powerful economy gives nations the wealth to control investment, build and maintain a powerful military, exploit natural resources and decide which countries receive aid in the world and under what conditions.
+ Military power: this is used through the threat of military action, or military force can be used to achieve geopolitical goals.
+ Political power: this is the ability to influence others through diplomacy or through international organisations.

- Cultural power: this includes how attractive a country's way of life, values and ideology are to others. Globalisation has led to a global culture spread via multimedia, such as Disney, News Corporation and Sony.
- Resources: these can be in the form of physical resources and human resources.

Skills activity

Constructing superpower indexes

- Data on superpower characteristics can be used to produce a superpower index (see Table 7.1).
- Countries can be ranked by order of strength and then their overall totals calculated. The larger the number, the more powerful a nation is considered to be.
- Weighting can be applied to the rank where the rank scores are multiplied to add greater or less importance.

Table 7.1 shows a superpower index using four quantitative measures:

- total PPP GDP (economic)
- total population (resources/demographic)
- nuclear warheads (military)
- TNCs (economic/cultural).

Table 7.1 Superpower index example, 2019/20

Country	Total PPP GDP US$ trillions	Total population, million	Nuclear warheads	Fortune Global 500 TNCs	Sum of ranks
China	27.8 (3)	1,402 (1)	290 (3)	119 (4)	11
Germany	4.2 (16.5)	83 (6)	0 (6.5)	29 (8)	37
India	11.3 (9)	1,362 (2)	140 (5)	7 (12)	28
Japan	5.4 (12)	126 (5)	0 (6.5)	52 (6)	29.5
Russia	4.2 (16.5)	146 (4)	6,500 (1)	4 (14)	35.5
UK	3.0 (21)	67 (7)	215 (4)	18 (10)	42
USA	21.4 (4)	329 (3)	6,185 (2)	121 (2)	11

1 Complete Table 7.2 to show which country is the most powerful if economic strength is deemed to be more important overall. Each measure is ranked, with 1 being the highest score. Total PPP GDP and TNCs are to be scaled (multiplied by three and two, respectively) to reflect their greater importance as measures of power.

Table 7.2 Superpower index, 2019/20: economic strength

Country	Total PPP GDP US$ trillions	Rank scaled x 3	Total population, million (rank)	Rank	Nuclear warheads	Rank	Fortune Global 500 TNCs	Rank scaled x 2	Sum of ranks
China	27.8 (3)	9	1,402 (1)	1	290 (3)	3	119 (4)		
Germany	4.2 (16.5)		83 (6)	6	0 (6.5)	6.5	29 (8)		
India	11.3 (9)		1,362 (2)	2	140 (5)	5	7 (12)	36	
Japan	5.4 (12)		126 (5)	5	0 (6.5)	6.5	52 (6)		
Russia	4.2 (16.5)		146 (4)	4	6,500 (1)	1	4 (14)		
UK	3.0 (21)		67 (7)	7	215 (4)	4	18 (10)		
USA	21.4 (4)	12	329 (3)	3	6,185 (2)	2	121 (2)	6	

Now test yourself

TESTED

1 Which of the five pillars of power do you think is the most important? Give your reasons.

Answer on p. 228

Typical mistake

Superpower status does not depend entirely on military capability.

Revision activity

Using the following criteria, rank which you think is the most important in creating the world's superpowers:

- Physical size
- Economic power
- Demographic factors
- Political factors
- Military strength
- Cultural strength
- Access to natural resources

Explain your rank order. Would you include any other criteria?

Mechanisms for maintaining power

Two forms of power maintenance have existed for centuries (see Figure 7.2):

- **Hard power** refers to the way that countries get their own way by using force.
- **Soft power** is the power of persuasion.

There are, however, few actions or policies that end in military action and therefore the difference between hard and soft power is best looked at along a spectrum. Economic power sits somewhere between hard and soft power and the most powerful countries use a combination of hard and soft mechanisms to get their own way. This is referred to as **smart power**.

The use of different types of power is necessary because invasions, war and conflict are very destructive. They do not always go to plan and fail to achieve the aims of those exercising hard power. Soft power on its own may not persuade one nation to do as another says, especially if they are culturally and ideologically very different.

Hard power	Economic power	Soft power
The spectrum of power		
• Military action and conquest, or the threat of it • The creation of alliances, both economic and military, to marginalise some nations • The use of economic sanctions to damage a nation's economy	• Economic or development aid from one nation to another • Signing favourable trade agreements to increase economic ties	• The cultural attractiveness of some nations, making it more likely that others will follow their lead • The values and ideology of some nations being seen as appealing • The moral authority of a nation's foreign policy

Figure 7.2 The power spectrum

Changing sources of power

The relative importance of superpower characteristics and mechanisms has changed over time.

- For example, the German bid for hyperpower status in the first half of the twentieth century was influenced by the ideas of a British geographer, Halford Mackinder, whose theory is known as geo-strategic location theory.
- It argued that whoever controlled Europe and Asia would control the world — Mackinder identified this area, protected by invasion from the sea, which reached from Russia to China and from the Himalayas to the Arctic, as the heartland.
- Mackinder argued that the heartland was the key geo-strategic location in the world because controlling it commanded a huge portion of the world's physical and human resources.

Now test yourself

2 Explain how superpowers use 'hard' and 'soft' power to maintain their position.

Answer on p. 228

TESTED

Geo-strategic location
A location that commands access to and control over a large territory and its resources.

Revision activity

1 Research Mackinder's geo-strategic location theory.
2 Complete Table 7.3 outlining the theory's strengths and weaknesses.
3 Is geo-strategic location still important today?

Table 7.3 Strengths and weaknesses of geo-strategic location theory

Strengths	Weaknesses

Changing patterns of power

REVISED

There are different distributions of global power (see Figure 7.3):

+ A unipolar world is dominated by one superpower, for example the USA following the fall of the Berlin Wall in 1989 and collapse of communist governments in Eastern Europe. In 1991 the USSR collapsed, leaving the USA as the sole superpower.
+ A bipolar world is one in which two superpowers, with opposing ideologies, try to establish themselves as global superpowers, for example the USA and the former USSR during the Cold War (1945–91).
+ A multipolar world is more complex, where several poles or centres compete for power in different regions. For example, during the time of colonial empires where there was no single superpower.

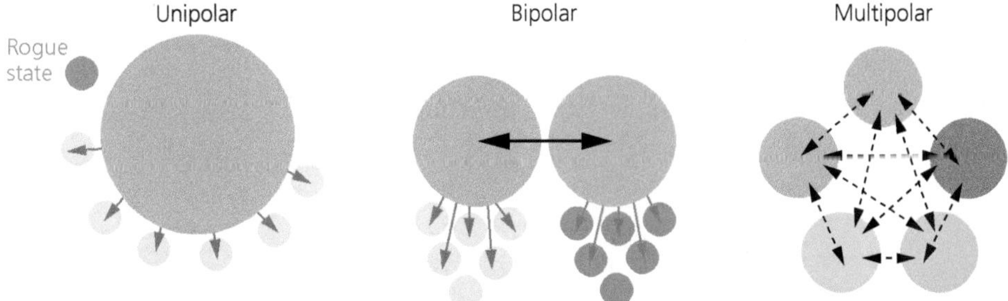

Figure 7.3 The different patterns of power

The colonial era

+ Superpower polarity was at its highest during the rule of the British Empire and colonialism.
+ Britain ruled over a global empire that contained over 20 per cent of the world's population and accounted for 25 per cent of the world's land area.
+ The colonies and the trade routes between the colonies and Britain were protected by the Royal Navy, which dominated the world's oceans during the two phases of empire building (see Table 7.4).

Colonialism Where an external nation takes direct control of a territory.

Table 7.4 The two phases of empire building

Mercantile phase, 1600–1850	Smaller colonies that could be easily defended by coastal forts were conquered, including New England (USA), Jamaica, Accra (Ghana) and Bombay (India). Coastal fringes and islands meant the navy and coastal forts could protect trade in raw materials (sugar, coffee, tea) and slaves. The British armed forces also protected private trading companies such as the Royal African Company, Hudson's Bay Company and East India Company for economic interests.
Imperial phase, 1850–1945	These coastal colonies began to extend inland, as Britain wanted to conquer vast territories. Religion, competitive sport (e.g. cricket) and the English language were introduced to the colonies. Complex trade developed, such as the export of UK-manufactured goods to new colonial markets. Colonialists argued that this brought stability and trade to these countries. Colonial power was maintained by populating colonies with British farmers, colonial administrators and military forces. Technology, such as railways and telegraph, was used to connect distant parts of the empire.

- After the end of the First World War (1918) the distribution of power became increasingly multipolar.
- Emerging powers, such as Japan, began threatening the traditional geographical spheres of influence of established superpowers and regional powers.
- Military power was increasingly important and, as war approached in 1939, an arms race took place, with countries strengthening their naval power.

The post-colonial era

The colonial era came to an end relatively quickly after the end of the Second World War in 1945. Most colonial powers had lost their colonies by 1970. Since then, neo-colonialism has emerged in many parts of the developing world.

- The **Cold War** dominated the era from 1945 to 1991.
- During this bipolar era the USA became an increasingly global superpower with worldwide military bases aimed at containing the USSR and halting the spread of communism (see Table 7.5).
- However, in recent years, China's rapid economic growth and the dependence of the USA on China for manufactured goods has meant that China has begun to challenge US **hegemony**.

Neo-colonialism An indirect form of control, which means that newly independent countries are not masters of their own destinies.

Key concepts

The **Cold War** was a period of geopolitical tension, between 1945 and 1991, between the superpowers of capitalist USA and communist USSR. During this period, nuclear weapons and the means to deliver them were perfected. Behind any political clash there was always the threat of nuclear war which prevented open conflict from breaking out.

Hegemony is the dominance of one superpower over other countries. Hegemony can be exercised in a range of ways and leads to a country being able to exert undue influence within a society. The sheer size and capability of the US military forces give it dominance over world affairs, which discourages others from acting against it.

Exam tip

It is important to be able to explain how world power fluctuated between unipolar power, bipolar and multipolar status at different times throughout the twentieth century.

Now test yourself

TESTED

3 What brought an end to the colonial era?

Answer on p. 228

Table 7.5 The Cold War superpowers compared: USA and USSR

	USA	USSR
Human resources	1989 — population of 287 million	1991 — population of 291 million
Physical resources	Self-sufficient in most raw materials; oil importer	Self-sufficient in most raw materials; oil exporter
Political system	Capitalist, free-market economy and global TNCs Investment in countries such as Japan, Singapore and Philippines to support economic growth and reduce the spread of communism Democracy with free elections held once every 4 years	Socialist, centrally planned economy; most businesses were state owned Single-party state with no free elections (dictatorship)
Allies	Countries supporting the USA formed NATO — mainly Western Europe Strong economic and military ties to Japan and South Korea	Eastern Europe (the Warsaw Pact countries) and alliances with Cuba among other developing nations Council for Mutual Economic Assistance (to support economic strength)
Military power	World's largest navy and most powerful air force, with a 'ring' of bases surrounding the USSR Global network of nuclear bases: large nuclear arsenal Extensive global intelligence gathering through the CIA	Very large army, and large, but often outdated, naval and air force capability Nuclear weapons Created a buffer of countries in Eastern and Central Europe to shield it from possible future attacks from the West KGB gathered extensive global intelligence
Cultural influence	Film, radio, television and music industry created a powerful vehicle for bringing a positive view of consumerism, family values, democracy and affluence to a global audience Propaganda between the two sides lasted for decades — Hollywood produced films designed to generate suspicion of communists	Exported a 'high' culture message focused on ballet, art and classical music in contrast to the 'popular' culture of the USA Strict censorship within the USSR

Geopolitical stability and risk

Different patterns of power bring varying degrees of geopolitical stability and risk.

- A **unipolar** world dominated by one **hyperpower** might appear stable, but the hyperpower is unlikely to be able to maintain control everywhere, all the time.
- A **bipolar** world has the potential to be stable, as it is divided into two opposing blocs. Stability depends on communication between the two blocs remaining open and each superpower having the capability to control countries within its bloc.
- **Multipolar** systems are complex, as there are multiple relationships between more or less equally powerful states. This creates more opportunities to misjudge the intentions of others, or fears over alliances creating more powerful blocs, which can increase the risk of conflict.

It could be argued that the period between 1910 and 1945 was a multipolar one and that this complex geopolitical situation contributed to two world wars.

Many observers believe that the twenty-first century could be multipolar, especially as we could see the rise of countries such as India and China while the power of the USA and the EU diminishes.

Revision activity

Study Table 7.5.

- Which is the most and least important factor in helping the USSR and USA to maintain power and influence?
- Are the factors different? If so, explain why this may be.

Now test yourself

4 Explain the difference between a unipolar, a bipolar and a multipolar world.

5 Explain how countries have maintained power over other countries. Give examples in your answer.

Answers on p. 228

TESTED

Emerging powers

REVISED

Who are the emerging powers?

Future superpowers are likely to emerge from two groups of countries, which overlap:

- The BRIC economies (Brazil, Russia, India and China) were identified as a group of emerging powers in 2001 (see Table 7.6).
- The G20 major economies, a group formed in 1999, is made up of 19 countries plus the EU. This includes several potential emerging powers, such as Indonesia, Mexico, Saudi Arabia, South Korea and Turkey.

Table 7.6 Strengths and weaknesses of the BRIC economies

	Economic	Political	Military	Cultural	Demographic	Environmental
Brazil	Produces half of South America's GDP (2015) Relies on primary product for export Self-sufficient in food and energy	Accusations of corruption in government Protests every year since 2013	Least significant of all the BRIC economies Spends over 60% of entire South American military budget	Global reputation for football Hosted football World Cup in 2014 Hosted Olympics and Paralympics in 2016	Contains half of South America's population Has a young population Population is also ageing	Contains 13% of all known flora and fauna species A leader in the use of biofuel Major environmental issues with the deforestation rates in the Amazon, illegal poaching and pollution caused by mining
Russia	Ninth largest global economy — large reliance on oil and gas exports Most unequal of emerging and developed nations — in 2018 the wealthiest 3% of Russians owned 89% of all financial assets	Political influence has declined globally since 1991 Still has a strong influence politically over many of its neighbours, all former republics of the USSR	Naval and aircraft stock is ageing Spending on military has increased	Russian is little spoken beyond the borders of former USSR nations Has a large tourist industry	Has a low fertility rate and its natural increase is tiny, at 0.02 per 1,000 Since 1991 the population has stayed low and declined in many of those years	Has a legacy of pollution
India	Economy quadrupled between 1997 and 2015 Has benefited from having an English-speaking education system, which has led to the growth of outsourced IT industries	The world's largest democracy A founding member of the UN	World's fourth largest military power in terms of its weaponry and personnel	Bollywood is the world's largest film industry but doesn't have Hollywood's global reach Birthplace of four world religions — Hinduism, Buddhism, Sikhism and Jainism	The world's second largest population Youthful population and large working population	One of the world's richest nations in terms of biodiversity Growing population and economic growth are a threat to the country's biodiversity

	Poor infrastructure, in particular, water and energy supplies and transport system A large proportion of citizens still live in poverty					Has one of the world's worst environmental problems with some of the world's worst slums and highest CO_2 emissions
China	Wealth is unevenly spread due to the country's huge population In 2018, 17% of the population had graduated from university, affecting the knowledge economy A major player in overseas investment	One-party government Rarely gets involved in global crises	World's largest army Has little global reach via its military but does have military strength	Has few global brands Has little influence on global entertainment	Has the largest population The one-child policy has created an ageing population Fairly isolated in regards to international migration — may need immigration in the future to support its shrinking working population	The world's largest CO_2 emitter Has started to commit to targets to reduce pollution levels

Strengths and weaknesses

All emerging powers have strengths and weaknesses. For example:

- China has a highly educated population, a strong economy and a vast military. Its weaknesses include an ageing population, environmental pollution and a heavy reliance on imported raw materials.
- Russia is a nuclear power with huge oil and gas reserves. The downsides are that it has an ageing and unhealthy population and strained relations with the EU and the USA.
- Indonesia has a youthful population and a large amount of natural resources. However, it has high levels of poverty and internal political instability.
- Turkey's economy is increasingly integrated with the EU. It has a youthful population with good levels of education. However, it has serious internal problems with its Kurdish minority and is suffering from the political instability.

Now test yourself

TESTED 

6 Why is a youthful population thought to be one of the prerequisites of emerging power?

7 Explain why China can claim to be a superpower.

Answers on p. 228

Theoretical explanations

Possible explanations of the changing global pattern of power are to be found in three development theories.

World systems theory, developed by Wallerstein in 1974, views development in regards to capitalist world systems and the development gap. The theory recognises three broad categories of economic development:

- core regions: drive the world economy and include the Organisation for Economic Co-operation and Development (OECD) countries and the USA and EU superpowers
- semi-periphery regions: the newly industrialised countries (NICs) of Latin America and Asia
- periphery regions: at the other extreme, these regions lie far away from the core and rely on the core regions to exploit their raw materials — the developing world.

Dependency theory describes how 'satellite' (periphery) countries provide a range of services to metropolitan (core) countries. The developing countries remain dependent on the developed countries and the developed countries control the development of developing nations. They do this by setting the prices paid for commodities, interfering in economies and using economic and military aid to 'buy' the loyalty of satellite states.

Modernisation theory views the way that countries develop as moving through five stages. It argues that pre-industrial societies develop very slowly until certain preconditions for economic take-off are met. Industrialisation and urbanisation follow and from these processes a country acquires wealth and political power.

Exam tip

Ensure you learn the three theoretical models of world systems theory, modernisation theory and dependency theory, as they can help to form the basis for a strong examination answer.

Now test yourself

8 Explain why and how superpowers can change over time.

Answer on p. 228

TESTED

Revision activity

Create a table showing the strengths and weaknesses of the three development theories.

The global impacts of superpowers

The global economic system

REVISED

Free trade and capitalism

The global economy is essentially a free-market, capitalist economic system operating throughout most of the world. There are few countries not involved. The promotion of the global economy and free trade are in the hands of powerful global inter-governmental organisations (IGOs). The four main players are shown in Table 7.7. The important point here is that the superpowers are able to influence the decisions and policies of those IGOs.

Table 7.7 Global organisations and capitalism

World Bank	Role is to finance development via development loans to developing countries. Done within a 'free-market' model that promotes exports, trade, industrialisation and private businesses, which benefits large developed-world TNCs.
International Monetary Fund (IMF)	To stabilise global currencies, and assist countries to reform their economies. This can lead to more open access to developing economies for TNCs.
World Economic Forum (WEF)	A Swiss non profit organisation that promotes globalisation and free trade via its annual meeting at Davos, which helps resolve disputes and promote global thinking.
World Trade Organization (WTO)	Established in 1995, it aims to free up global trade and reduce trade barriers. Previously known as the General Agreement on Tariffs and Trade.

Free trade When a government does not discriminate, and the trade of goods is free of import/export taxes and tariffs.

Inter-governmental organisations (IGOs) Mainly global organisations whose members are nation states that uphold treaties and international law.

Revision activity

Note the four IGOs in Table 7.7 and how each influences the world economy. How do each of their roles differ?

Skills activity

Understanding world trade growth using linear and logarithmic scales

Study Figures 7.4 and 7.5.

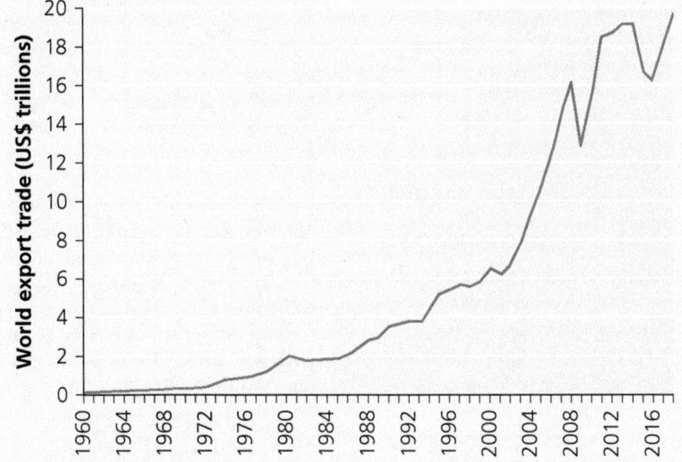

Figure 7.4 Growth in world export trade 1960–2018 using a linear scale

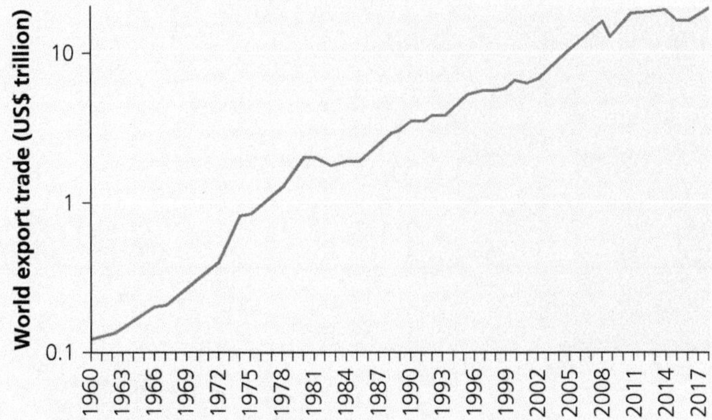

Figure 7.5 Growth in world export trade 1960–2018 using a logarithmic scale

1 What are the advantages and disadvantages of using each scale?

2 Which scale is more effective in showing world export trade?

3 Complete Table 7.8.

Table 7.8 The advantages and disadvantages of using linear and logarithmic scales

	Advantages	Disadvantages
Linear graph		
Logarithmic graph		

TNCs and their roles

TNCs come in two distinct forms:

- publicly traded TNCs with shareholders who receive dividends based on the company profits every year
- state-owned TNCs where all profits return to the state.

In many cases, state-owned TNCs are large but not well known, as their brands are not global.

The dominance of TNCs in the global economy is the outcome of:

- their economies of scale, which mean that they can outcompete smaller companies
- large bank balances where they have the ability to borrow money to invest in new ventures
- their ability to expand globally and open up new markets.

A further role of TNCs is their promotion of new technology whereby they invest huge sums in research and development (R&D) to develop new products. These new developments are protected through intellectual property law by:

- patents: new inventions, technologies and systems
- copyright: artistic works, e.g. music, books and artworks
- trademarks: protecting designs, e.g. logos.

Typical mistake

Do not underestimate the economic and political clout of TNCs compared with that of superpower governments.

Global cultural influences

TNCs bring to the countries in which they operate influences from their country of origin. They can have a cultural impact on their global consumers.

Western culture is global due to the dominance of the USA and the economic power of the EU, which have led some people to identify the spread of cultural globalisation. Its characteristics include:

- a culture of capitalism
- a culture of consumerism
- a white, Anglo-Saxon culture
- a culture that 'cherry picks' and adapts parts of other world cultures and absorbs them.

Making links

Because of their size, global reach and economic influence, TNCs exercise considerable leverage as players, particularly as far as FDI (see page xxx) is concerned. Critics argue that TNCs have too much power and abuse it in the form of worker exploitation, low wages and environmental pollution.

Now test yourself

9 Explain how TNCs have become dominant economic forces in the global economy.

10 Explain how global cultural influence is an important aspect of power.

Answers on p. 228

International decision making

Global action

The UN Security Council is the primary global mechanism for preventing conflict. It aims to maintain international law by:

- applying sanctions to countries that are considered a security risk, harbouring terrorists, threatening or invading another state or breaching human rights
- permitting the use of military force against a country where necessary
- authorising a UN peacekeeping force, e.g. Bosnia (1998) and Kuwait (1990–94).

Sanctions Political or economic decisions that aim to persuade a country back to the negotiating table without the use of military force.

A standout feature over the past 40 years is how many times the USA has intervened militarily in overseas countries. It has done this through:

- UN Security Council action
- working together with allied countries as a coalition, but outside a UN remit
- unilaterally — that is, with no support from another country.

Making links

Some countries, notably the USA, the UK and France, regularly act as 'global police'. They deploy forces overseas to intervene in the affairs of other countries, either unilaterally or as part of a coalition.

Alliances

Military alliances are a key element of superpower status. Two international military alliances dominate following the dissolution of the Warsaw Pact in 1991:

- North Atlantic Treaty Organisation (NATO)
- ANZUS, a treaty involving Australia, New Zealand and the USA.

Other military alliances include a Russian military alliance, the Collective Security Treaty Organization (CSTO), which includes only former USSR republics bordering Russia, and the Shanghai Cooperation Organisation (SCO) involving Russia, China and some of the former USSR republics.

Now test yourself

11 Why is there a need for global police?

Answer on p. 228

TESTED

Economic alliances strengthen interdependence between nations. Examples include the following:

- European Union
- North American Free Trade Agreement (NAFTA)
- Association of Southeast Asian Nations (ASEAN)

Those countries that are members of military alliances are also often involved in economic alliances, for example many NATO members are EU countries.

Figure 7.6 illustrates part of the alliance networks. This creates a powerful axis of economic and military security that reflects the ideology of each bloc.

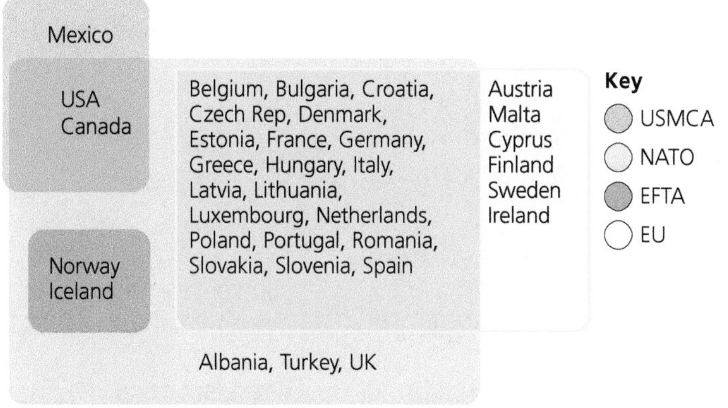

Figure 7.6 Western economic and military alliances, 2020

Geopolitical stability

Figure 7.7 shows the main pillars supporting global security. The system that is in place to help maintain global security was set up in 1945 and revolves around the United Nations (UN). However, this post-war system is under strain:

- The USA, the UK and France (the leaders) are not as powerful as they were.
- As emerging powers, there is a strong case for countries such as India and Brazil to have more of a say in global security.
- Neither Africa nor Latin America has a say.
- The global financial crisis of 2007–08 strained the IMF and global financial system to the limit and many economists wonder whether there is a 'better way'.
- The continuing threat of global terrorism from al-Qaeda, Islamic State and the Taliban might suggest that global security co-operation is not as strong as it could be.

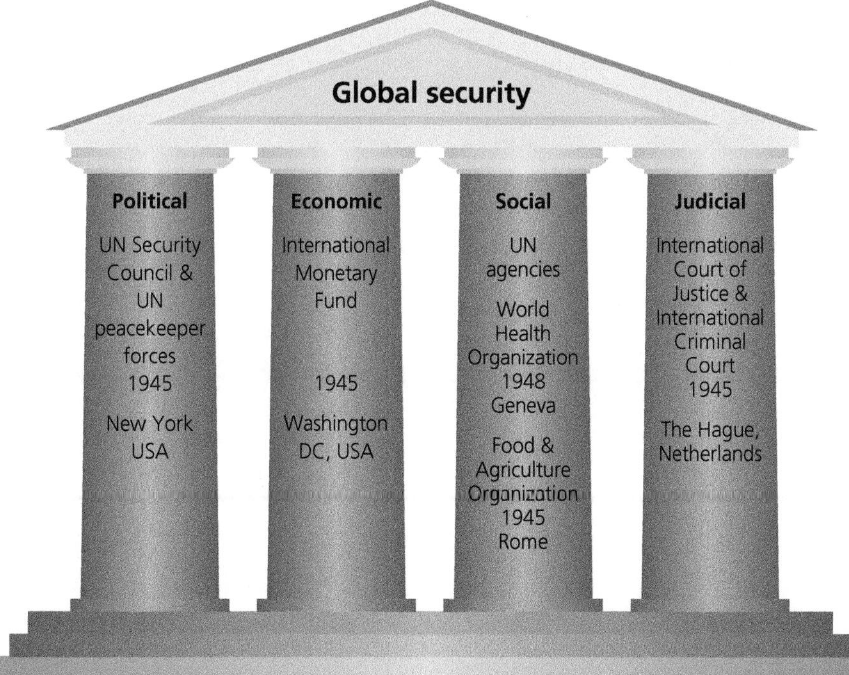

Figure 7.7 Pillars of global security

Revision activity

Find out about three organisations shown in Figure 7.6. Create a table to show the strengths and weaknesses of each organisation.

Exam tip

Make sure that you are aware of how the overlap of military and economic alliances supports and enhances superpower status.

Making links

The actions and attitudes of IGOs towards maintaining aspects of global security are largely determined by the willingness of member states to act. In most cases, this requires strong political will from a group of countries with support from a superpower or emerging power.

Now test yourself

TESTED

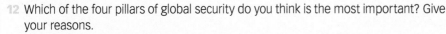

12 Which of the four pillars of global security do you think is the most important? Give your reasons.

13 How effective are superpowers and emerging nations in international decision making?

Answers on p. 228

Global environmental concerns

REVISED

Resource demands

Maintaining a large economy, a military machine with global reach and a wealthy population requires energy, mineral, land and water resources. This has resulted in superpowers having very large carbon footprints.

The high resource consumption of superpowers and emerging powers creates a number of environmental issues:

- Urban air quality in emerging cities is low due to heavy reliance on fossil fuels.
- The volume of global trade is such that shipping now makes a significant contribution to global CO_2 emissions. China's immense demand on raw materials means it accounts for 90 per cent of the global growth in sea traffic in the twenty-first century.
- Deforestation and land degradation are issues in some emerging powers as they convert more land into farmland, continue to urbanise, and the demand for water and use of chemicals in farming grows in order to increase yields.

Revision activity

- Produce a list showing the ways in which superpower resource demands can cause environmental degradation and contribute to global warming.
- How are the USA, China, Russia and the EU committed to reducing their CO_2 emissions? Give examples.

Reducing carbon emissions

- Recognition of the link between carbon emissions and global warming is increasing.
- However, there are national differences in the willingness to take action and to sign up to global agreements, such as the Climate Change Conference in Paris (2015).
- One aspect of the problem is that developed countries are in a better position to turn to renewable sources of energy. The emerging countries are heavily dependent on fossil fuels and are less willing to sign international agreements.

Making links

Attitudes and actions over the threat of global warming are influenced by many factors, including a country's level of development and the government's viewpoint, as illustrated in the following examples:

- On the whole, Europeans trust the science behind global warming more than Americans do.
- Often economic growth and personal wealth are given a higher priority than environmental issues in emerging countries.

Exam tip

A range of energy alternatives is now available that can help reduce emissions. But access to those alternatives is far from equal.

Now test yourself

TESTED

14 Why might emerging countries be less willing to reduce carbon emissions and sign global agreements on environmental issues?

Answer on p. 228

Growth in middle-class consumption

The growth of the middle class is a significant concern in relation to the growth of the BRICS and other emerging powers. While rising affluence in these countries is positive in regards to development, the rise in demand will result in a huge strain on the availability and cost of key resources, such as rare earth minerals, oil and gas, staple grains and water (see Table 7.9).

Middle class The global middle class can be defined as people with an annual income of over US$10,000. They are significant consumer spenders.

Table 7.9 Pressure on resources from rising consumption

Food	Water
As countries develop, the pressure on food supply increases due to demands for new food types, e.g. China's growing affluence has led to a 99% increase in the consumption of meat. Land previously used for staple food grains has been converted to produce meat and dairy products.	Some emerging powers, such as India, already have water supply problems. By 2030 India's situation could be critical, with 60% of areas facing water scarcity. The World Bank believes that economic growth could lead to China having a water crisis due to uneven distribution of water resources and increased consumption of water for industry, agriculture and domestic use.
Energy	**Resources**
In 2015, global oil demand was approximately 95 million barrels per day. The use of fossil fuels is likely to rise by as much as 30% by 2030. This could affect supply and lead to price increases and/or supply shortages. Countries such as Russia and Brazil could be in a stronger position as they have their own domestic supplies, as opposed to those relying on imports (e.g. India).	Demand for rare earth minerals — used in LCD screens and numerous other hi-tech gadgets — could raise prices. The demand for lithium-based batteries is very high and may be hard to meet in the future. These are also toxic if buried in landfill. Even more basic metals, such as copper, tin, nickel and platinum, are at risk of supply shortages and dramatic price changes, especially as they are used in mobile phone production.

Now test yourself TESTED

15 Explain why the growth of a middle class is thought to be good for development.

16 Why might the growth of the middle class be a threat to the planet?

Answers on p. 228

Contested spheres of influence

The contest for global influence

REVISED 

Tensions over resources

Superpowers are huge consumers of resources. Securing access to physical resources is of great importance for both governments and TNCs. This can lead to resources being contested in a number of different economic, environmental and political spheres. This could be because:

- the land border separating two countries is in dispute, e.g. tensions between Russia and many Eastern European countries
- the ownership of a landmass is being disputed, e.g. the Arctic region
- the extent of a nation's offshore exclusive economic zone is in dispute or claimed by another nation, e.g. international disputes over islands in the South China Sea.

Making links

If a country and/or TNC deems a resource indispensable then tensions are more likely to arise. If players consider investing in alternatives or give conservation a higher priority than exploitation then tensions are likely to reduce.

Exam tip

Make sure that you have real-life examples with which you can support your answers in the exam, such as a case study of tensions over the Arctic oil and gas, and over territory and physical resources such as the South and East China Seas.

Revision activity

Create a mind map to show the reasons why:

- IP rights are not always defended by consumers
- some countries are unwilling to promote IP agreements, e.g. Thailand.

Intellectual property

A possible source of tension is intellectual property (IP) rights. A worldwide system of IP has been in place since 1967, run by the World Intellectual Property Organization (WIPO), a part of the UN. Countries must reach an agreement on intellectual rights to protect those businesses and individuals who have developed new inventions, trademarks, artistic works and trade secrets from use by others.

Abuse of the international IP treaties and counterfeiting of goods can affect relations between countries, especially between the USA and China:

- TNCs may be reluctant to invest in China, due to the fact that counterfeiting will affect their profits.
- Lack of action by Chinese authorities on IP issues might suggest its government is less likely to co-operate on other issues relating to international law.
- If one government believes another will not follow international rules, this could result in limited trade agreements.

Intellectual property (IP) Intangible property that is the result of a person's creativity, such as patents, copyrights, trademarks, etc.

Sphere of influence A physical region over which a country believes it has economic, military, cultural or political rights.

Now test yourself

17 Explain how global influence may be tested politically.

Answer on p. 229

TESTED

Political spheres of influence

Political spheres of influence can be contested, leading to tensions over territory and physical resources.

A prime example is China's **Island Chain Strategy** in the South and East China Seas. This involves claiming sovereignty over small, uninhabited islands as well as artificial islands, and so extending its maritime border and economic exclusion zone. So the aims are both military and economic. This strategy is creating tensions with Brunei, Japan, Malaysia, the Philippines, South Korea, Taiwan and Vietnam.

Other current examples include the long-running dispute between India and Pakistan over the ownership of Kashmir, and between Japan and Russia over the Kuril Islands.

Revision activity

Make sure that you have case study examples of both a place where there is tension over territory and physical resources and a place where tension has resulted in open conflict.

Changing relations with developing countries

Developing economic ties

Existing superpowers, such as the EU and the USA, are often accused of having unfair relationships with developing countries. This means relationships based on:

- neo-colonialism: superpowers having unfair economic and political influence over developing countries, despite not ruling them directly
- unfair terms of trade: developing countries' reliance on cheap commodity exports (coffee, cocoa, copper, oil) set against expensive manufactured imports from developed countries

- skilled workers from developing countries migrating to developed countries to boost developed world economies, leading to a brain drain
- inequalities in developing countries resulting in local wealthy elites who control imports and exports, benefiting from the neo-colonial relationship.

Table 7.10 illustrates how the economic relationship between China and Africa can be interpreted in two contrasting ways: neo-colonialism or development opportunity.

Typical mistake

Do not think that superpowers are magnanimous and generous in their dealings with less powerful countries.

Table 7.10 China's relationship with Africa

Neo-colonial challenge?	Development opportunity?
Infrastructure investments, particularly in improving transport links, ensure China can export raw materials.	China has invested heavily in ports, railways and roads to export raw materials — infrastructure that can be used by African citizens themselves.
Skilled and technical jobs are often filled by Chinese migrant workers — nearly 1 million Chinese citizens have moved to African countries since 2005.	China has trained 40,000 African personnel, which has helped to create vital jobs and also modernise the economy.
Local producers are undercut by cheap Chinese imports, which have forced them out of business.	Chinese mines and factories have brought modern working practices and technology to Africa.
FDI may only bring temporary construction jobs.	Finance from China has funded 17 major HEP projects since 2000. Hydropower accounts for 15% of total electricity share in Africa.
Aid from China is tied to FDI, meaning that if investment is allowed then China will provide some aid.	Investment deals are often accompanied by aid, so the benefits from Chinese money are more widely spread.

Making links

The role of emerging powers in Africa could mirror that of the USA in Taiwan, South Korea and Singapore in the 1960s. The USA's support, both political and economic, helped these states achieve their NIC status. Or could it be that the interest in Africa was simply a repeat of the colonial and imperial model of previous centuries?

Exam tip

The shifting balance of power can be difficult to predict, and although Asian dominance looks set to happen, Japan's experience (economic recession in the 1990s) is a warning that economic growth is not always promised.

The rising importance of Asia

Asia is, or soon will be, the dominant global region. It hosts a number of powerful countries, including those with emerging and existing superpower credentials, such as China, India, Indonesia, Japan, Malaysia and South Korea. It is expected that by 2050 Asia will be the world's most populous continent as well as the world's largest by GDP.

The rise of Asia is also creating economic and political tensions within the continent:

- The rivalry between China and India as superpowers will intensify.
- Other countries, such as Indonesia, Japan, Malaysia and the Philippines are jostling for the third, fourth and fifth places in the regional power rankings.
- Japan feels threatened, particularly by China.
- North Korea poses a security threat, particularly to Japan and South Korea, but even to the USA.

Now test yourself

18 Give examples of the economic ties between emerging powers and the developing world.

19 Suggest how the rising importance of certain Asian countries might create economic and political tensions within the region.

Answers on p. 229

TESTED

Tensions in the Middle East

The Middle East is one global region that has proved troublesome over the past 50 years. It has various characteristics that make it a frequent location of conflict and tension, including the complexities of Middle Eastern politics, religions, ethnic differences and territorial disputes (see Table 7.11). In addition, there are the tensions created by **Islamic terrorism**.

Key concept

Islamic terrorism involves a number of extremist terrorist organisations fighting jihad (holy war) against the non-Islamic West. Their motives are complex and include religious motivation and the ideological belief that war should be fought against all non-Muslims. Many are fighting against what they regard as long-term interference by the West in the Middle East. Justification of the attacks against civilians comes from extreme interpretations of Islamic holy books.

Table 7.11 Sources of instability in the Middle East

Religion	Oil and gas	Governance
Most of the region is Muslim, but Sunni (Saudi Arabia, United Arab Emirates) and Shia (Iran) sects are in conflict with each other.	The region has the largest proven oil reserves with 65% of the world's crude oil exports originating in the region.	Many of the countries are relatively new states where democracy is either weak or non-existent. Religious and ethnic allegiances are often stronger than those of national identity.
Resources	**Youth**	**History**
Although rich in fossil fuels, the region has limited access to fresh water and farmland, meaning territorial conflict over natural resources is more probable.	Many countries in the region have young populations with high unemployment and relatively low education levels; the potential for youths to become disaffected is high.	Many international borders in the region are arbitrary; they were drawn on a map by colonial powers and do not represent the actual geography of religious or cultural groups.

Making links

The differences between Western ideology and culture (capitalism, democracy, individual freedom, gender equality and perhaps Christianity) and Islamic ideology and culture (primacy of religion, strict laws and gender discrimination) are hard to reconcile.

Ideology A system of ideas, beliefs and values that forms the basis of economic or political theory and policy.

Now test yourself

TESTED

20 Why are tensions in the Middle East an ongoing challenge to superpowers and emerging powers?

Answer on p. 229

Revision activity

Create a mind map to show the changing relationships between existing and emerging superpowers. Use examples to show evidence.

Ongoing economic restructuring

REVISED

Economic problems

The EU and the USA are the two largest economic powers, but this is not guaranteed. Both face considerable economic challenges that may prevent them from maintaining their global importance. The USA is believed to be in a better position than the EU for two reasons:

- Although the USA consists of 50 states, which have their own rights and laws, the differences between states are minor and they are not sovereign, unlike the 27 countries that make up the EU. The former are much more likely to agree on policy.
- The USA is not ageing as quickly as the EU.

Both superpowers face the ongoing costs of economic restructuring, whereby employment has shifted from predominantly primary and secondary industries into tertiary and quaternary. This has resulted in traditional manufacturing cities seeing a rise in unemployment levels and requiring major investment in regeneration.

Economic restructuring The shift from primary activities and secondary industry towards tertiary and quaternary industry as a result of deindustrialisation. It has heavy social and economic costs.

The economic costs of maintaining superpower status

The USA has the largest military budget in the world, spending more than US$900 billion annually on maintaining its global supremacy. This includes all military spending and intelligence services, as well as NASA and foreign aid.

In comparison, the UK has the fifth largest defence budget and spends US$56 billion. China spends around US$200 billion. To some extent emerging powers can increase their military power by copying technologies that were initially developed (at huge cost) by the USA.

Exam tip

The UK may not be considered a superpower, but it is one of only a handful of countries that has the technology to launch nuclear warheads almost anywhere in the world.

The future balance of global power

Figure 7.8 sets out four different superpower futures. Each has different implications:

- US hegemony (unipolar): would China be prepared to go along with this? China has military power and is keen to extend its maritime limits, but its economy is faltering.
- Regional mosaic (multipolar): this structure is inherently unstable as countries of equal power make complex and competing alliances, with no single country acting as the 'global police'.
- New Cold War (bipolar): China rises to become equal in power to the USA; countries align themselves with one or the other ideology.
- Asian century (unipolar): this would cause a fundamental shift in the centre of gravity of the global economy, but the demand for resources in the West would continue.

Making links

For the governments of superpowers and emerging superpowers, preparation is key. Geopolitical futures are so uncertain that governments and TNCs undertake 'what if' thinking. The goal is for these key players to have a range of policy options at the ready depending on which 'future' becomes reality.

US hegemony (unipolar)	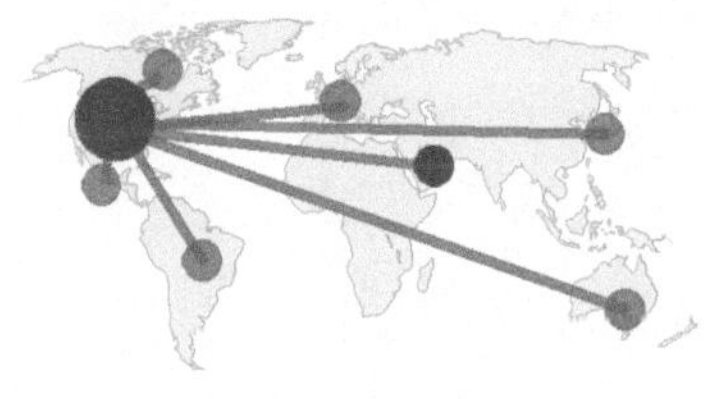	US dominance, and economic and military alliances, continue in a unipolar world. China faces an economic crisis, similar to Japan's in the early 1990s, and ceases to grow rapidly.
Regional mosaic (multipolar)	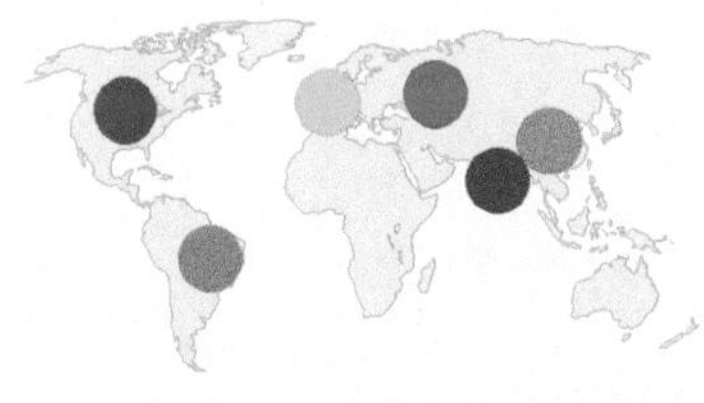	Emerging powers continue to grow while the EU and USA decline in relative terms, creating a multipolar world of broadly equal powers with regional but not global influence.
New Cold War (bipolar)	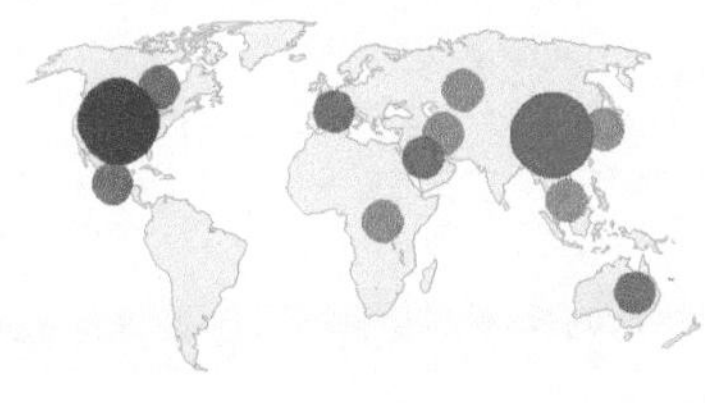	China rises to become equal in power to the USA, and many nations align themselves with one or other ideology, creating a bipolar world similar to the 1945–90 Cold War period.
Asian century (unipolar)	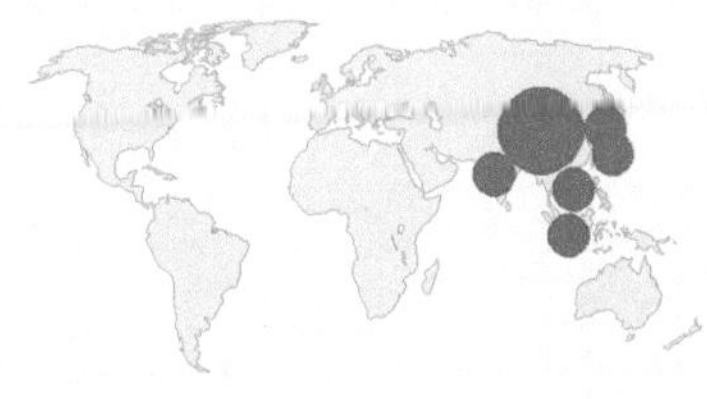	Economic, social and political problems reduce the power of the EU and USA; economic and political power shifts to the emerging powers in Asia, led by China.

Figure 7.8 Alternative superpower futures in 2030

Revision activity

Produce a list of the reasons why the future balance of global power is uncertain.

Now test yourself

21 Suggest reasons why some people in the USA might question the cost of maintaining global military power.

22 Explain the ways in which the economic restructuring of existing superpowers challenges their power in the future.

23 To what extent is the position of the USA as a unipolar superpower being questioned?

Answers on p. 229

TESTED

Exam skills

You should be familiar with the following skills and techniques used throughout the topic of superpowers:

- constructing power indexes using complex data sets, including ranking and scaling
- mapping past, present and future sphere of influence and alliances using world maps
- interpreting graphs of world trade growth using linear and logarithmic scales
- mapping emissions and resource consumption using proportional symbols
- plotting the changing location of the world's economic centre of gravity on world maps
- analysing future gross domestic product (GDP) using data from different sources.

Exam practice

A-level

1 a) Explain the influence that superpowers have on the global economy. (4)

b) Assess the consequences for people and the environment in developing countries as a result of their relationships with superpowers. (12)

2 a) Explain how geopolitical power stems from a range of characteristics. (4)

b) Assess the extent to which TNCs have influence over the global economic system. (12)

3 a) Explain how countries have maintained power and influence over other countries. (4)

b) Assess the importance of superpowers in playing key roles in international decision making. (12)

Answers and quick quiz 7 online

Summary

You should now have an understanding of:

- how geopolitical power stems from a range of human and physical characteristics of superpowers
- how patterns of power change over time and can be unipolar, bipolar or multipolar
- how emerging powers vary in their influence on people and the physical environment, which can change rapidly over time
- the way in which superpowers have a significant influence over the global economic system
- how superpowers and emerging nations play a key role in international decision making concerning people and the physical environment
- global concerns about the physical environment, which are disproportionately influenced by superpower actions
- how global influence is contested in a number of different economic, environmental and political spheres
- the way in which developing nations have changing relationships with superpowers, with consequences for people and the physical environment
- how existing superpowers face ongoing economic restructuring, which challenges their power.

8 Global development and connections

8A Health, human rights and intervention

- Traditional definitions of development have equated it with economic development.
- Today, however, while it is still widely acknowledged that economic growth provides much of the driving force behind development, there is a preference for a broader view.
- This acknowledges that development also has important social, political and environmental dimensions.
- Development studies now tend to assess progress in terms of human welfare and human rights.
- The course of global development today is guided by the decisions and geopolitical interventions of national governments and international organisations.
- These interventions range from development aid to military campaigns and can have a considerable impact on health, well-being and human rights.

Human welfare The health, happiness, good fortune, prosperity, etc. of a person or a group.

Human development

Concepts of human development

REVISED

While most agree that **human development** has an important economic dimension, it is clear that there are some very different perceptions of what it is all about. Those perceptions are coloured by the beliefs, values, morals and codes of conduct of different societies.

A prevailing view today is that human development should focus on:

- health
- life expectancy
- access to education
- human rights.

Key concept

Human development has been defined as the process of enlarging people's freedoms and opportunities, and improving their well-being and quality of life. The economic dimension of development is highlighted because it is recognised to be the main driver of the entire human development process. It provides the core of what has been described as the 'development cable'. However, human development is multi-faceted, involving a range of different strands that all contribute to well-being and quality of life (see Figure 8.1).

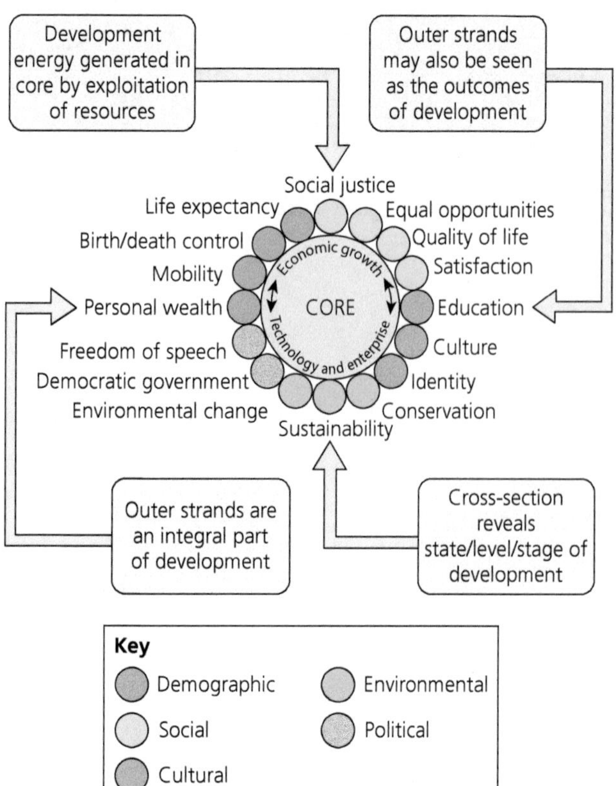

Figure 8.1 The core and strands of the human development cable

Exam tip

Be sure you are aware of the five different strands of human development shown in Figure 8.1.

Measuring human development

Geographers are interested in how the level of development varies spatially — between countries and within countries. The traditional measures used for investigating development in this way have been the economic ones of per capita gross domestic product (GDP) and per capita GNI. But given the increasing recognition of a broader definition of human development, two relatively new measures are being used:

- The human development index (HDI): based on life expectancy, years of schooling and per capita income (see Figure 8.2).
- The happy planet index (HPI): based on sustainable well-being, life expectancy and ecological footprint (Figure 8.3).

However, these indexes are not the only way of assessing improvements in human development. Other approaches contest these models, for example Sharia Law or Bolivia under the rule of Evo Morales.

Gross domestic product (GDP) The total value of all goods and services produced by a country in a year. It is calculated by taking the value of all produced goods plus the value of all services.

Ecological footprint A measurement of the area of land or water required to provide a person (or society) with the energy, food and resources they consume and the waste they produce.

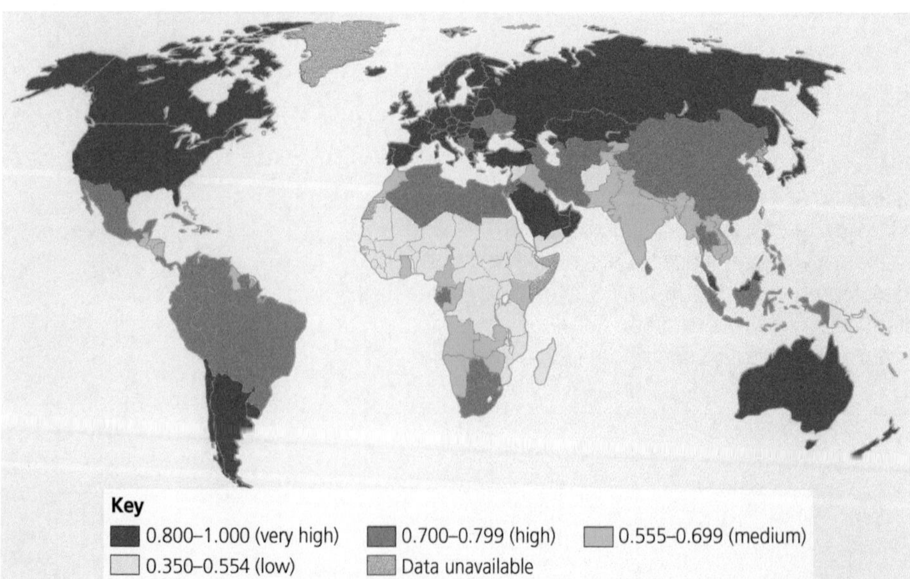

Figure 8.2 HDI scores in 2018

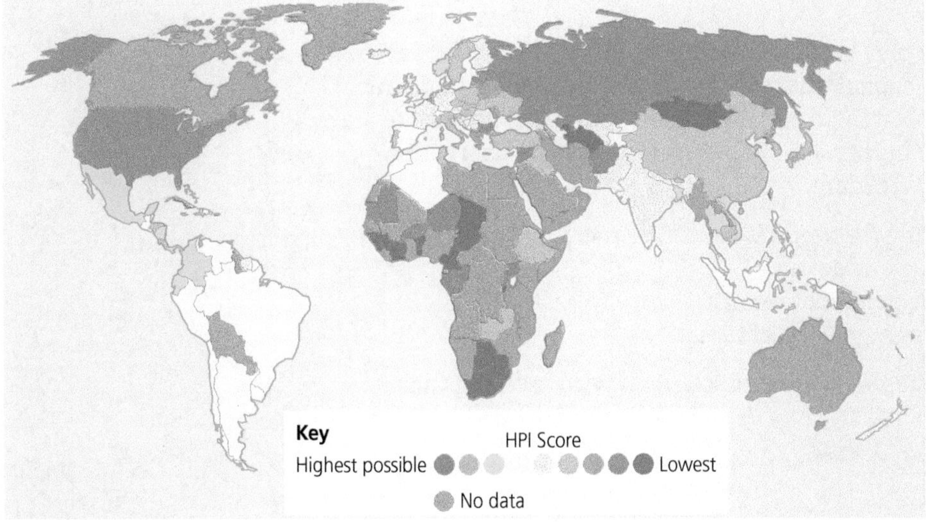

Figure 8.3 Global variation in HPI scores

Revision activity

Makes notes on the different pictures conveyed by Figures 8.2 and 8.3. The former basically reveals the conventional distinction between developing, emerging and developed countries. The latter, while highlighting many of the developing countries in Africa as low scoring, also shows in the same category developed countries in Europe, North America and Australasia. There are some surprise 'winners', for example in Central America.

Exam tip

Remember that these two measures (HDI and HPI) are each based on three variables, and that they have one variable in common: life expectancy.

Typical mistake

Do not assume that economic development and human development are one and the same thing.

Now test yourself

1 Which do you think provides a better measure of human development, the HDI or the HPI? Give your reasons.

Answer on p. 229

TESTED

Skills activity

Understanding scatter graphs to show the relationship between GDP and HDI

Study Table 8.1.

Table 8.1 Information on the HDI and GDP for 10 countries

Country	Sierra Leone	Namibia	Paraguay	Sri Lanka	New Zealand	Poland	Canada	Brazil	Germany	USA
GDP (US$ trillion)	0.0038	0.0132	0.0397	0.0874	0.204	0.526	1.647	2.054	3.693	19.485
HDI	0.452	0.646	0.728	0.782	0.931	0.880	0.929	0.765	0.947	0.926

Sources: HDI figures from http://hdr.undp.org/en/content/latest-human-development-index-ranking. GDP figures from https://www.worldometers.info/gdp/gdp-by-country/

1 Produce a scatter graph to show the 10 countries in Table 8.1.
2 Draw a best-fit line to show their relationship.
3 What is the relationship between GDP and HDI? Suggest reasons for the relationship shown.

Development goals

While the importance of economic growth as a means of delivering development is widely accepted, observers increasingly argue that human development should be targeted at specific non-economic goals, such as improving:

+ environmental quality
+ life expectancy
+ health
+ human rights.

Now test yourself

2 What are the links between the first three of the four bullet points opposite?

Answer on p. 229

TESTED

Education

Education has a pivotal role to play in the development process. It enriches the human resources (human capital) that are vital in the context of economic development. At the same time, it increases awareness of the importance of a whole range of concepts, from human rights to environmental conservation, from healthy living to political freedom.

The educational challenge facing the world is to ensure that:

- everyone has access to a basic education
- access is not conditioned by gender or wealth
- every country is able to offer higher education.

UNESCO is the intergovernmental organisation charged with meeting this global challenge.

Now test yourself TESTED

3 Why is education so important to a country's well-being?

4 Why do levels of education vary both in and between countries?

Answers on p. 229

Revision activity

Start a glossary of key terms and add to it as you make your way through this topic. You could make a start with:

- HDI
- HPI
- GDP
- Gini coefficient

Human health and life expectancy

REVISED

The most widely used measure of the general health of a population is life expectancy. Data on life expectancy is readily available worldwide. Happily, life expectancies have been rising almost everywhere, but they vary enormously from country to country, for example from 84 years in Japan to 46 years in Sierra Leone.

Life expectancy The average number of years a person might be expected to live based on the year of their birth.

Now test yourself TESTED

5 Name two other widely used measures of health at a national level.

Answer on p. 229

Exam tip

Remember that there are gender differences in life expectancy. In nearly all countries, female life expectancy is greater. The gender difference is greater in developed countries (by as much as 5 years). It is in only a handful of countries that male life expectancy is greater.

Variations within the developing world

- These variations in life expectancy and health are explained by differential access to the basic survival needs of food, clean water, sanitation and healthcare.
- A shortfall in any of these necessities immediately increases the risk of disease, ill health and premature death.
- Any shortfall tends to impact most on rates of infant and maternal mortality.

Variations within the developed world

- Variations here are more to do with differences in lifestyles, levels of poverty and deprivation, and the accessibility and quality of healthcare.
- It is ironic that the lifestyles of the better-off carry health risks such as obesity, smoking and alcoholism.
- In countries lacking a national health service, access to healthcare is impeded by poverty.

Variations within countries

- Such variations occur within almost all countries and are largely explained in terms of wealth and poverty.
- In multi-ethnic countries, such as the UK, there are life expectancy differences between ethnic groups.
- Indigenous people, such as the Aborigine people of Australia, often show life expectancies that are up to 10 years less than those of the non-indigenous population.
- A number of factors play a part here, such as the marginalisation of indigenous people, low incomes and lifestyles.

Now test yourself

TESTED 

6 (a) What are the main factors responsible for the variations in human health and life expectancy at the three spatial scales:
   - globally: between developing and developed countries
   - internationally: between countries falling in either of the above categories
   - internally: within countries?

(b) To what extent are the factors the same for developed and developing countries?

Answers on p. 229

Revision activity

Create a mind map to explain how different determinants, e.g. occupation and income, can impact on people's health outcomes.

Defining development targets and policies

REVISED

The relationship between economic and social development

- Given that in most countries economic development provides the means that drive and sustain human development, the link between the two is critical.
- The link is, in effect, in the hands of government. It is government that determines how much of a country's economic wealth should be spent on providing those vital components of human development, such as education, health and other social services.
- This, in turn, depends very much on governmental attitudes towards **social progress**.

Key concept

Social progress derives from the idea that societies have the power to improve their ability to meet basic human needs and to create opportunities for people to improve their lot within society. The pace of improvement is often very slow, but it can be accelerated by:

- government intervention, for example setting up healthcare, providing housing and free education
- social enterprise on the part of responsible businesses
- social activism at a grassroots level that presses for change.

Government attitudes towards social progress are largely conditioned by the type of government or political regime.

- For example, democratic governments are likely to spend more of the national budget on the welfare of the people.
- They contrast with totalitarian governments, which have a reputation for a low level of spending on health and education.

Democratic government A system of government through elected representatives of the people.

Totalitarian government A system of government that is centralised and dictatorial. It requires complete subservience to the state, with political control in the hands of elites.

The leading IGOs

Intergovernmental organisations (IGOs) are among the major players in the promotion of global development.

- Three are particularly influential in the encouragement of economic development: the World Bank, the World Trade Organization (WTO) and the International Monetary Fund (IMF).
- The UN Educational, Scientific and Cultural Organization (UNESCO) and the Organisation for Economic Co-operation and Development (OECD) are more focused on broader aspects of development, such as health, education and human rights.

Now test yourself TESTED

7 Check that you know the specific roles of each of the five IGOs listed.

Answer on p. 229

The Millennium Development Goals (MDGs)

Since 2000, global development has been focused on achieving specific targets. The MDGs defined eight goals with a deadline of 2015 (see Table 8.2).

Table 8.2 The Millennium Development Goals

Goal	Details	Progress made
Goal 1: Eradicate extreme poverty	Reduce poverty by half Create productive and decent employment Reduce hunger by half	Extreme poverty rate has reduced from 47% in 1990 to 14% in 2015.
Goal 2: Achieve universal primary education	Universal primary schooling	In 2015, 91% of primary-aged children were enrolled in school.
Goal 3: Promote gender equality and empower women	Equal girls' enrolment in primary school Women's share of paid employment Women's equal employment in national parliaments	In 2015, 41% of woman were paid workers outside of agriculture compared to 35% in 1990.
Goal 4: Reduce child mortality	Reduce mortality of under-fives by two-thirds	Infant mortality has reduced from 12.7 million in 1990 to 6 million in 2015.
Goal 5: Improve maternal health	Reduce maternal mortality by three-quarters Access to reproductive healthcare	In 2015, maternal mortality was at 210 per 100,000 live births compared to 380 per 100,000 in 1990.
Goal 6: Combat HIV/AIDS, malaria and other diseases	Halt and begin to reverse the spread of HIV/AIDS Halt and reverse the spread of tuberculosis	In 2015, 13.6 million had access to antiretroviral therapy treatment.
Goal 7: Ensure environmental sustainability	Halve proportion of population without improved drinking water Halve proportion of population without sanitation Improve the lives of slum dwellers	In 2015, 4.2 billion had access to piped drinking water compared to 2.3 billion in 1990.
Goal 8: Develop a global partnership for development	Use of internet	In 2015, 43% of the population had access to the internet.

The Agenda for Sustainable Development

Since 2015, world leaders have agreed a 2030 Agenda for Sustainable Development. The 17 goals of sustainable development (SDGs) go much further than the MDGs and are more focused on the root causes of poverty and the universal need for a style of development that works for all people and is more aware of the environment (see Figure 8.4).

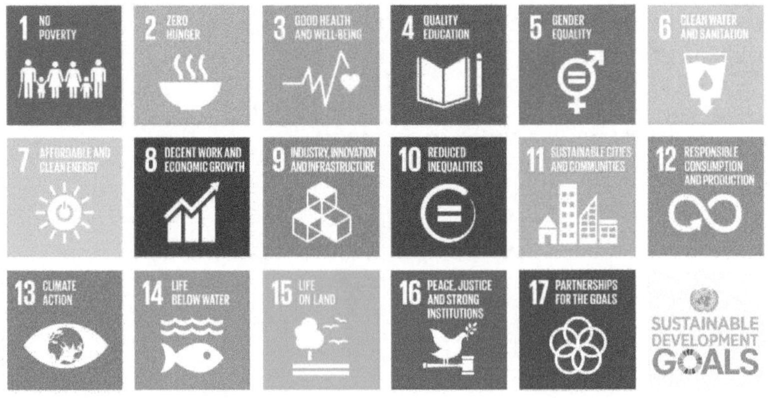

Figure 8.4 Sustainable development goals for 2030

Sustainable development
Development that meets the economic, social and environmental needs of today's population without compromising the ability of future generations to meet their own needs.

Revision activity

Create a mind map to show how the SDGs differ from the MDGs.

Now test yourself TESTED

8 What is meant by sustainable development?

9 What is the significance of governments and IGOs in defining development targets and policies?

Answers on p. 229

Human rights

The importance of human rights

REVISED

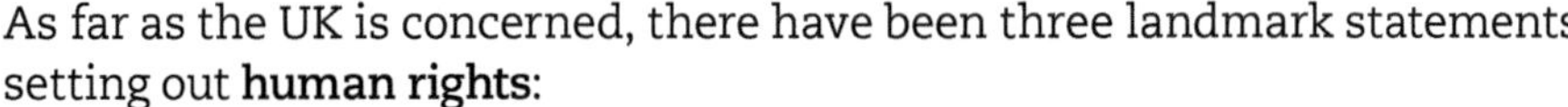

As far as the UK is concerned, there have been three landmark statements setting out **human rights**:

- Universal Declaration of Human Rights (UDHR) (1945): not really binding and therefore unenforceable at an international level.
- European Convention on Human Rights (ECHR) (1953): this and the UDHR were a response to human rights violations that had occurred during the Second World War (1939–45).
- UK Human Rights Act (1998): recognises the human rights set out in the ECHR and comprises 14 articles.

Also, be aware of the Geneva Convention (1949), which set out rules that apply only in times of armed conflict.

Key concept

Human rights are the moral principles that underlie standards of human behaviour, the inalienable rights to which a person is entitled regardless of their nationality, language, religion, ethnicity or gender. The rights include liberty, freedom of movement and speech, personal security and access to education and justice.

Now test yourself TESTED

10 Explain the importance of international law and agreements in protecting human rights.

Answer on p. 229

Revision activity

You should be aware of the range of human rights and the background to the three landmark statements listed above.

Differences between countries

REVISED

Human rights vs economic development

- If pressed, most governments would give the promotion of economic development a higher priority than that of human rights.
- But that is not to suggest a disregard for human rights.
- The difficulty here is that there are few, if any, widely recognised measures relating to the status of human rights, measures that might be used to compare countries.

Figure 8.5 is a map produced by Freedom House. Countries are rated in terms of political freedom and respect for civil liberties. Each country is broadly classified as free, partially free or not free.

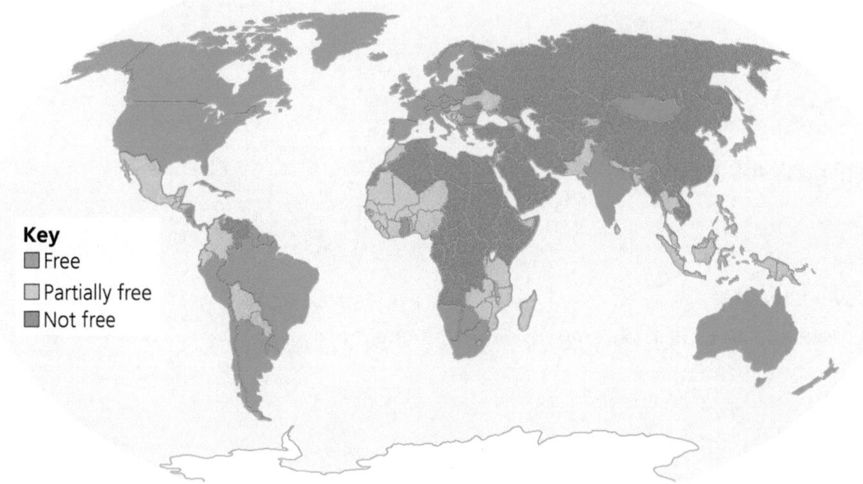

Figure 8.5 Levels of freedom in the world, 2019

Source: Freedom House

Revision activity

- Make summary notes about the 'free' and 'not free' distributions shown in Figure 8.5.
- Can you detect any possible correlations between these distributions and types of political regime?

Exam tip

To illustrate contrasting attitudes to economic development and human rights, compare North and South Korea.

The transition to democracy

The question arises: is there any correlation between economic development and human rights? Of the ten top economic superpowers, six are long-established democracies with a satisfactory current human rights record: the USA, Japan, Germany, France, the UK and Italy. As for the other four, China and Russia have poor records, while the situation is rather better in the fledgling democracies of Brazil and India.

Political corruption

The term political corruption immediately suggests election rigging, but it can take other forms, such as:

- allowing private or business interests to dictate government policy
- taking decisions at a government level that benefit those who are 'funding' the politicians
- diverting foreign aid and scarce resources into the private pockets of politicians.

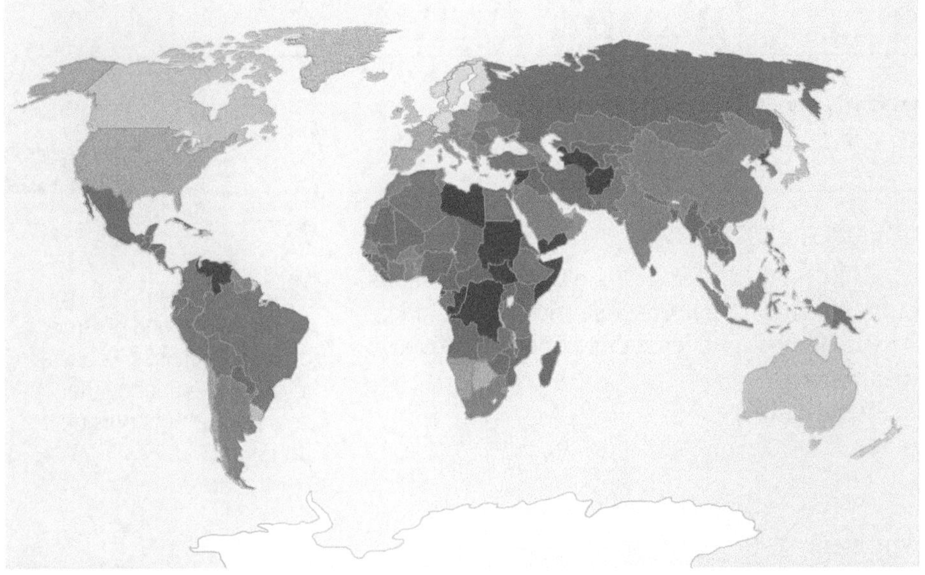

Key
Score
Highly corrupt  Very clean
0–9 10–19 20–29 30–39 40–49 50–59 60–69 70–79 80–89 90–100 No data

Figure 8.6 A global view of corruption based on the corruption perceptions index, 2019

Source: adapted from www.transparency.org

A comparison of Figures 8.5 and 8.6 suggests a partial correlation between high levels of corruption and 'not free' government.

Now test yourself

11 Explain how high levels of corruption could threaten human rights.

Answer on p. 230

TESTED

Revision activity

For examples of corrupt regimes, you might look at the cases of Myanmar or Zimbabwe.

Differences within countries

REVISED

Discrimination

The most widespread discrimination within countries is that based on gender and ethnicity.

- It was only in the twentieth century that women in most parts of the developed world gained equal rights and opportunities.
- Some would argue that even in these countries the employment playing field is not an even one when it comes to wages, salaries and promotion.
- However, there remain many parts of the world where women continue to be regarded as second-class citizens.
- Although discrimination on the basis of ethnicity has been legislated against in many countries, it still occurs, as for example in the housing market, at the workplace and in terms of access to education.

Differences in health and education

- There is plenty of evidence to suggest a broad correlation between human rights on the one hand and access to health and education on the other.
- The UDHR stated that such access is one of the most basic human rights.
- However, it is clear that the failure to respect the human rights of indigenous people has impeded their access to education and healthcare.
- However, the correlation also works in the other direction, as education might be expected to lead to a greater respect for human rights.

The growing demand for equality

Within many countries, the struggles of women and minority ethnic groups for equality have been long and not always successful. Compare the struggles of women in Afghanistan or Bolivia with those in European countries. Compare the struggles of indigenous people within Brazil with those in Australia.

A possible verdict on today's world: all people may be equal, but they are not so when it comes to wealth, freedom and opportunity.

Now test yourself

12 Explain why there are differences in human rights between countries.

Answer on p. 230

TESTED

Revision activity

Create a case study on variations in human rights within a country, e.g. Australia.

- What are the issues?
- What is being done to reduce discrimination?

Interventions and human rights

Geopolitical intervention in defence of human rights

Range and motives of intervention

It needs to be understood that geopolitical interventions are made for reasons other than just the defence of human rights. Other possible motives include:
- extending development aid to the poorest and least developed countries
- encouraging education and healthcare
- strengthening security and political stability
- promoting and protecting trade
- accessing resources
- offering low-interest loans
- increasing global or regional influence.

Geopolitical intervention Occurs when and where a country exercises its power in order to influence the course of events outside its borders.

From these motives, it can be seen that there are three main intervention mechanisms or pathways:
- development aid (see below)
- economic support: for example, trade, investment
- military support: for example, training and equipping the armed forces of another country, helping to deal with insurgents and terrorists, all-out military occupation.

The specification focuses on the first and last of these inventions (see below).

International intervention players

The main players are:
- individual governments, often those of the superpowers
- IGOs such as the UN, the EU, the World Bank and the WTO
- NGOs such as Amnesty International, Human Rights Watch, Oxfam and Médicins Sans Frontières (Table 8.3).

Table 8.3 Major NGOs in human development

Organisation	Founded	Mission
Amnesty International	1961	Founded in the UK and focused on the investigation and exposure of human rights abuses around the world. Takes on both governments and powerful bodies, such as major companies. Today it combines its considerable international reputation with the voices of grassroots activists on the spot to ensure that the UDHR is fully implemented. It also provides education and training so that people are made aware of their rights.
Human Rights Watch	1978	Founded under the name of Helsinki Watch to monitor the former Soviet Union's compliance with the Helsinki Accord (aimed at reducing Cold War tensions). Like Amnesty International it is constantly on the lookout for violations of the UDHR. It is not frightened to name and shame non-compliant governments through media coverage and direct exchanges with policymakers.

Organisation	Founded	Mission
Oxfam	1942	Founded in the UK to help deal with the hunger and starvation that prevailed during the Second World War. Today it has three main targets: development work aimed at lifting people out of poverty and improving health (safe water and sanitation); assisting those affected by conflicts and natural disasters; and campaigning on a range of issues, from women's rights to the resolution of conflicts.
Médicins Sans Frontières (Doctors Without Borders, or MSF)	1971	Founded in France with the belief that all people have the right to medical care regardless of race, religion or political persuasion. Today it provides healthcare and medical training in around 70 countries and has a reputation for providing emergency aid in conflict zones. It remains independent of any economic, political or religious influences.

Conditional intervention

Some Western governments have sought to combat violations of human rights elsewhere in the world. They have done so through various offers of 'help' in exchange for undertakings to respect human rights. The 'help' has been mainly in the form of:

- humanitarian aid
- preferential trade deals
- investment
- military protection.

However, such interventions can easily be interpreted by other governments as threatening the sovereignty of the countries receiving such 'help'. A primary example would be the intervention by the Russians in Ukraine, undertaken because Russia felt the need to protect the human rights of the ethnic Russians living there.

But there are other kinds of conditions, for example the development aid offered by China to African countries in exchange for preferred access to resources that it needs.

Now test yourself

13 Explain the advantages and disadvantages of a country/IGO intervening in the affairs of another.

Answer on p. 230

Revision activity

Create a table with the points for and against different forms of geopolitical intervention in defence of human rights.

Development aid

REVISED

The range of development aid

Development aid can vary in scale, from installing a village well to constructing a large irrigation project; financially, from a small charitable gift to a global appeal raising millions of pounds; in time scale, from short term (for example, emergency aid) to long term (for example, disease eradication programmes); and in the mix of providers, from local charities to major IGO and NGO players.

Key concept

Development aid is aimed at development in the broader sense, such as safeguarding human rights and improving human welfare and quality of life. Nonetheless, in many instances it also has an economic dimension.

Development aid has three main delivery routes:

- bilaterally: directly from one country to another
- multilaterally: indirectly through donations by individual governments to the major IGO players
- charitably: through individual donations to NGOs and their emergency appeals.

Basically the global flow of development aid is from developed to developing countries. Figure 8.7 shows the main donors in terms of official development assistance (ODA).

Official development assistance (ODA) A measure used by the OECD as an indicator of the flows of international aid. Flows are transfers of resources, either in cash or in the form of commodities and services.

A growing concern in some countries is the size of their aid budgets. Should they be donating more or less?

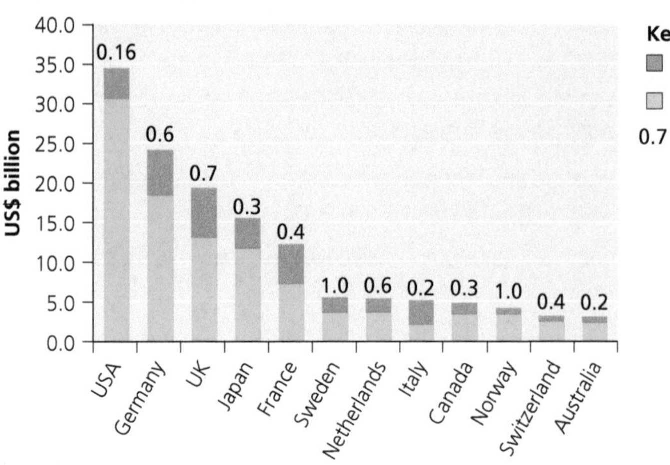

Figure 8.7 The largest ODA donors in 2019

Skills activity

Understanding proportional flow-line maps to show both direction and level of aid from the UK to its recipients

Study Table 8.4

Table 8.4 The ten largest recipients of official development assistance from the UK, 2019

Rank	Country	Official Development Assistance (ODA) (in UK£ million)
1	Pakistan	305
2	Ethiopia	300
3	Afghanistan	292
4	Yemen	260
5	Nigeria	258
6	Bangladesh	256
7	Syria	223
8	South Sudan	207
9	Dem. Rep. Congo	185
10	Somalia	176

Source: https://assets.publishing.service.gov.uk/government/uploads/system/uploads/attachment_data/file/927135/Statistics_on_International_Development_Final_UK_Aid_Spend_2019.pdf

1. Get a blank map of the world.
2. Draw proportional flow lines on the world map to show the direction and level of aid from the UK to its recipients.

How to draw a flow line map:

1. The width of each arrow represents a different amount of money, using a set scale (e.g. 1 cm = £100 million).
2. The arrowhead on each flow line indicates the direction of movement.
3. Look at the data and identify the highest number that you have to show, e.g. Pakistan £305 million.
4. Decide on an appropriate width of flow line to show this e.g. 1 cm = £100 million.
5. Check that the value works with the minimum value.
6. Plot the flows on a world map, with arrows leading from the UK.

Positive impacts

Development aid has been fairly successful in a number of different contexts:

- in dealing with life-threatening and often contagious diseases, eliminating some (for example, polio) or limiting others (for example, malaria and Ebola)
- in fighting extreme poverty
- in improving access to education and healthcare
- in providing safe water and sanitation
- in stimulating the creation of jobs.

Negative impacts and concerns

Development aid is coming in for criticism on a number of different counts:

- Aid in the form of capital grants and loans is now thought to be inappropriate, if only because it tends to increase the receiving country's indebtedness. Technical assistance and skills training are now preferred.
- There are concerns about the recipients of development aid. For example, large amounts of UK aid are still going to India, which is an emerging country on the up. Are other countries more deserving of aid?
- A worrying amount of development aid is being siphoned off by corrupt officials and also used as bribes (for example, land grabbing in Kenya).
- It encourages the growth of elites in receiving countries, who disregard the human rights of others.
- It is thought to encourage dependency rather than stimulate enterprise and self-advancement.
- The involvement of TNCs is questioned because of profit leakage and the considerable environmental damage they can cause in the exploitation of resources (for example, the exploitation of oil in the Niger Delta).

Exam tip

Be sure that you are able to cite at least three of the concerns about development aid.

Revision activity

Create a table showing the positive and negative aspects of loans and donations as a way of providing development aid.

Land grabbing The acquisition of large areas of land in developing countries by domestic and transnational companies, governments and individuals. In many instances, the land is simply taken over and not paid for.

Profit leakage The process whereby the profits made by a business are not retained and reinvested in the country where they were made.

Now test yourself

14 Which do you think is the better indicator of a country's generosity: total ODA or ODA as a percentage of GNI? Give your reasons.

15 Why can some development have negative environmental and cultural impacts?

Answers on p. 230

Military interventions

REVISED

Defending human rights

Protecting human rights has often been used as an excuse for military intervention. But whether or not that excuse was valid is another matter. All too often, there have been ulterior motives.

Providing military aid

Military aid can take various forms:

- providing military equipment and armaments
- training the military forces of the receiving country
- providing military personnel: troops, advisors.

A contentious aspect of selling arms to another country is that the supplying country often has little control over how those arms are used and against whom. The UK's supply of military equipment to Saudi Arabia has recently been questioned as some of that equipment has been used in attacks on Yemen.

Waging war on terror and torture

- The world today is troubled by Islamist terrorist organisations such as the Taliban, al-Qaeda and Daesh (also known as IS).
- The last is causing much trouble in the Middle East, particularly in Iraq and Syria, and has mounted occasional attacks in other parts of the world (see Figure 8.8).
- As a result, the Western superpowers find themselves increasingly embroiled in a war on terror.

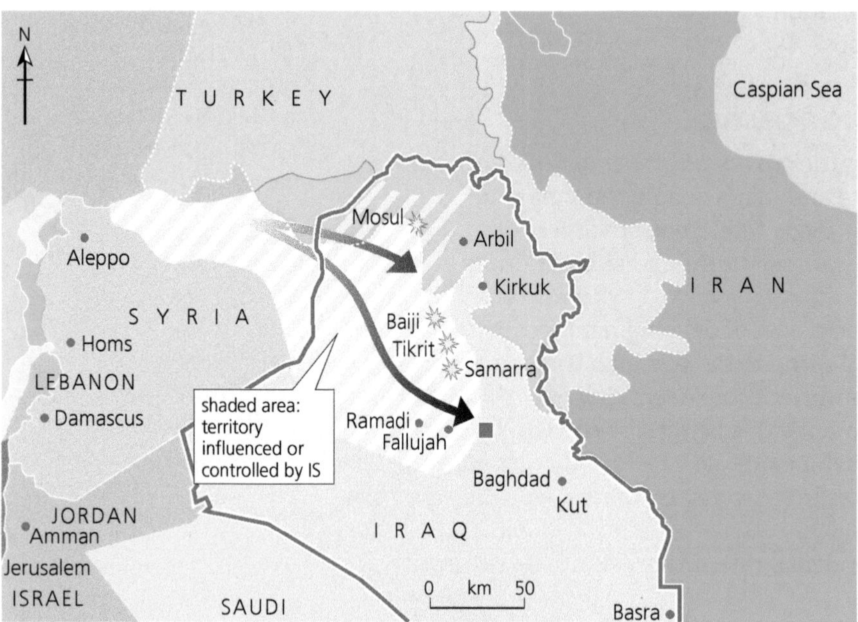

Figure 8.8 The spread of IS in the Middle East

The West's feeling that it should intervene militarily against IS is motivated by three main concerns:

- the political stability of the entire Middle East region, and possibly that of North Africa
- protecting access to the oil reserves in both regions
- the atrocious abuse of human rights.

The military intervention needed here is more one of covert surveillance and intelligence gathering, as well as providing Arab troops on the ground with aerial cover, equipment and training.

The issue of rendition and torture as a means of obtaining vital information is a challenging one:

- Whose rights are more important, those of the terrorist not to be tortured or the right to life of those innocent people who could become the victims of terrorism?
- Should there be limits to the actions a government undertakes to ensure national security and territorial integrity?

Rendition The practice of sending international criminal or terrorist suspects covertly to be interrogated in a country where there is less concern about the humane treatment of terrorist suspects.

Now test yourself TESTED

16 What are the three motives for taking action against IS?
17 What is the link between rendition and torture?
18 What is the justification for military intervention in the defence of human rights?

Answers on p. 230

Revision activity

What does the UN Convention against Torture (1987) stipulate about the treatment of suspects?

The outcomes of geopolitical interventions

Evaluating interventions

REVISED

Measures

- Given the diversity of interventions (military, development aid, etc.) and hoped-for outcomes, there are many possible ways of assessing whether or not a particular intervention has had a successful outcome.
- Table 8.5 shows some possible measures for evaluating interventions aimed at human development and human rights.
- Among these measures, there is no single 'silver bullet'.
- However, there is much to be said for using more than one measure.
- It is also the case that progress in human development is easier to measure than progress in human rights.

Table 8.5 Possible measures for evaluating intervention outcomes

Intervention target	Possible measure
Human development	Life expectancy
	Provision of healthcare (doctors per 100,000)
	Literacy rate (percentage of population)
	Quality of physical infrastructure (percentage with access to safe water and sanitation)
	Per capita GDP or GNI
Human rights	Freedom of speech
	Gender equality (gender index)
	Democratic elections
	Respect for minorities
	Recognition of refugee status

Human rights indicators

- Potential indicators of progress in human rights tend to be qualitative rather than quantitative.
- Of the five measures shown in Table 8.5, possibly the incidence of democratic elections is the most telling.
- The fact is that broad respect for human rights is more likely to flourish in a democracy than in a one-party state without any form of opposition.

Now test yourself TESTED

19 Explain why it is more difficult to measure progress in human rights.

Answer on p. 230

Importance of economic growth

- It is accepted that economic growth provides the fuel needed to drive human development.
- But equally it is recognised that serious tensions often exist between economic growth and human rights.
- This is particularly the case if a country is keen to fast-track that growth. This is well exemplified by today's China.
- Even so, many countries today prefer to have their development status assessed by economic measures such as per capita GDP and per capita GNI.

Revision activity

Create a mind map showing the arguments for and against the success of geopolitical intervention. Use examples in your answer.

Evaluating development aid

REVISED

Development, health and human rights

- Given that much development aid is targeted at improving health and human rights, this should make evaluation rather more straightforward.
- What becomes clear is a mixed record, with successes in countries such as Botswana and Guyana contrasting with serious shortcomings in, for example, Haiti and Ivory Coast.
- Corruption is one factor helping to explain the lack of achievement.

Typical mistake

Not all development aid is provided for humanitarian reasons.

Impact on inequalities

- Recipients of development aid also differ in terms of the impact of that aid on internal economic inequalities.
- In some countries, such inequalities have been reduced (for example, in Cambodia and Senegal) while in others, such as Bangladesh and Honduras, they have increased (with corruption again being a factor).
- But any increase in inequalities almost inevitably hits the poor and least powerful sectors of society. For them, there are negative impacts on health, life expectancy and human rights.

Superpower objectives

There is no doubt that the superpowers have become donors of development aid for a range of motives that is not necessarily humanitarian. They may be motivated by the need to secure:

- a presence in strategic locations
- future supplies of resources
- military alliances with other countries
- a global sphere of influence.

Now test yourself

20 What statistical measure is widely used in the assessment of spatial inequalities?

21 Which are today's leading superpowers?

Answers on p. 230

TESTED

Revision activity

Research why the tackling of the Ebola (2014) outbreak in West Africa was a relative success.

Evaluating military interventions

REVISED

The most obvious evaluation criterion is whether or not the intervention has been successful — has it achieved its objectives? Whichever way the verdict goes, there will be costs.

Costs

The costs of military intervention fall into two broad categories:

- military: e.g. troops killed and wounded; ammunition spent; equipment lost
- civilian: e.g. number of deaths, injured and displaced (refugees); damage to, and destruction of homes, public buildings and infrastructure; disruption of business, etc.

The above are essentially short-term costs. There will also be longer-term costs in the area of intervention, depending in part on whether or not the intervention has been successful, such as:

- the time taken to fully recover and get back to normal after the cessation of hostilities
- the impacts of a loss of territory or sovereignty
- a possible loss of human rights.

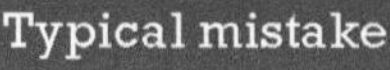

Typical mistake

Do not think of military intervention as just the movement of troops into a conflict zone.

Costs of non-military intervention

What the specification has in mind here is peacekeeping interventions by the UN, as for example in Kosovo. It is a possible way of navigating away from a conflict situation.

UN peacekeeping is guided by three basic principles:

- all parties in a conflict must consent to the intervention
- the need to show impartiality
- non-use of force except in self-defence and defence of the mandate.

The UN is involved in several different peacekeeping operations (see Figure 8.9). The record of past operations has not been entirely successful, particularly when it has come to putting in place the right sort of governance.

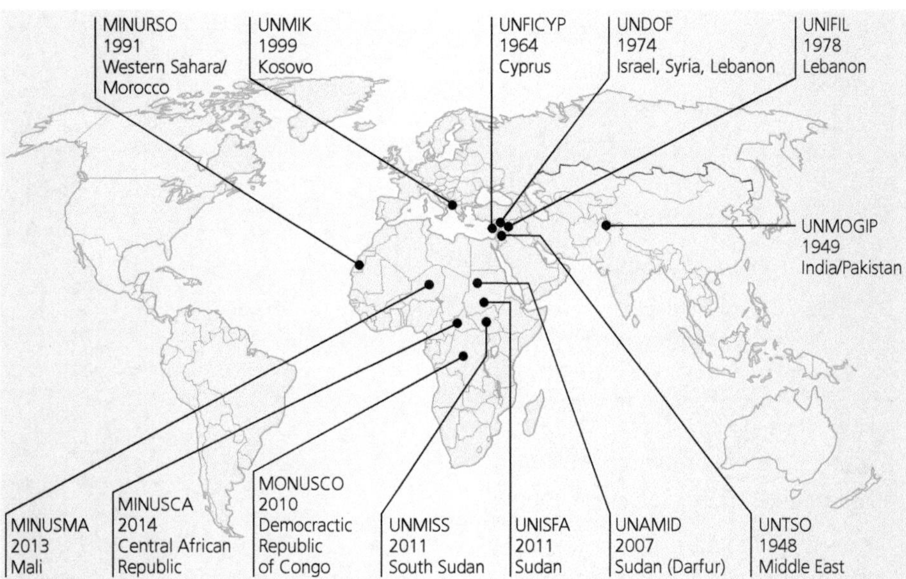

Figure 8.9 The locations of UN peacekeeping operations

Revision activity

Make notes about the distribution of the peacekeeping operations shown in Figure 8.9.

Exam tip

It would be worthwhile to have some details of one of the peacekeeping operations shown in Figure 8.9.

Costs of no action

Faced with threats to human well-being and human rights, the global community has three options:

- turn a blind eye and do nothing — this may well mean that the threats worsen
- make a limited intervention (military) to deal with the short-term threat — this may bring short-term relief but longer-term costs
- make an extended intervention — but not necessarily a military one.

Now test yourself TESTED

22 Why might it be better not to make an extended military intervention?

23 What is the difference between direct and indirect military intervention?

Answers on p. 230

Exam skills

You should be familiar with the skills and techniques used in the following investigations of health, human rights and interventions:

- comparing different measures of development using ranked data
- analysing the relationship between health and life expectancy using scattergraphs and correlation techniques
- using proportional symbols to show government spending on health, welfare and education
- correlating the distribution of the corruption index with types of government
- using flow-line maps to show the global movements of aid between donors and recipients
- evaluating source materials used to determine the impact of development aid
- interpreting images to evaluate the impacts of economic development on the environment where minority groups live
- using the Gini coefficient to investigate inequalities between and within countries
- critical analysis of data that might be used in the assessment of interventions.

Exam practice

A-level

1 a) Study Figure 1.

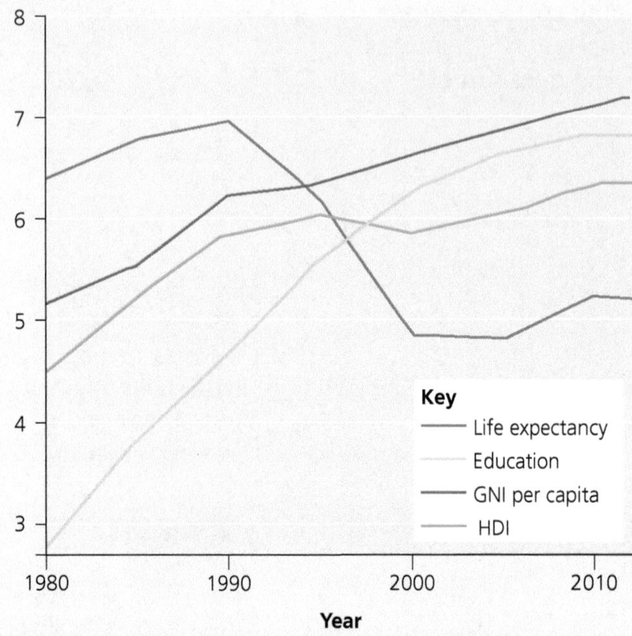

Figure 1 Botswana's HDI and its three components since 1980

i) By how much did the HDI increase since 1980? (1)

ii) Suggest one reason for the change in life expectancy in Botswana. (3)

b) Suggest reasons why access to education is such an important human right. (6)

c) Explain why indigenous peoples are often the target of discrimination. (8)

d) Evaluate the view that development aid is the best intervention the developed world can make in the developing world. (20)

2 a) Explain one reason why health and life expectancy varies greatly between countries. (4)

b) Study Figure 2.

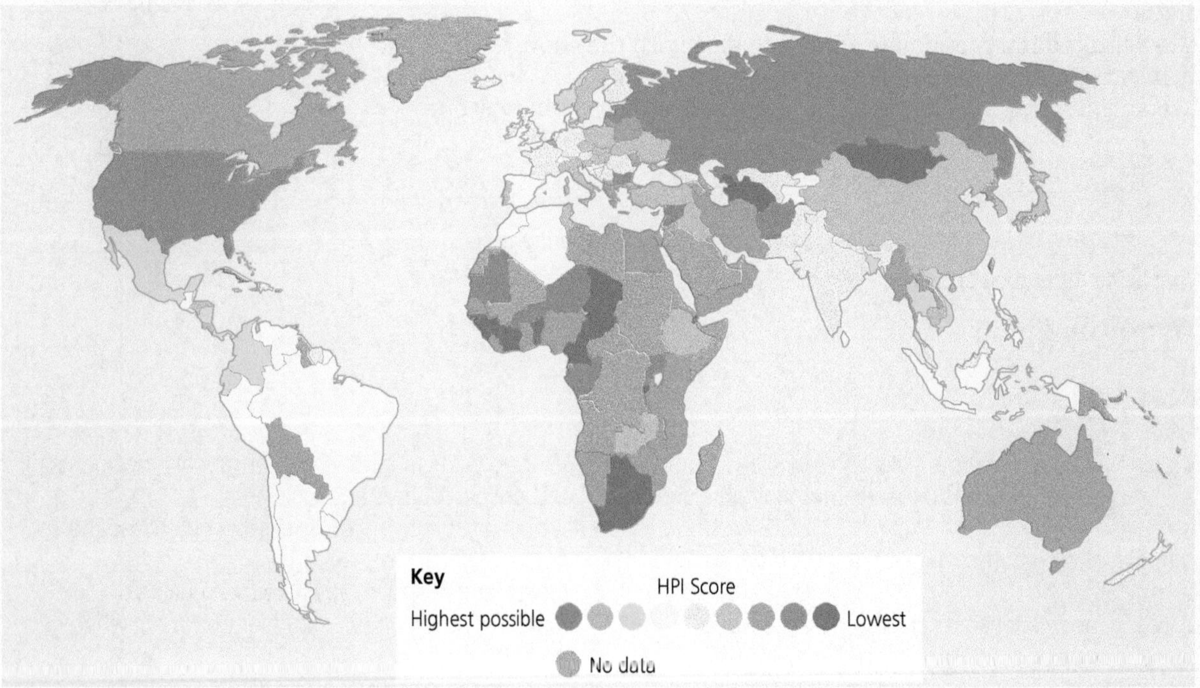

Figure 2 Global variation in HPI scores

Suggest reasons why HPI shows little correlation with levels of economic development. (6)

c) Explain why human rights vary from place to place. (8)

d) Evaluate the success of geopolitical intervention in the defence of human rights. (20)

Answers and quick quiz 8A online

Summary

You should now have an understanding of:

- the nature of human development
- the importance of economic growth to the delivery of human development
- the importance of education
- the links between health and life expectancy
- variations in health and life expectancy, both between and within countries
- the major players in global development
- setting development goals
- human rights and international agreements
- how countries vary in both their definition and protection of human rights
- significant variations in human rights within countries, which are reflected in different levels of social development
- discrimination based on gender and ethnicity, which remains widespread
- geopolitical intervention in defence of human rights, which can take several different forms
- how development aid takes many different forms, but it does have its critics and the record is a mixed one
- how development aid of an economic kind can have serious negative impacts on the environment and culture
- how military interventions are frequently justified in terms of human rights
- evaluating the outcomes of different types of geopolitical intervention.

8B Migration, identity and sovereignty

Globalisation means increasing movements of capital, goods and people. Migration flows, both internal and international, are now running at record levels. International migration not only alters the ethnic composition of populations, it also changes national identity and attitudes to it. Indeed, the growing interdependence of countries that has come with globalisation is creating tensions, most notably with ideas of national identity and sovereignty.

The fact that the governance of globalisation lies in the hands of a small number of global organisations is another factor encouraging the rise of nationalism in some countries.

Impacts of globalisation on international migration

Internal and international migration

REVISED 

A new pattern of labour demand

- Globalisation has generated much new employment and it is the locations of this employment that are determining the directions and destinations of economic migrant flows.
- This new pattern makes sense if considered in terms of the core–periphery model.
- National core–periphery systems have been strengthened not only by rural-urban migration but also by flows converging on leading cities, as has happened so spectacularly in China.
- Core–periphery structures also exist at an international scale.
- In the EU a core region encompasses northern France, Belgium and much of western Germany.
- Because the EU is founded on a belief in the free movement of labour, huge volumes of economic migrants are leaving the periphery in southern and eastern Europe and heading for the 'cores of the core', leading cities such as Berlin, Brussels and Paris.

Economic migrant A person who travels from one country to another in order to find work and improve their standard of living.

Core–periphery model Relates to the uneven distribution of population and wealth between two regions and the resulting flows of migrants, trade and investment from a lagging peripheral region to a prospering core region.

Figure 8.10 illustrates global migration patterns.

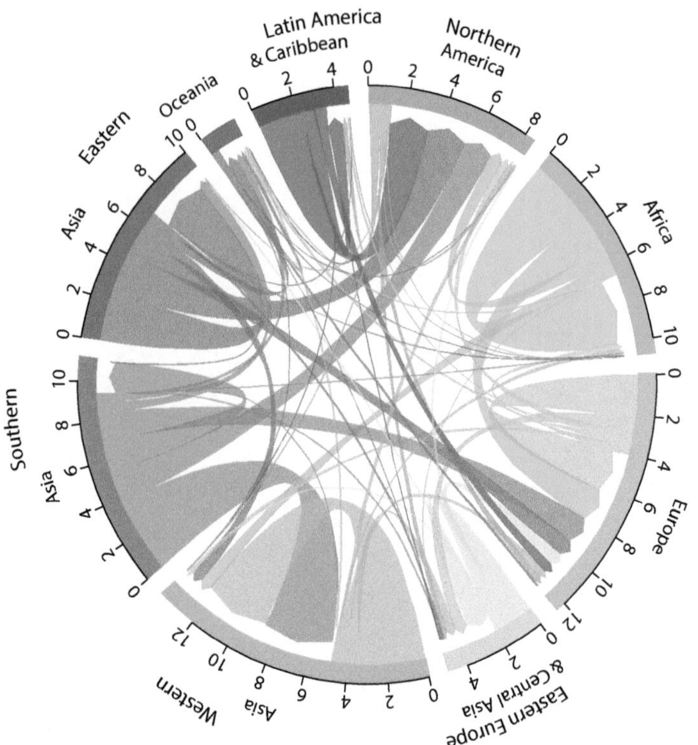

Visualising migration flows

Movements of people are often shown as lines or arrows on a world map, but circular chord diagrams can be a more effective way of visualising global migration flows. This diagram shows migration between 196 countries between 2010 and 2015, broken down into nine world regions. The colour shows which region each flow came from, the width of a flow shows its size, and numbers indicate the total migration in and out of a region, in millions.

Figure 8.10 A chord diagram of global migration patterns

Exam tip

Be sure to have specific examples of both rural-urban migration within a country, e.g. China, and international migration between countries, e.g. EU Schengen Agreement.

Now test yourself

1 Define an economic migrant.

Answer on p. 230

TESTED

Immigration policies

- As an indicator of the rising tide of international migration, it is estimated that around 4 per cent of the global population lives outside its country of birth.
- This proportion varies between countries because of different policies relating to international migration — compare the strict **immigration** policies of Australia and Japan (see Table 8.6) with the more liberal policies of Singapore and Sweden.

Table 8.6 Different attitudes to immigration

Australia	Japan
Skills-based immigration policy	Closed-door policy to immigration
70% of immigrants are accepted based on skills shortages where there are insufficient Australian workers	Has a cultural aversion to immigration due to the belief they are 'homogenous' people
Australia's immigrants contribute more to Australia's GDP than non-migrants Australia's immigrants offset an ageing population, as 88% are under 40	Little change in political mindset despite Japan having: + an ageing population + a sluggish economy due to decreasing working population and increasing elderly population UN estimates Japan needs 17 million migrants by 2050 to maintain its population

Revision activity

- Using Figure 8.10, make notes about the international migration characteristics of (i) Northern America, and (ii) Southern Asia.
- Describe the pattern shown.
- Explain the pattern.

Key concept

Immigration can be controversial and resentments can arise in host countries where people may sense a threat to their national identities.

Non-economic migration

Employment is not the only factor in international migration, with unemployment providing the 'push' and job opportunities the 'pull'.

There are two other noticeable groups of international migrant, both responses to push forces and therefore forced:

+ political refugees and asylum seekers: the victims of persecution and conflict, such as prevail in the Middle East today
+ environmental refugees: where life support systems are collapsing due to climate change and overpopulation, as for example in the Sahel.

Refugee Defined by the UN as someone whose reasons for migrating are genuinely to do with fear of persecution or death.

Asylum seeker A person who seeks to gain entry to another country by claiming to be a victim of persecution, hardship or some other compelling circumstance.

Now test yourself TESTED

2 How does a political refugee differ from an environmental refugee?

3 What are the links between globalisation and increased international migration?

Answers on p. 230

The complex causes of migration

Main causes

The causes of international **migration** fall into three main groups:

+ economic — move for work
+ political — to escape persecution and conflict
+ environmental — due to sudden or long-term changes to the local environment.

The risks that huge numbers of migrants have been taking to cross the Aegean and Mediterranean into Europe testifies to the desperation of both economic migrants and refugees to escape from their homelands.

Most international migrants move for work-related reasons and are often driven by poverty and deprivation. For these people, the dividing line between voluntary and forced migration may not always be clear.

Also involved in these work-related migrations are the families, who often follow an economic migrant once they have established themselves in the destination country.

Key concept

Migration is the outcome of push–pull forces — 'pushing' in the place of origin and 'pulling' in the destination. In the case of voluntary migration, it is likely that the pull factors are stronger, while in the case of forced migration the balance is reversed.

Economic theory

The core–periphery model states that with development, there is a backwash (flows) of migrants to the core from the periphery — a response to the greater opportunities available in the core (see Figure 8.11).

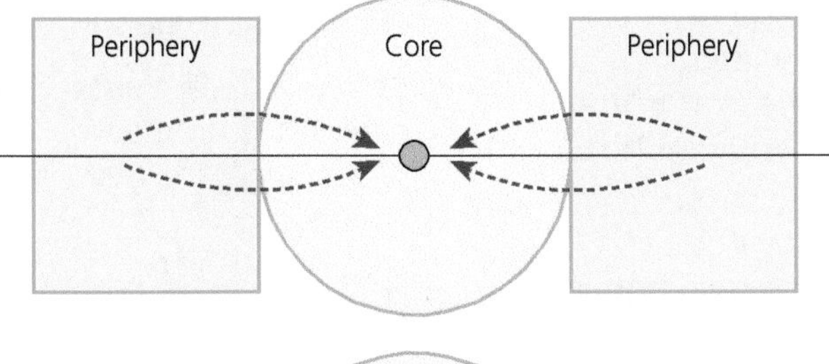

A strong economic core develops fuelled by the in-migration of people (workers and investors) from the peripheral regions of a state.

Time

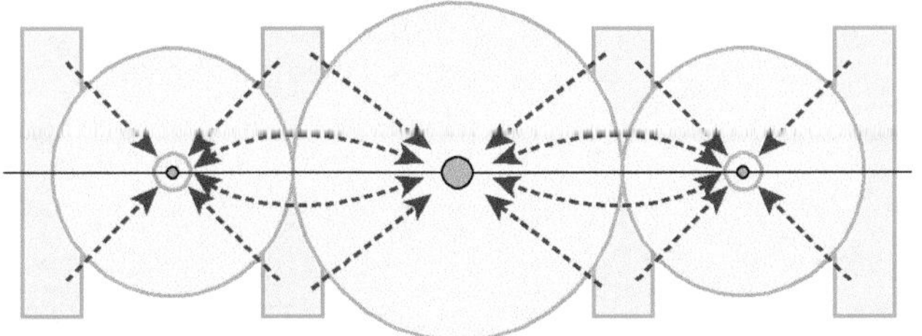

In economic theory, additional core regions form as part of the development process over time. The growth of these cores is fuelled by flows of raw materials and workers from neighbouring areas.

Figure 8.11 Backwash processes in the core–periphery model

- The periphery to core migration within a country is not without its problems.
- For example, there is likely to be some political concern for the plight of the periphery.
- But when cores reach such a scale and magnetism that they draw in large numbers of migrants, the situation can become much more serious.
- Such flows can seriously challenge national identity and **sovereignty** (see below).

Key concept

Sovereignty is the ability of a place and its people to self-govern without any outside interference.

Now test yourself 

4 What are the downsides to the growth of large core regions?

5 What are the causes of international migration? Categorise the causes into social, economic and political.

Answers on p. 230

Exam tip

The core–periphery model is important in understanding the geographical distribution of development and other aspects of globalisation. Be sure you have a sound grasp of the basics.

Unrestricted flows

- There is a strong divergence of opinion about the unrestricted movement of migrants.
- There are those who argue that economic efficiency is maximised when goods, capital and labour can move freely across international borders.
- But there are those who focus on the problems of overheated and overgrown cores and the plight of debilitated peripheries.
- The free movement of labour is one of the founding principles of the EU and made even easier by the Schengen Agreement.
- However, some member states are concerned about the great influx of migrants involving both economic migrants and refugees.
- Immigration can reach such a pitch that countries feel they can no longer cope in providing housing, education and healthcare.

Revision activity

Research a source country of migration into Europe, e.g. Syria. Then create a mind map, showing:

- economic, environmental and social causes of migration
- the pathways used to reach the ports
- migrant experiences on arrival in Europe.

Making links

A government's attitude to migration plays a large role in deciding migration policies, which impact the flow in and out of the country.

The consequences of international migration

REVISED

Migration has become one of the key issues of the twenty-first century. Because of globalisation and a shrinking world, migration is now a major management challenge for most governments, particularly in those developed countries that are attracting large flows of immigrants.

- Immigration changes the culture and ethnicity of a population.
- The history of the UK since around 1950 provides a prime example.
- Despite much assimilation, the UK today has become a truly multicultural and multi-ethnic society.
- Less than 75 years ago it was overwhelmingly a white society.
- Diversity, as such, was generated by influxes of Jewish and Irish immigrants.

Culture Ideas, beliefs, social behaviour and customs of a group.

Ethnicity Social groups distinguished by their religion, language, customs and heritage.

Assimilation The process by which persons of diverse cultural and ethnic backgrounds interact and intermix in the life of the larger community or nation.

Now test yourself

TESTED

6 What is meant by a shrinking world and what is its significance?

Answer on p. 230

Political tensions

The paradox of globalisation is that while it promotes a global way of thinking, it simultaneously ignites fears that immigrants pose a destabilising threat. The perception prevails that immigrants are taking over housing, jobs and businesses. Almost inevitably, tensions are generated between the natives and the newcomers. Most times, the tensions take the form of ill-informed suspicions rather than outright conflict or hostility.

A good example is provided by the USA and labour flows from neighbouring Mexico. Heightening the tension here is the fact that huge numbers of Mexican people are entering the USA illegally. This raises the suspicion that they are not paying taxes on their earnings.

Exam tip

Some people assume that culture is part of ethnicity; others believe otherwise. The specification seems to separate the two. So be warned!

An uneven playing field

- The world's poorest people are sometimes the least likely to become economic migrants. This is because those countries with tight border controls are in a position to select whom they admit. They can choose on the basis of skills that the country needs, as with Australia.
- Money and influence can also be used to acquire necessary visas. This sort of discrimination is often used as an argument in favour of open national borders.
- In the case even of refugees, there is also an element of discrimination.
- The first to escape tend to be the healthy, young adults and the better informed.
- The sick, elderly and those with a poor awareness of the threat hanging over them tend to lag behind, and often until it is too late.

Skills activity

Constructing proportional circles

Using the data in Table 8.7, construct proportional circles showing the proportion of migrant remittances for a range of African nations.

Table 8.7 World Bank remittance data for selected African nations, 2018

African nation	Migrant remittance inflows as a share of GDP in 2018 (%)	Diameter of circle (cm)
Rwanda	2.7	2
Malawi	2.6	
Djibouti	2	
Mozambique	2	
Burundi	1.6	
Sierra Leone	1.5	
Benin	1.4	1.0
Sudan	1.1	
Algeria	1	
Cameroon	0.8	

How to construct proportional circles

1. Find a blank map of Africa, which shows national borders. Print it off on an A4 piece of paper.
2. Work out the size of your largest circle. On an A4 piece of paper the maximum diameter you would want is 2 cm.

3 Calculate the diameter of the remaining circles using the following equation:
Symbol size = maximum symbol size × (the value/maximum value)
Example: Benin is 2 × (1.4/2.7) = 1.0 cm
4 Draw the symbols on the map using a pair of compasses and a ruler. Alternatively, you can draw it digitally.
5 Draw a key.
6 Using the data, explain the costs and benefits of international migration to source countries.

Typical mistake

It is wrong to assume that all migrants are poor people. Make sure you understand different types of migrants and the reasons why they move.

Now test yourself TESTED

7 What are the main arguments for and against open borders?
8 Why can immigration cause political tensions? Use specific examples in your answer.

Answers on p. 230

Revision activity

Create a mind map on the impacts of migration. Categorise impacts into social, economic and cultural. Be sure to use specific examples.

The evolution of nation states in a globalised world

- A state is a territory over which no other holds power or sovereignty.
- The term 'nation' refers to a group of people who live in a particular territory but who may lack sovereignty.
- So, the Scottish and Welsh nations are part of the UK, which is a sovereign state.

States and the processes that shape them

REVISED

Nation states vary in their unity

- Some states pride themselves on their homogenous culture, which they have maintained over long periods of time. Examples would include Iceland (because of its isolated location – see Table 8.8) and North Korea (because of its self-imposed political isolation).
- Other states boast about their cultural and ethnic diversity, for example Australia and the USA.
- In virtually all democracies, national unity is tested by political parties, particularly at the time of general elections.

Table 8.8 Iceland's national characteristics

	Characteristics
Geographic	Extremely isolated — island in the Atlantic Ocean hundreds of kilometres from its nearest neighbour, Greenland
	Dependent on the sea and difficult geographical terrain
Political	It is a 'young' republic — Iceland gained independence from Denmark in 1944, becoming a sovereign state
	The Althing, established around AD 930, claims to be the longest-running parliament in the world
	Iceland does not have a standing military and has a long history of neutrality
Cultural	Iceland's law and society ensure citizens' national identity and ancestry are protected
	By law, being born in Iceland does not give you automatic citizenship — you must be born of Icelandic parents
	The Icelandic language has remained unchanged since AD 870
	74% of Icelanders are part of the Evangelical Lutheran Church of Iceland
	All children's names must come from an approved list
	The phonebook is ordered by first name, as surnames are predominantly the father's name followed by *son* for a boy and *dottir* for a girl

Skills activity

Constructing divided bar charts

Use the data in Table 8.9 to create a divided bar chart of the ethnic composition of Iceland and Singapore.

Table 8.9 Ethnic composition of Iceland and Singapore

Iceland	Singapore
91% Icelandic	74% Chinese
9% Other	13% Malay
	9% Indian
	4% Other

How to construct a divided bar chart

1. Divided bars may be drawn horizontally or vertically.
2. Two bars should be drawn — one to represent Iceland and one Singapore — and a scale should be added to the longer side, marked from 0% to 100%.
3. Starting from the largest proportion draw a line across the bar and then shade this section in.
4. Then show the next proportion by adding this to the first, for example for Singapore the first line is drawn at 74% and the second at 87% (74 + 13).
5. Shading should be used to show the different ethnicities of the population and sections can be labelled or a key can be constructed.

The nature of borders

- Prior to today's shrinking world, mountain ranges and sometimes rivers formed natural barriers to population movements. As a consequence, they provided obvious and quite effective borders.
- Inside those borders, the isolated conditions favoured the gradual development of a distinctive culture and national identity.
- Of course, the sea has always been an effective border. This explains why many national states are island states.
- In Europe, the political map corresponds broadly with the cultural and linguistic maps.
- In Africa, however, the situation is very different. Political boundaries, drawn centuries ago by colonial powers, do not correspond well with the distributions of different cultural (tribal) and ethnic groups. This explains much of the conflict that has characterised post-colonial Africa.

Exam tip

It is worth knowing one or two examples of states where the boundaries cut across cultural or ethnic divides, e.g. Rwanda.

Contested borders

It is not only in Africa where those contested borders dissecting tribal homelands have generated conflict. Other examples include:

- the wars that have waged in Europe between France and Germany over possession of the mineral-rich regions of Alsace and Lorraine
- the boundaries of Russia, which have changed a great deal, most recently in 2014 when Russia annexed part of Ukraine
- the unhappy fit in the Middle East between state borders and the distribution of strongly opposed ethnic groups.

Another issue here is that of non-recognition. The UN recognises 196 states, but two are not universally recognised as sovereign states:

- Kosovo: it broke away from Serbia in 2008, but it is not recognised by Russia and Serbia.
- Taiwan: before the Second World War Taiwan was part of China. It broke away when the communists overran mainland China. Needless to say, China does not recognise Taiwan as a sovereign state.

Revision activity

Research Crimea and why there are contested borders. Use the following to help your research:
- Summarise a brief history of Crimea (e.g. religious background, links to Russia and Ukraine)
- Find out what happened in 2014 and why Crimea is so important to Russia.
- Research the current states of relations between those who contest sovereignty.

Now test yourself

9 What factors may lead to national borders becoming contested?

Answer on p. 231

TESTED

Nationalism

REVISED

Many people today view themselves as being global in their outlook. In contrast, others continue to view nationalism as a better option.

Nationalism The belief held by people belonging to a particular nation that their own interests are much more important than those of people belonging to other nations.

Empires

In the nineteenth century nationalism was important in:
- the maintenance of empires that European powers had built up earlier as they colonised Africa, Asia and Latin America. The British Empire was the greatest in extent.
- fuelling conflict in Europe, as between France, Germany and Britain.

New nation states

- The era of empire started to come to a close in the late nineteenth century and between 1945 and 1970 the remaining colonised nations gained their freedom. A raft of new nation states emerged.
- The rapidity of decolonisation often left a 'power vacuum' and the transition to independence did not bring development.
- In many countries, power was seized by the army, for example in the Congo, Indonesia and Nigeria.
- In others, old tribal and ethnic tensions resurfaced, as in Rwanda and Uganda. In yet others, the handover of power was distinctly messy, as in Kenya, Malaysia and Vietnam.

The fight for an independent Vietnam

- Vietnam was colonised by France in the nineteenth century.
- Much of the agricultural land was converted into plantations for coffee, tea and rubber, which were exported to the colonial powers.
- Society was divided into those who owned land and those that were landless.
- The landless classes often worked on the plantations under very poor conditions.
- Post-1945, Vietnamese nationalists challenged the French rule.
- In 1954 the French military were defeated and this paved the way for nationalist Ho Chi Minh to reclaim Vietnam from France.
- The USA was worried about the spread of communism as many Asian countries gained independence. This led to Vietnam being split into North and South along the 17N line of latitude.
- The North was run by Vietnamese nationalists supported by communist China, and the South by independent non-communists supported by USA troops.
- From 1961 there was a war between the North and South, as the north Vietnamese nationalists fought to reunify the whole country. A total of 1.4 million Vietnamese people were killed.
- In 1975 the South was defeated and an independent, unified Vietnam emerged.

Colonial heritage

- Interesting relics or reminders of the former colonial empires are the flows of migrants between former colonies and the imperial core country.
- Figure 8.12 illustrates important post-colonial migrations into the UK between 1950 and 1980.
- Most, but not all, the flows of migrants into the UK were primarily 'pulled' by the job opportunities created by the UK's booming economy and its labour shortage.
- Noteworthy exceptions were the refugee movements 'pushed' by persecution and discrimination.
- These immigrations were the start of a process that has profoundly affected the ethnicity and cultural diversity of the UK's population today.

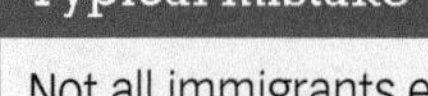

Typical mistake

Not all immigrants entering the UK are economic migrants.

Revision activity

Make notes on the origins of the immigration flows shown in Figure 8.12. What economic and social impacts would this immigration have?

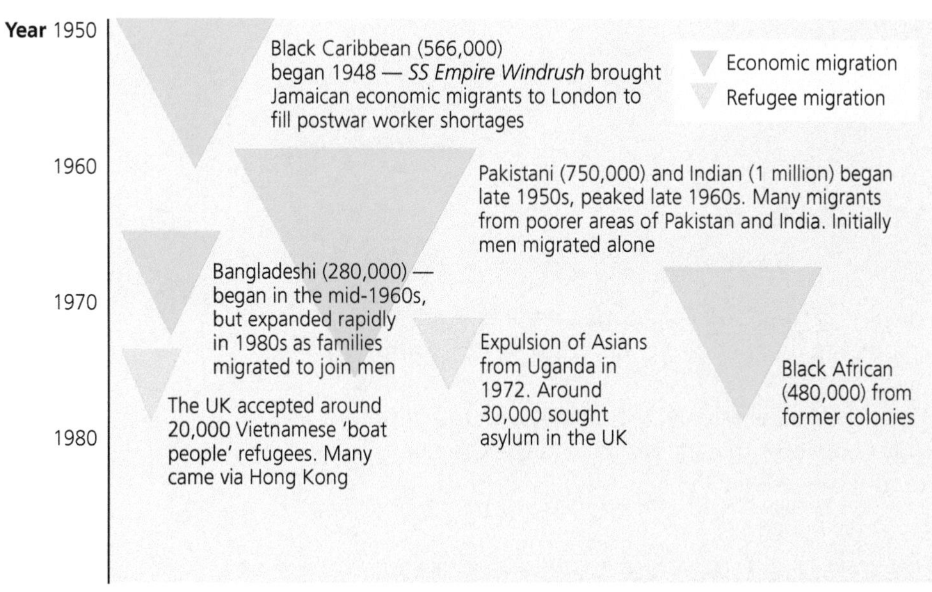

Figure 8.12 UK post-colonial immigration, 1950–80

Now test yourself TESTED

10 Has nationalism played a role in the development of the modern world?

11 Why were there labour shortages in the UK and what type of jobs did the immigrants fill?

Answers on p. 231

Globalisation and new state forms

Globalisation has led to the deregulation of capital markets. This has created an opportunity for some of the newcomers to the ranks of global states to adopt a new way of making a living.

Tax havens

- One of the most popular of these opportunities has been the creation of low-tax regimes, which have provided havens for business corporations and wealthy people.
- Such states include the Bahamas, Bermuda, the British Virgin Islands, the Cayman Islands, the Channel Islands, Gibraltar, Hong Kong, Switzerland and some US states (for example, Nevada and Wyoming).
- These havens are popular with wealthy expatriates and TNCs that are in the business of transfer pricing.

Expatriate Someone who has migrated to live in another state but remains a citizen of the state where they were born.

Transfer pricing A financial flow occurring when one division of a TNC charges a division of the same company based in another country for the supply of a product or service. It can lead to less corporation tax being paid.

- The latter can be elusive when it comes to paying tax. It is difficult to identify in which country or countries they should pay the taxes on their profits (see Figure 8.13).

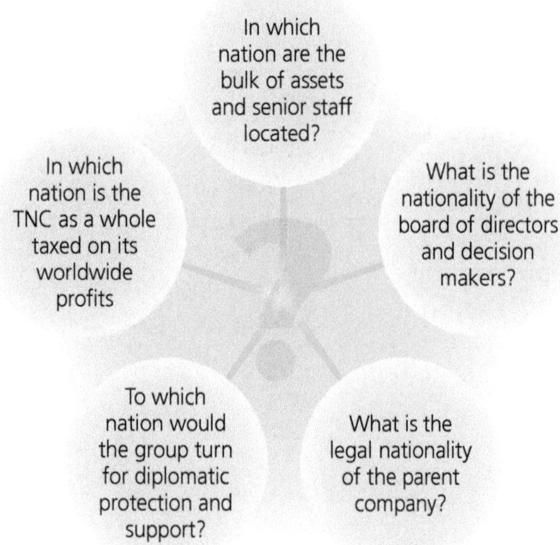

In which nation are the bulk of assets and senior staff located?
What is the nationality of the board of directors and decision makers?
What is the legal nationality of the parent company?
To which nation would the group turn for diplomatic protection and support?
In which nation is the TNC as a whole taxed on its worldwide profits

Figure 8.13 Investigating where a TNC is domiciled

Objections

- The existence of these tax havens is unpopular with many states that desperately need all the corporate taxes they can collect. These taxes are vital sources of revenue.
- Non-government organisations (NGOs), particularly those connected with aid, are among those unhappy about the havens. NGOs also see the tax havens as depriving them of much-needed funding.

Alternative models

- The present global economic system is far from perfect.
- It has been frequently criticised by the leaders of several Asian, African and South American nations.
- Their criticisms centre on the inequalities in the system whereby some states gain disproportionately from global trade at the expense of others.
- Unfortunately, the governments that most criticise global capitalism have yet to persuade the rest of the world that there exists a viable alternative form of state building. The question arises: would you like to live in those countries with alternative models, such as Bolivia, Ecuador, North Korea or Zimbabwe?

Exam tip

It is important to realise that globalisation is not the silver bullet solution to creating a more even global spread of development.

Now test yourself TESTED

12 In which of the four countries with alternative models of government — Bolivia, Ecuador, North Korea or Zimbabwe — would you prefer to live? Give your reasons.

13 How do global inequalities threaten the global economic system?

Answers on p. 231

Revision activities

Why have the tax affairs of companies such as Apple, Google and Amazon been criticised?

Create a table with the arguments for and against tax havens. Include examples to support your argument.

The impacts of global organisations

Since the end of the Second World War in 1945, there has been an acceleration towards **global governance**. Leading the way was the United Nations (UN), an umbrella organisation for many global agencies, treaties and agreements.

> **Key concept**
>
> **Global governance** is based on the notion of steering or piloting rather than the direct form of control associated with national government. So global governance sets the rules, norms and codes that regulate human activity at an international level. Enforcement of this sort of steering is no easy matter.

Global organisations in the post-1945 world

REVISED

The UN

- The UN was the first post-war IGO to be set up. Since 1945, its remit has grown to include a whole range of different areas of governance.
- The **Security Council** is an important and powerful arm of the UN.
- Its primary responsibility is the maintenance of international peace and security.
- Five powerful states (China, France, Russia, the UK and the USA) are permanent members.
- In recent years, its record of successful interventions has been somewhat tarnished by divisions within the Council (see below).

> **Revision activity**
>
> Note the names and responsibilities of at least four UN agencies or organisations.

> **Making links**
>
> Attitudes to national identity, particularly in light of globalisation, can be affected by IGOs such as the UN. This is especially dependent on how involved IGOs become in the affairs of other countries.

Interventions

The main interventions made by the UN fall under the following headings:

- direct military intervention, e.g. in the Congo
- peacekeeping in states with internal conflicts, e.g. Ivory Coast
- economic sanctions against countries stepping out of line, e.g. trade embargo on Iran
- defending human rights, e.g. setting up war crime trials
- protecting refugees, e.g. in Syria
- promoting development, especially in agriculture, education and healthcare, e.g. over much of Africa.

> **Typical mistake**
>
> It is wrong to think that the power of the UN is paramount.

Unilateral actions

Despite recognising the UN, some member states have conducted their own military campaigns and contrary to the resolutions of the Security Council. Examples include:

- the UK's war in 1982 with Argentina over the sovereignty of the Falkland Islands
- the US invasion of Iraq in 2003, with the support of the UK
- Russia's annexation of Crimea in 2014.

Since 1945, there have been many examples of unilateral military interventions in so-called failed states, such as Yemen, Somalia and Syria, or on the pretext of waging a war on international terrorism, for example in Afghanistan.

> **Failed state** A country whose government has lost political control and is unable to fulfil the basic responsibilities of a sovereign state, with severe adverse impacts on some or all of its population.

Figure 8.14 illustrates some of the major US interventions abroad since the end of the Second World War.

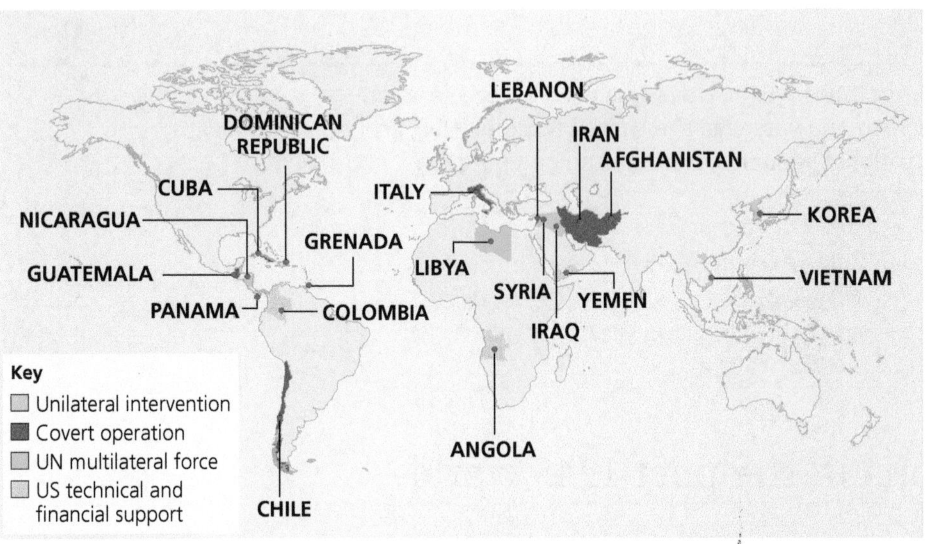

Figure 8.14 Some major US interventions abroad since 1945

Now test yourself

TESTED

14 Why has it been necessary for the UN and its agencies to make interventions?

15 How does the UN contribute to achieving world peace?

Answers on p. 231

Revision activity

Research the trade embargo in Iran. How successful was the UN's intervention?

The control of trade and financial flows

REVISED

Three powerful players

Three IGOs — the World Bank, the International Monetary Fund (IMF) and the World Trade Organization (WTO) — have been particularly significant in the management of the global economy (see Table 8.10). Established in the 1940s, they have since become important in maintaining the global dominance of Western capitalism.

Table 8.10 Three global players in economic development

Organisation	Founding date	Headquarters	Role in world trade
World Bank	1944	Washington, DC, USA	To give advice, loans and grants for the reduction of poverty and the promotion of economic development. Main role is to offer long-term assistance rather than crisis support.
IMF	1944	Washington, DC, USA	To monitor the economic and financial development of countries and to lend money when they are facing financial difficulty. Help is provided to countries across the development spectrum. Between 2010 and 2015, almost US$40 billion was lent to Greece, for instance.
WTO (previously GATT)	1995 (previously 1947)	Geneva, Switzerland	To formulate trade policy and agreements, and to settle disputes. Overall, aims to promote free trade on a global scale. Unfortunately, a round of negotiations that began in 2001 stalled for 14 years. Trying to get 162 member states to agree anything can be challenging. Difficult problems for the WTO to deal with include: + wealthy countries failing to agree on how far trade in agriculture should be liberalised + the fast growth of emerging economies including China (which makes it harder to agree on fair policies for so-called developing countries).

Controversial borrowing rules

Lending by the IMF and the WTO has undoubtedly helped many lagging states to develop economically. However, since 1970 tougher conditions have been attached to large-scale lending. For states experiencing severe financial difficulties, there are two forms of help available:

+ structural adjustment programmes (SAPs): providing loans but with strict conditions attached
+ heavily indebted poor countries (HIPC) policies: aimed at ensuring that no poor country faces a debt burden it cannot manage.

Critics of these two concessions say that they tend to increase rather than decrease poverty. The concessions also undermine the economic sovereignty of borrowing states. They are regarded by some as a neo-colonial strategy used by developed countries to maintain influence over how peripheral countries develop.

Regional groupings

As a reaction to the failure of the WTO to deliver free trade globally, states have organised themselves into regional groupings or trade blocs (see Figure 8.15).

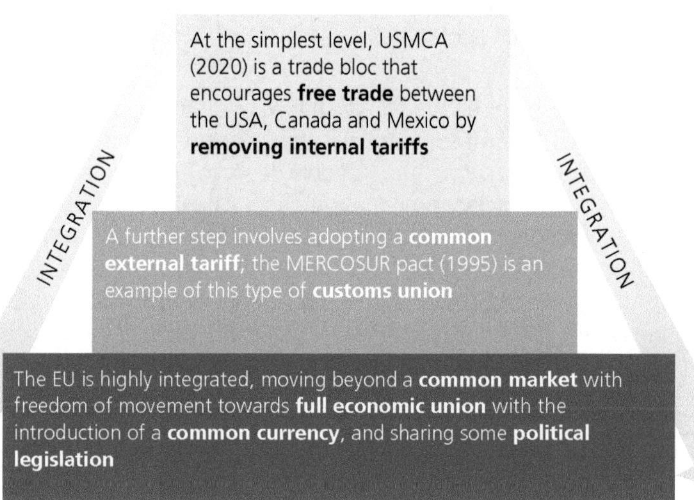

At the simplest level, USMCA (2020) is a trade bloc that encourages **free trade** between the USA, Canada and Mexico by **removing internal tariffs**
INTEGRATION
INTEGRATION
A further step involves adopting a **common external tariff**; the MERCOSUR pact (1995) is an example of this type of **customs union**
The EU is highly integrated, moving beyond a **common market** with freedom of movement towards **full economic union** with the introduction of a **common currency**, and sharing some **political legislation**

Figure 8.15 Different types of trade bloc and their degree of integration

Exam tip

It is recommended that you have some information about USMCA, MERCOSUR and the EU.

Revision activity

Find out the difference between the roles of the IMF, World Bank and WTO.

Now test yourself

TESTED

16 How have IGOs impacted economic sovereignty?

Answer on p. 231

Global environmental governance

REVISED

The UN has attempted to manage some of the world's pressing environmental issues, but with varying degrees of success.

Global environmental issues

The atmosphere and the biosphere are important resources shared by all members of the global community. Table 8.11 sets out important agreements for each of the two spheres.

Table 8.11 Global agreements and actions on the atmosphere and the biosphere

Agreements and actions	UN aims	Evaluation
Montreal Protocol on Substances that Deplete the Ozone Layer	In the late 1960s the UN Environment Programme first called for an international response to the issue of ozone depletion caused by worldwide use of a group of chemicals called chlorofluorocarbons (CFCs) in fridges and aerosol sprays. The 1977 'World Plan of Action on the Ozone Layer' gave the UN Environment Programme (UNEP) responsibility for promoting and co-ordinating international research and data-gathering activities.	The Montreal Protocol was signed in 1987. It was a remarkable agreement on account of the number of individual governments that were prepared to back an important global goal ahead of narrower economic self-interest. Within a decade, irrefutable proof that CFCs were to blame led to all UN states ratifying the treaty. CFC use was phased out rapidly as a result of this exceptional milestone in international co-operation.
Climate change agreement	Climate change was first raised as an urgent issue in 1992 at the UN Earth Summit conference. Many of the meetings that followed were plagued by uncertainty over the evidence and also wrangling over which nations should be held responsible for the majority of the anthropogenic carbon stock that has been added to Earth's atmosphere.	On the whole, international co-operation on climate change has been slow, which many people deem to be a failure of international governance. Although a new international agreement on action was reached in Paris in 2015, critics say the pledges that were made to reduce carbon emissions do not go far enough. These pledges cannot be enforced either.
Convention on International Trade in Endangered Species of Wild Fauna and Flora (CITES)	CITES entered into force in 1975. It banned trade in threatened species and their products. Now adopted by 181 countries, it has effectively saved some species but not others. Success stories include recovery of the Hawaiian nēnē bird and the Arabian oryx.	Rising wealth in China, Indonesia and South Korea has actually increased illegal trade in some prohibited substances such as ivory and rhino horn. The problem can be summed up as 'new money, old values'. Without a cultural shift away from the use of these products, CITES will not be able to protect some species.
Millennium Ecosystem Assessment	This international collaboration helped popularise the 'ecosystem services' approach to biodiversity management. A financial value is calculated for threatened biomes and species, thereby strengthening the rationale for their preservation.	'Ecosystem services' is a philosophy that fits well with the capitalist values of the global economic system and is a pragmatic approach for the UN to have adopted and helped promote globally.

Now test yourself TESTED

17 Which of the four aims in Table 8.11 do you think is the most challenging? Give your reasons.

18 How successful have IGOs been in managing environmental problems?

Answers on p. 231

Revision activity

Make brief revision notes about all four agreements in Table 8.11.

Global waters

IGOs have been developing laws for managing the world's oceans and international rivers:

- UN Convention on the Law of the Sea (launched 1982): defines a nation's territorial limit as extending to 12 miles and its exclusive economic zone to 200 miles from the coastline. It recognises that oceans are global commons and sets out laws to protect the marine environment from pollution.
- Helsinki Water Convention (launched 1992): aims to protect and ensure the quantity, quality and sustainable use of transboundary water resources by facilitating co-operation between the states involved.

Global commons Global resources so large in scale that they lie outside of the political reach of any one state. International law identifies four global commons: the oceans, the atmosphere, Antarctica and outer space.

Managing Antarctica

- Antarctica, the world's fifth largest continent, is now universally recognised as a 'continent of peace and science'.
- Recognition of its intrinsic value and special status started in 1959 with the Antarctic Treaty.
- Nobody owns Antarctica, but 20 nations have scientific bases there.
- For the immediate future, Antarctica is safe from exploitation, except possibly by tourism. Around 40,000 tourists visit Antarctica each year.
- However, as non-renewable resources run out elsewhere in the world, it might be foreseen that there will be increasing calls to exploit the continent's coal, oil, copper, silver, gold and titanium.

Now test yourself

TESTED

19 What are 'international rivers'? Give three examples.

20 How do tourist cruises threaten Antarctica?

21 Why is the Antarctic Treaty important for Antarctica's future?

Answers on p. 231

Threats of national sovereignty in a more globalised world

- The theory of hyperglobalisation (see page 72) states that globalisation will, over time, reduce the power of individual countries.
- Global flows of commodities, ideas and people in a shrinking and borderless world will gradually lead to the creation of a global village, in which group identities will give way to global citizenship.
- In contrast, others argue that there is a growing resistance to globalisation and membership of regional groupings.

Global citizenship A way of living in which a person identifies strongly with global-scale issues, values and culture rather than with a particular nation, state or place.

The elusive concept of national identity

In many nations, attempts are being made to reassert **national identity**.

Key concept

National identity is an elusive concept in terms of trying to identify and measure it. Part of the problem lies in the fact that it is the product of many interacting things, some of which are intangible (for example, the fact that national identity means different things to different people adds to the difficulty). Perception is paramount. What we can be certain about is that national identity, and the wish to preserve it, lies at the heart of nationalism.

Factors reinforcing nationalism

- There is evidence in some countries of a growing concern about the amount of sovereignty being surrendered as a result of participation in globalisation.
- The feeling is that governance is now in the hands of the UN and IGOs rather than national government.
- Some of the support for the UK's decision to exit from the EU came from the conviction that the UK was being increasingly ruled from Brussels and not London.
- Loss of autonomy equals loss of sovereignty.

There are other non-political factors encouraging nationalism, for example:

- education that informs about national heritage
- prestige in the world of sport — for example, the Olympic medal tables
- the pride that individuals feel in belonging to a particular nation
- pride that comes when citizens are renowned internationally, from pop stars to Nobel prize winners
- pride that so many overseas tourists wish to visit.

Identity, loyalty and national character

Clearly, the three aspects of identity, loyalty and national character are closely interrelated and important components of national identity. Aspects of national identity are sometimes tied to distinctive legal systems and methods of government. For example:

- Americans' belief in the First Amendment
- the legacy of the French Revolution and the importance of *liberté* in modern France
- the UK's Magna Carta and its foundation of British laws and liberties.

Multinational countries

One of the features of most countries today is their increasingly multinational nature. This is largely the outcome of recent international migration flows. The UK is a prime example. The question here is this: does this increasing ethnic and cultural diversity strengthen or weaken national identity (see Table 8.12)?

Table 8.12 How English national identity has changed over the past 100 years

	Early twentieth century	Twenty-first century
Religious beliefs	Generally widespread, with high levels of Anglican or Catholic church attendance	Largely secular and non-religious, although some minority faiths are prospering
Food	Locally sourced seasonal food; native herbs preferred to international spices	Global, varied tastes in food; strong spices are widely used in cooking
Identity	People had a strong sense of local belonging (either to a town or county); regional dialects were stronger than today; most were also extremely patriotic and would fight for their country	Many would be less willing to fight for their country, although they are often strong supporters of national sports teams; younger people may see themselves as 'global citizens'
Roots of vocabulary	Celtic, Saxon, Scandinavian (Norse), Roman, Greek, French	Additional Indian subcontinent, Caribbean and American influences (due to migration and TV/film/media)

Now test yourself TESTED

22 Can you suggest one explanation for the changes in English national identity shown in Table 8.12?

23 Does it matter if national identity is diluted by globalisation?

Answers on p. 231

Challenges to national identity

REVISED

National identity is being challenged in a number of ways. Three examples follow.

Foreign-owned companies

Many UK companies are foreign-owned, for example, EDF, Jaguar and Land Rover. For this reason, the term 'Made in Britain' does not quite mean the same as it did say 50 years ago. So a possible ingredient of national identity has been lost.

Westernisation

Westernisation, which is spreading as a sort of global culture, is dominated by US cultural values. Some would say that Westernisation equals 'Americanisation'. No matter what it is called, it is being promoted by large retailing and media corporations (see Table 8.13).

There can be little doubt that the spread of Westernisation, with its particular set of capitalist values, is eroding some of the differences between nation identities.

Table 8.13 Three large entertainment companies that contribute to 'Westernisation'

Walt Disney Company	This is the world's largest entertainment TNC, with annual earnings that exceed US$50 billion. Disney's films, TV channels and resorts have been accused of portraying an unreal version of the 'American Dream' that erodes and replaces traditional cultures as a consequence. Disney is not deliberately trying to alter cultures in other countries, but this is often what happens.
McDonald's	McDonald's helps to spread the idea of American fast food across the world, with 37,000 restaurants across 100 countries that serve 70 million customers per day. McDonald's advertising is regularly attached to major sporting events as well as movie franchises. Connections such as these reinforce American and Western culture.
Apple	Apple helps to connect people in very similar ways across the world regardless of where they are. This indirectly helps to spread Western music, news and other media, and as a consequence contributes to Westernisation.

Ownership of property

The non-national ownership of property is perceived by many as threatening to national identity. Examples include:

- Russian and other overseas investment in the London property market as a safe haven for unstable currencies — note the particular concentrations in Figure 8.16
- UK retirees in Spain purchasing residential properties
- US and Indian ownership of TNCs.

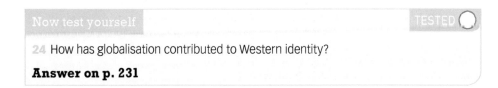

Now test yourself TESTED

24 How has globalisation contributed to Western identity?

Answer on p. 231

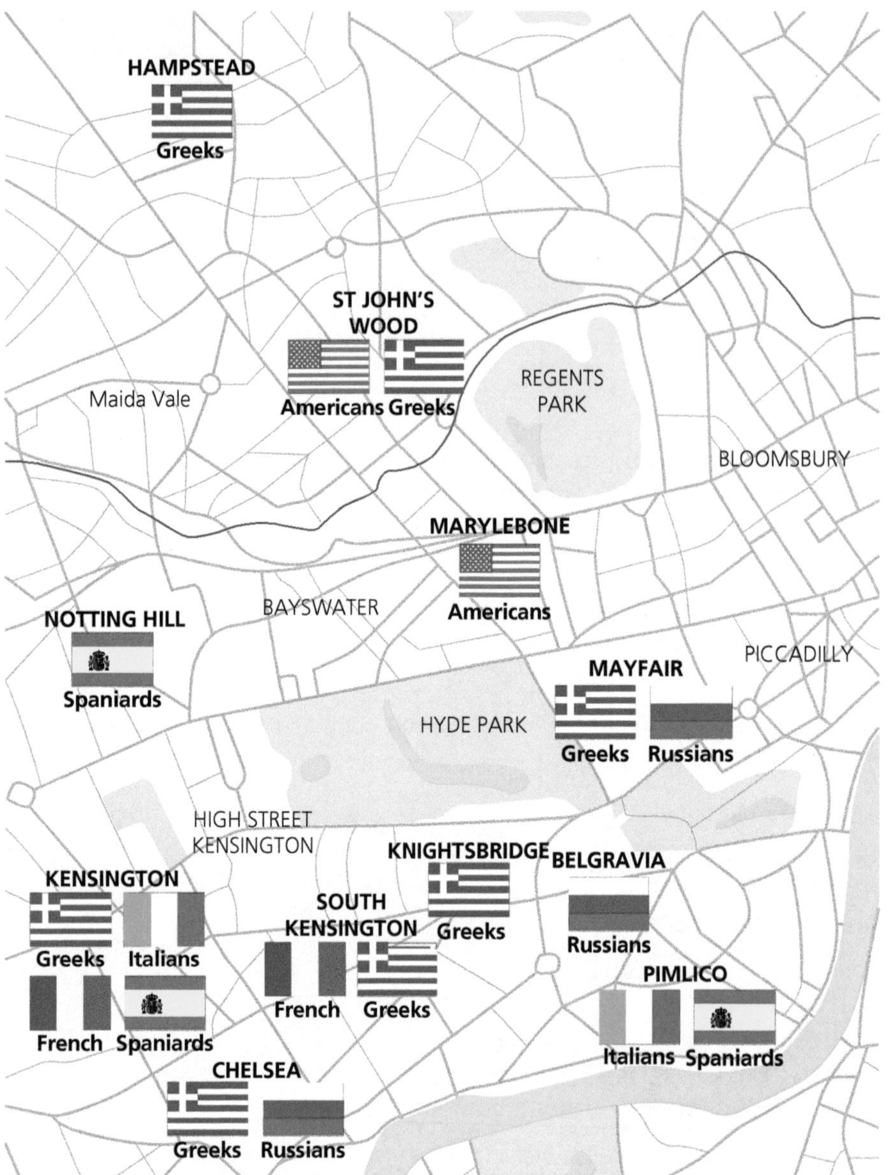

Figure 8.16 Some London neighbourhoods developing a strong non-British identity

Consequences of disunity within nations

REVISED

The distinction between nation and state has already been drawn (see page 190). The term nationalism is used in two different contexts:

- the promotion of sovereign states, such as the UK
- the promotion of small nations that lack sovereignty but are part of a sovereign state, such as Scotland and Wales in the UK.

Nationalist movements

In recent years there has been an upsurge in nationalism of the latter type. Examples include:

- the Scottish National Party's bid for independence from the UK
- the Catalonian wish to gain full independence from Spain
- the break-up of former Yugoslavia in the early 1990s, which gave many small nations, such as Croatia, Montenegro and Slovenia, the chance to assert their sovereignty.

An interesting question is this: is the upsurge in nationalism really a reaction to globalisation? Certainly there are those who see globalisation as eroding nationalism and promoting global citizenship.

Exam tip

With this sort of contentious question, you will impress the examiner if you show that you are capable of seeing both sides of the argument.

Political tensions within the BRICs

Many emerging economies, including members of the BRIC (Brazil, Russia, India, China) group, are experiencing significant internal political tensions:

- Brazil: polarisation of rich and poor; corruption
- Russia: ethnic discrimination; power of political elite
- India: polarisation of rich and poor
- China: tensions between rural and urban communities; human rights issues.

Failed states

National unity and national identity are most challenged in the so-called failed states. Examples include Rwanda, Somalia and Sudan in Africa, and Afghanistan, Iraq and Yemen in Asia. Here governments have lost control and political and economic power lies with wealthy elites, international investors, terrorist organisations or a dominant and powerful ethnic or cultural group.

The issue raised by the failed states is a thorny one. Should they be left alone to heal their own disunities? Or should there be some form of intervention? If the latter, who should intervene — the UN or a superpower with its own agenda?

Making links

Decisions that are taken by stakeholders today will have a massive impact on future generations. For example, when it comes to globalisation, does it provide benefits for all or are there losers that suffer as a result? Furthermore, as seen with Brexit, is the EU beneficial or does it challenge the national identity of a country?

Many countries see nationalism as the preferred option, but others see it as a challenge to governmental authority.

Now test yourself

TESTED

25 Give some possible causes of disunity within nations.

26 Explain the consequences of disunity within nations.

Answers on p. 231

Exam skills

You should be familiar with the skills and techniques used in the following investigations of migration, identity and sovereignty:

- using flow lines on global maps to show migration movements
- interpreting migrant oral accounts
- interpreting a range of opinions about the contributions made by migrants
- using divided bar graphs to compare the ethnic diversity of countries
- comparing maps of languages and colonial histories
- using the Gini coefficient and other techniques to measure inequalities of income and wealth
- evaluating sources to determine the impact of those IGOs managing global environmental issues
- using proportional circles to show the output and level of foreign ownership of different economic sectors
- critical analysis of source materials used in the assessment of the costs and benefits of foreign ownership
- critical analysis of source materials used in the assessment of attempts to promote national identity.

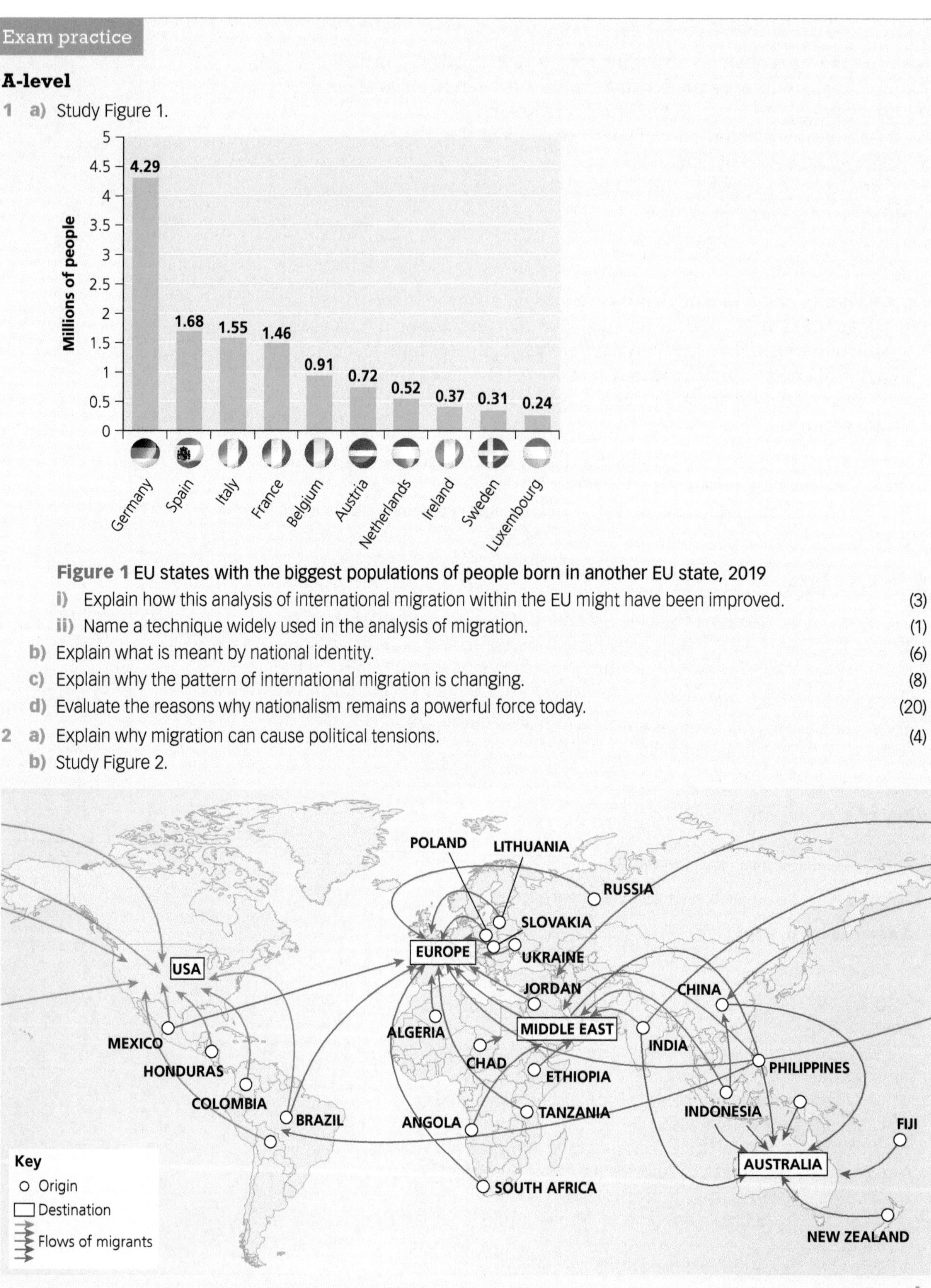

Exam practice

A-level

1 a) Study Figure 1.

Figure 1 EU states with the biggest populations of people born in another EU state, 2019

i) Explain how this analysis of international migration within the EU might have been improved. (3)

ii) Name a technique widely used in the analysis of migration. (1)

b) Explain what is meant by national identity. (6)

c) Explain why the pattern of international migration is changing. (8)

d) Evaluate the reasons why nationalism remains a powerful force today. (20)

2 a) Explain why migration can cause political tensions. (4)

b) Study Figure 2.

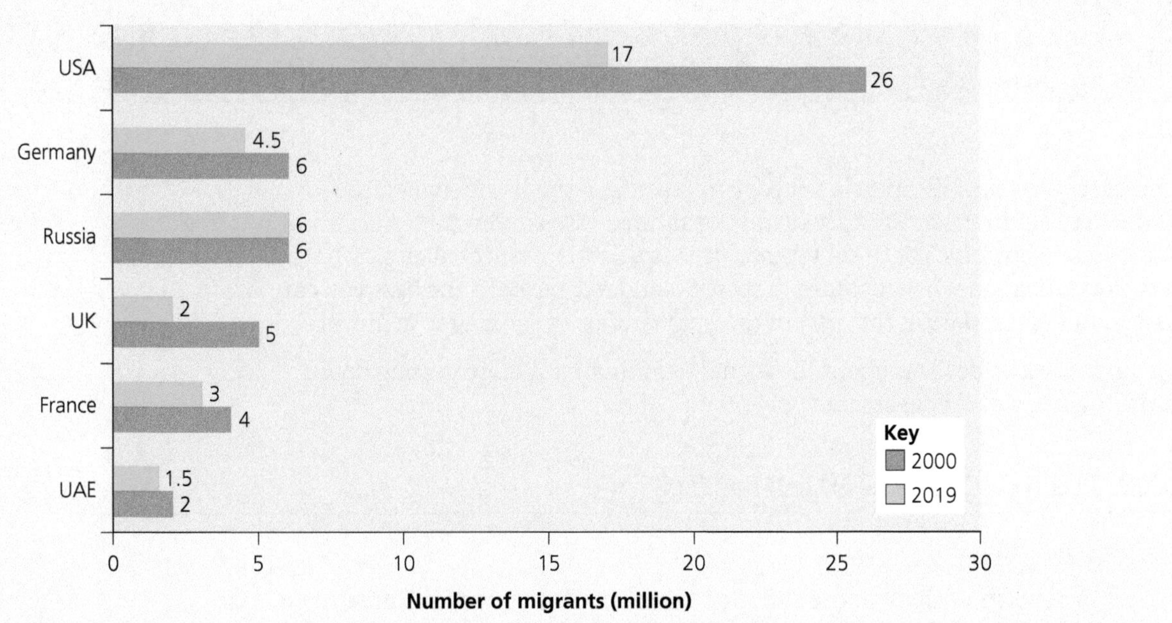

Figure 2 Major world migration flows and the number of international migrants living in selected countries, 2019

Explain why there are variations in the ability for people to migrate across national and international borders. (6)

c) Explain the factors that may lead to national borders being contested. (8)

d) 'IGOs lead to an erosion of national economic sovereignty.' Evaluate the extent to which you agree with this statement. (20)

Answers and quick quiz 8B online

Summary

You should now have an understanding of:

- globalisation and the changing pattern of labour demand and migration
- the causes of migration
- the varied consequences of international migration
- the challenges posed by migration
- nation states and their diverse histories
- nationalism and its role in the modern world
- the deregulation of capital markets and the emergence of new state forms
- the importance of IGOs in today's world
- the IMF, the World Bank and the WTO, and the maintenance of Western capitalism
- IGOs and the management of environmental problems
- the elusive nature of national identity
- the challenges to national unity
- the consequences of disunity within nations
- failed states and their internal tensions.

9 Synoptic themes

This part of the specification is not easy to grasp and the link between the content set out in the specification and the sample assessment materials is not immediately obvious. For these reasons, this is a particularly challenging part of the qualification when it comes to revision. Indeed, perhaps the best you can do is to understand what this part of the qualification is seeking to achieve.

It is important to be clear about the actual examination before getting down to the business of preparing for it.

Examination requirements

REVISED

A-level examination

Paper 3 is devoted entirely to the synoptic themes. All six questions are linked to maps, diagrams, tables, etc. provided in a resource booklet.

This paper accounts for 20 per cent of the qualification.

Synoptic themes (PAF)

REVISED

The term is misleading in that what the specification sets out are not linking topics and concepts but rather three different perspectives from which to look at some of the important issues of today.

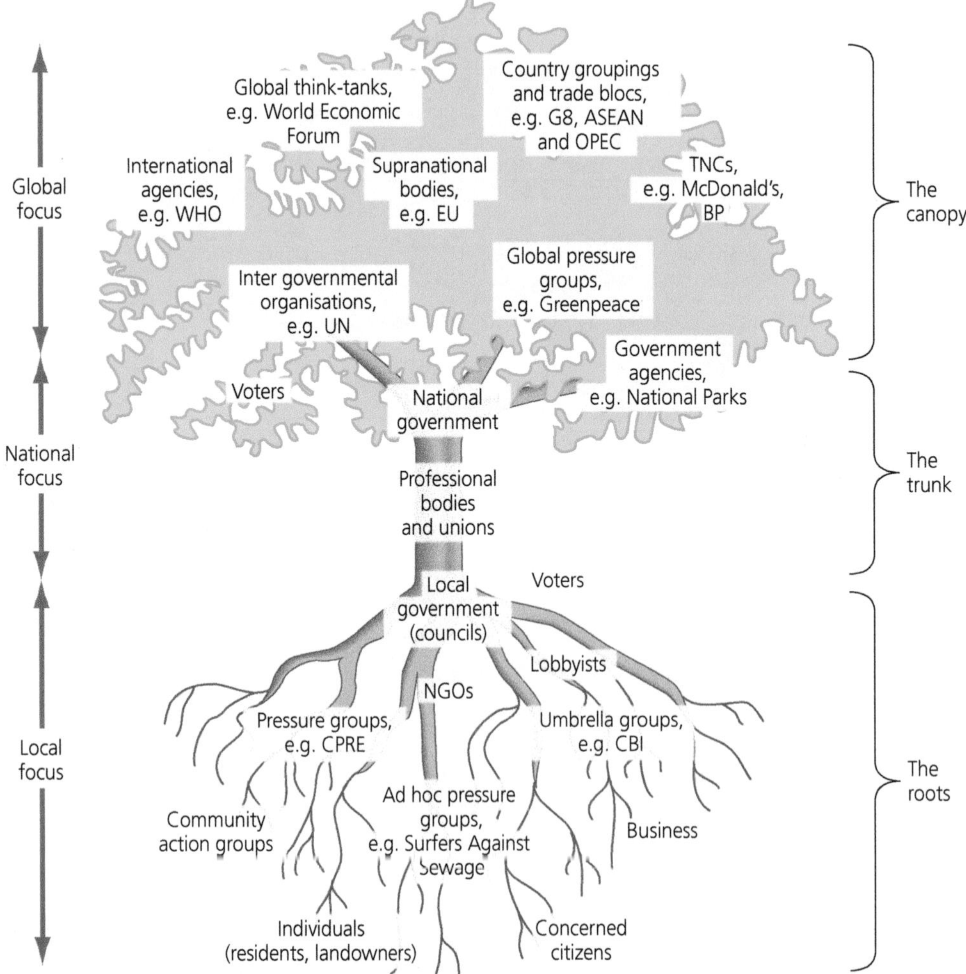

Figure 9.1 Players in decision making

Players (P): they are individuals, groups, organisations, governments, businesses and other stakeholders involved in geographical issues and decisions in a number of different ways:

+ as **causes** of or contributors to the issue, as for example the role of refugees in the so-called migrant crisis
+ as the **victims**, both directly and indirectly, of the issue, as for example the role of local residents affected by an airport expansion
+ as **managers** making decisions about how best to deal with the issue, as for example governments dealing with energy security.

It is important to note that some players have greater influence than others. Figure 9.1 shows the range of potential decision makers on three spatial scales: global, national and local.

Attitudes and actions (A): all the players involved in an issue have their own attitudes or views on it. The attitudes of different players may well clash. In some instances, they may have taken actions that they thought were appropriate or necessary. Influences on attitudes include identity, political and religious views and profits, as well as importance attached to social justice, equality and the natural environment.

Futures and uncertainties (F): this concerns rather more those players who make decisions that are likely to affect people in the future. Basically, there are three different approaches to management of the future:

+ opting for business as usual
+ seeking sustainable strategies
+ devising radical alternatives (mitigation and adaptation).

The links between the three standpoints are shown in Figure 9.2.

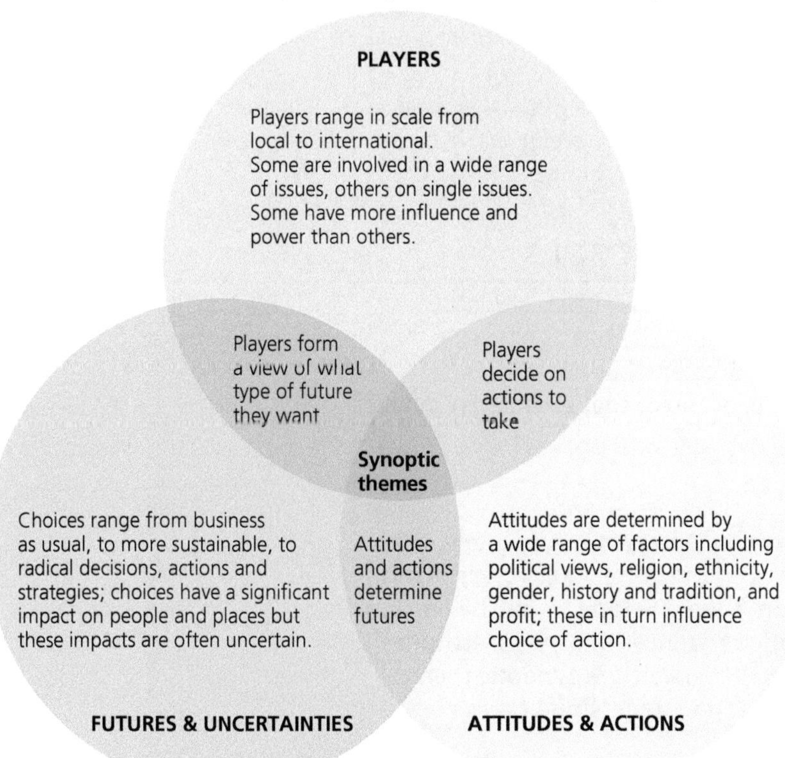

Figure 9.2 The links between the three synoptic standpoints

Whatever strategy is adopted, there will be positive and negative impacts, winners and losers. There will also be consequences that relate to risk, resilience and thresholds. But can we really be sure that what we decide now will have the desired outcome tomorrow? So in anticipating any futures, there are uncertainties: scientific, demographic, economic and political. These uncertainties relate to:

+ the accuracy or otherwise of forecasts and predictions
+ unforeseen events and changes
+ possible outcomes.

Unfortunately, experience teaches us that forecasts and predictions can go horribly wrong.

Table 9.1 gives examples from each of the topics.

Table 9.1 Some examples of the so-called synoptic themes

Topic	Theme	Sample players (other than national and local government)	Attitudes and actions	Futures and uncertainties
1	The hazard management cycle	IGO, NGOs, local residents	Contrasting attitude to hazard risks	Forecasting disaster impacts
2A	Exploiting glaciated landscapes	Tourism, TNCs, conservationists	Actions and their impact on natural systems	Impacts of climate change
2B	Exploiting coastal landscapes	Developers, tourism, conservationists	Actions and their impact on natural systems	Impacts of climate change
3	Promoting globalisation	IGOs, NGOs, TNCs	Contrasting attitudes to globalisation	Patterns of resource consumption
4A	Regeneration	Investors, businesses, local groups	Contrasting attitudes to regeneration	Success of regeneration plans
4B	Cultural diversity	Local groups, political parties	Contrasting attitudes to places	Change and legacies
5	Shared waters	IGOs, major water users, scientists	Contrasting attitudes to water supply	Rising global water demand
6	Securing energy supplies and pathways	TNCs, IGOs, green parties	Energy consumers' attitudes to the environment	Rising global energy demand
7	Global policing	IGOs, NGOs, green parties	Contrasting attitudes to the reduction of carbon emissions	Future power structures

Synoptic thinking and core concepts

REVISED

Synoptic thinking means thinking like a geographer. In other words:

+ seeing the links between the topics you have studied
+ making connections between different places and people
+ recognising that there are some concepts or topics that link across most of what you have studied.

Core concepts

The specification sets out a number of core concepts that are, in effect, bridging or truly synoptic topics. These concepts require you to bring together your knowledge and understanding of more than one topic. It will be one or more of these concepts, rather than specific references to players, attitudes or futures, that provides the overarching structure for the synoptic theme questions. For this reason, it is important that you understand these core concepts and the topics they might touch on.

The core concepts are now explained and illustrated. You should note that the core concepts do not necessarily have links with all the topics. Hence the blanks in the following tables.

Causality

Causality is the relationship between something that happens or exists and the thing or circumstances that cause it (see Table 9.2). Two obvious examples are provided by climate change and globalisation. The causes of climate change are largely physical, with a possible human input (e.g. carbon emissions), while the causes of globalisation are largely human, with a possible physical input (e.g. the availability of resources).

Table 9.2 Some examples of 'causality' topics

Topic	Example of causality topic
1	Tectonic hazards
2A	The fragility of glaciated landscapes
2B	Coastal flooding
3	International migration
4A	The need for regeneration
4B	Changing population structures
5	Water insecurity
6	Energy insecurity
7	The rise of superpowers

Systems and feedback

Systems have two different contexts in this specification:

- as a set of interrelated physical parts with inputs and outputs
- as a three-step decision-making mechanism (see Figure 9.3).

Feedback occurs when outputs of a system are returned to the system as inputs. The resulting loop creates a cause-and-effect chain, sometimes referred to as a feedback loop (see Table 9.3).

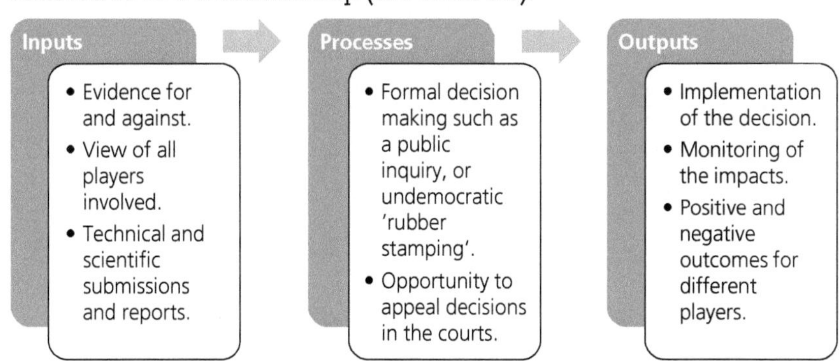

Inputs
• Evidence for and against.
• View of all players involved.
• Technical and scientific submissions and reports.
Processes
• Formal decision making such as a public inquiry, or undemocratic 'rubber stamping'.
• Opportunity to appeal decisions in the courts.
Outputs
• Implementation of the decision.
• Monitoring of the impacts.
• Positive and negative outcomes for different players.

Figure 9.3 The decision-making system

Table 9.3 Some examples of 'systems and feedback' topics

Topic	Example of systems and feedback topic
1	
2A	The glacial mass balance
2B	The coastal cell
3	
4A	
4B	
5	The hydrological cycle
6	The carbon cycle
7	

Inequality

Inequality (see Table 9.4) exists at all geographical scales. It relates to differences in opportunity, access to resources, wealth and influence between different groups. In terms of players, decision-making power and influence tend to be in the hands of people with political power, financial resources or cultural influence.

Table 9.4 Some examples of 'inequality' topics

Topic	Example of inequality topic
1	
2A	
2B	
3	Development gaps
4A	Multiple deprivation
4B	Rural and urban poverty
5	Access to water
6	Access to energy
7	Superpowers and the minnows

Identity

Identity (see Table 9.5) refers to the beliefs, perceptions and characteristics that make one group different from another. This is strongly related to place. Differences in identity mean that we should expect contrasting groups of people not to share the same fears, desires, ambitions or concerns. Identity and a strong attachment to places shape the attitudes of traditional and tribal societies.

Table 9.5 Some examples of 'identity' topics

Topic	Example of identity topic
1	
2A	Indigenous people
2B	
3	
4A	Attachment to place
4B	Attachment to place
5	
6	
7	Political groups

Globalisation and interdependence

Globalisation (see Table 9.6) is the increasing integration of the world's economies, societies and cultures. It is creating a more interconnected and interdependent world. For example, the economic success of one country is tied to and dependent on economic success in other countries.

Table 9.6 Some examples of 'globalisation and interdependence' topics

Topic	Example of globalisation and interdependence topic
1	
2A	Pressure on cold environment resources
2B	Development of coastal zone
3	
4A	Impact on regeneration
4B	Impact on places
5	Impact on water demand
6	Impact on carbon cycle and energy
7	

Mitigation and adaptation

Mitigation is preventing something from happening, such as preventing a natural event from becoming a hazard. Adaptation is dealing with the impacts of something, such as building flood defences. See Table 9.7.

Table 9.7 Some examples of 'mitigation and adaptation' topics

Topic	Example of mitigation and adaptation topic
1	Strategies to modify hazard vulnerability
2A	Protecting permafrost
2B	Coping with future coastal threats
3	
4A	
4B	
5	Desertification
6	Rebalancing the carbon cycle
7	

Sustainability

Sustainability (see Table 9.8) has been a geographical buzzword for some time. More recently, the term 'environmental stability' has entered geographical speak. It implies that people need to reduce their overall impact on the Earth to a sustainable level, one that prevents irreversible environmental damage.

Resources are a major focus in concerns over sustainability. It is here that contrasting attitudes are most obvious, with exploitation at one end of the spectrum and conservation at the other (see Figure 9.4).

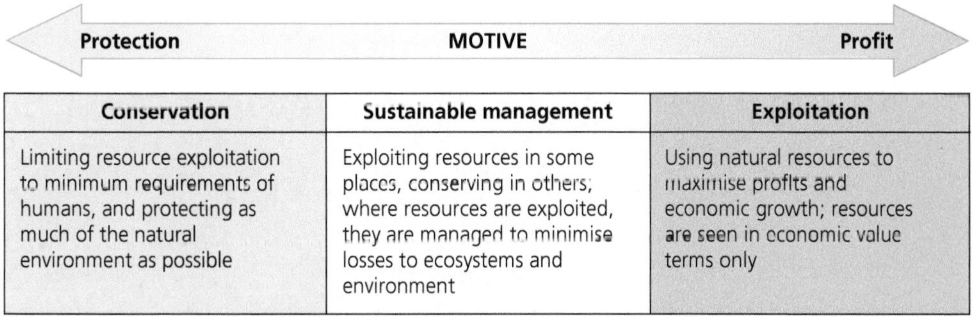

Conservation	Sustainable management	Exploitation
Limiting resource exploitation to minimum requirements of humans, and protecting as much of the natural environment as possible	Exploiting resources in some places, conserving in others; where resources are exploited, they are managed to minimise losses to ecosystems and environment	Using natural resources to maximise profits and economic growth; resources are seen in economic value terms only

Figure 9.4 Contrasting attitudes to resources

Table 9.8 Some examples of 'sustainability' topics

Topic	Example of sustainability topic
1	
2A	Threats facing glaciated landscapes
2B	Wetland reclamation
3	Impact on resources
4A	
4B	
5	Managing water resources
6	Managing energy resources
7	Tensions over resources

Risks and thresholds

Risk is best thought of as degree of exposure to harm or the chances of something happening. Risk can relate to avoiding hazards, running out of food, water or energy, contracting a disease or sinking into poverty. Thresholds mark the boundaries between safety and risk. They are tipping points (see Figure 9.5). See Table 9.9 for examples.

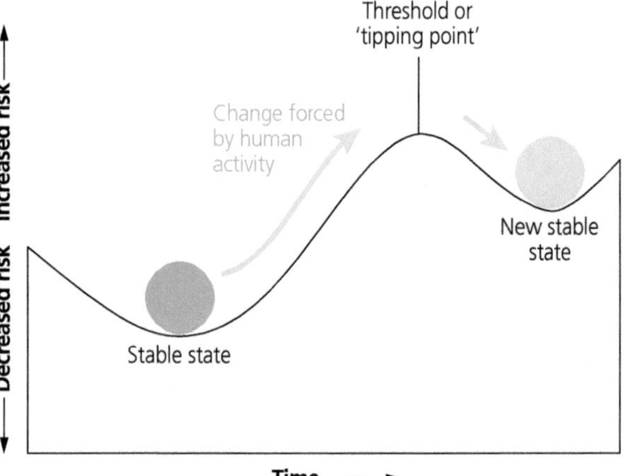

Figure 9.5 The threshold concept

Table 9.9 Some examples of 'risks and thresholds' topics

Topic	Example of risks and thresholds topic
1	The hazard risk equation
2A	Climate warming and unique landscapes
2B	Climate warming and coastal flooding
3	
4A	
4B	
5	Future droughts and floods
6	Continuing use of fossil fuels
7	Geopolitical stability

Resilience

Resilience (see Table 9.10) is the ability to cope with change or stress. Resilient communities can cope with change, both known and unforeseen.

Table 9.10 Some examples of 'resilience' topics

Topic	Example of resilience topic
1	Coping with tectonic hazard events
2A	Coping with resource exploitation
2B	Coping with rising sea level
3	Coping with international migration
4A	Coping with change
4B	Coping with change
5	Coping with water insecurity
6	Coping with energy insecurity
7	Coping with superpowers

Skills

REVISED

The examination questions' synoptic themes often involve interpreting resources, such as maps, diagrams, photographs, etc., and using what they show to support your arguments. So be sure to brush up on your ability to 'read' resources.

Having reminded you of what is involved in those parts of the specification concerned with the synoptic themes, it has to be said that there is little you can do by way of swotting up. The brutal truth is that the examiner can pick any one of the core concepts and pitch their questions at any one or more of your topics. So the possible questions are almost endless.

Whatever they turn out to be, it will then be up to you to apply the knowledge and understanding you have gained from those topics. You need to show an ability to address the questions as directly and explicitly as possible. For many of you, this will be the most challenging part of the whole qualification.

Question practice

REVISED

It is recommended that you try your hand at the relevant questions given in the sample assessment materials. Ask your teacher to print off the questions and the relevant parts of the resource booklets. Good luck!

Glossary

Term	Definition	Page
Abrasion	The process by which solid rock is eroded by rock fragments being transported by glaciers.	35
Adaptation	In the present context, any action that reacts and adjusts to changing climate conditions.	145
Assimilation	The process by which different groups within a community intermingle and become more alike. The process particularly applies to the integration of immigrant minority ethnic groups.	110
Assimilation	The process by which persons of diverse cultural and ethnic backgrounds interact and intermix in the life of the larger community or nation.	188
Asylum seeker	A person who seeks to gain entry to another country by claiming to be a victim of persecution, hardship or some other compelling circumstance.	187
Backwash	The return flow of seawater back down the beach to meet the next incoming wave.	50
Basal sliding	Occurs where ice temperatures are at or close to 0ºC and a layer of basal meltwater forms between the ice and the bedrock.	32
Beach morphology	The shape of a beach, including its width and slope (the beach profile) and features such as berms, ridges and runnels. Also includes the type of sediment (shingle, sand, mud) forming the beach.	50
Blue water	Freshwater stored in rivers, streams and lakes — the visible part of the hydrological cycle.	118
Carbon cycle pumps	The processes operating in oceans that circulate and store carbon.	135
Carbon pathways	The routes taken by flows of carbon between stores.	137
Colonialism	Where an external nation takes direct control of a territory.	151
Connection	Any type of physical, social or online link between places. Places may keep some of their characteristics or change them as a result.	80
Core–periphery model	Relates to the uneven distribution of population and wealth between two regions and the resulting flows of migrants, trade and investment from a lagging peripheral region to a prospering core region.	185
Cost–benefit analysis	Before a project can go ahead, economic, social and environmental costs are weighed up against the benefits of the project.	90
Cryosphere	Water frozen into ice and snow.	118
Cultural diffusion	The gradual spread of culture from an influential civilisation.	71
Cultural imperialism	Promoting the culture/language of a nation in another nation, which usually occurs with a larger, more powerful nation having influence over a less powerful nation.	71
Culture	Ideas, beliefs, social behaviour and customs of a group.	188
Currents	Flows of seawater in a particular direction driven by wind, tides and differences in density, salinity and temperature.	52
Deindustrialisation	The process of economic and social change due to a reduction in the industrial capacity or activity of a country, region or city. This process is widely experienced in the developed world with the global shift of manufacturing to emerging economies.	80
Deindustrialisation:	The decline of regionally important manufacturing industries in terms of either workforce numbers or output and production measures.	67
Democratic government	A system of government through elected representatives of the people.	171
Dependency ratio	The ratio of dependants (children 0–15 and over 65s) to the working-age population (16 64).	101
Deprivation	A condition in which a person's well-being falls below a level generally regarded as a reasonable minimum. Measuring deprivation usually relies on indicators relating to employment, housing, health and education.	84
Desalinisation	The conversion of salt water into freshwater through the partial or complete extraction of dissolved solids.	129

Desertification	The degradation of land in arid and semi-arid areas resulting from various factors including climatic variations and human activities.	122
Diaspora	The distribution of people from their original country around the world.	75
Disasters	Occur when hazards have a significant impact on vulnerable populations. Officially, a hazard becomes a disaster when 100 or more people are killed and/or 100 or more people are affected.	14
Dynamic equilibrium	The balanced state of a system when inputs and outputs balance over time. If one of the inputs changes, then the internal equilibrium of the system is upset. By a process of feedback, the system adjusts to the change and the equilibrium is regained.	46
Ecological footprint	A measurement of the area of land or water required to provide a person (or society) with the energy, food and resources they consume and the waste they produce.	168
Economic migrant	A person who moves from one country to another in order to find work and improve their standard of living and quality of life.	76
Economic migrant	A person who travels from one country to another in order to find work and improve their standard of living.	185
Economic restructuring	The shift from primary activities and secondary industry towards tertiary and quaternary industry as a result of deindustrialisation. It has heavy social and economic costs.	164
Ecosystem resilience	The capacity of an ecosystem to recover from disturbance or to withstand an ongoing pressure, such as drought.	123
Ecosystem stress	Constraints on the development or survival of ecosystems. The constraints can be physical (drought), chemical (pollution) and biological (diseases).	123
Enclave	A group of people surrounded by a group or groups of entirely different people in terms of ethnicity, culture or wealth.	109
Energy mix	The combination of different available energy sources used to meet a country's total energy demand.	138
Entrainment	Process by which surface sediment is incorporated into a fluid flow (e.g. air, water or ice) as part of the process of erosion.	34
Environmental refugees	People and communities forced to abandon their homes due to natural processes. These processes may be sudden, as with landslides or volcanic eruptions, or gradual, such as coastal erosion and rising sea levels.	57
Epicentre	The point directly above the hypocentre (focus) on the Earth's surface.	12
Ethical consumption	Occurs when the consumer takes into account the costs (economic, social and environmental) of producing food and goods, and of providing services.	76
Ethnic segregation	Voluntary or enforced separation of people of different cultures or nationalities.	109
Ethnicity	Social groups distinguished by their religion, language, customs and heritage.	188
Expatriate	Someone who has migrated to live in another state but remains a citizen of the state where they were born.	193
Failed state	A country whose government has lost political control and is unable to fulfil the basic responsibilities of a sovereign state, with severe adverse impacts on some or all of its population.	195
Fetch	The uninterrupted distance across water over which the wind blows. It is the distance over which waves are able to grow in size.	49
Foreign direct investment (FDI)	A financial injection made by a business into another country's economy, either to build new facilities (factories or shops) or to acquire, or merge with, a firm already based there.	65
Fossil water	Ancient, deep groundwater from former pluvial (wetter) climatic periods.	118
Free trade	When a government does not discriminate, and the trade of goods is free of import/export taxes and tariffs.	156
Free-trade blocs	Voluntary international agreements that encourage the free flow of goods and capital between member countries.	64
Gentrification	The movement of middle-class people back into rundown inner-urban areas, resulting in an improvement of the housing stock and image.	84
Gentrification	The movement of middle-class people into rundown, inner-urban areas and the associated improvement of the housing stock and area image.	110
Geo-strategic location	A location that commands access to and control over a large territory and its resources.	150

Geopolitical intervention	Occurs when and where a country exercises its power in order to influence the course of events outside its borders.	176
Glaciation	The modification of landscapes while covered by ice sheets or glaciers.	25
Global citizenship	A way of living in which a person identifies strongly with global-scale issues, values and culture rather than with a particular nation, state or place.	199
Global commons	Global resources so large in scale that they lie outside of the political reach of any one state. International law identifies four global commons: the oceans, the atmosphere, Antarctica and outer space.	199
Global shift	The international relocation of different types of industrial activity, especially manufacturing industries.	67
Glocalisation	The changing of the design of products to meet local tastes or laws.	62
Green water	Freshwater stored in the soil and vegetation — the invisible part of the hydrological cycle.	118
Gross domestic product (GDP)	The total value of all goods and services produced by a country in a year. It is calculated by taking the value of all produced goods plus the value of all services.	168
Halophytes	Plants that can tolerate salt water, be it around their roots, being submerged at high tide or being exposed to sea spray.	48
Hazard hotspots	Locations that are extremely disaster prone for a number of reasons. Notable is the fact that they experience more than one type of natural hazard.	19
Hazards	Natural events that threaten or actually cause injury and death, as well as damage and destruction to property.	14
Hotspot volcanoes	Volcanoes found in the middle of tectonic plates and thought to be fed from the underlying mantle (a thick layer of high-density rocks lying between the Earth's crust and its molten core). These volcanoes occur where the mantle is unusually thin and hot. The summits of the Hawaiian Islands are classic examples.	10
Human welfare	The health, happiness, good fortune, prosperity, etc. of a person or a group.	167
Hyperpower	A state that is dominant in all aspects of power (economic, political, cultural, military) — for example, the USA from 1990 to 2010.	148
Hypocentre (focus)	The point of origin where the pressure is released inside the Earth.	12
Ideology	A system of ideas, beliefs and values that forms the basis of economic or political theory and policy.	164
Infrastructure	The basic services and systems needed for a place to work effectively, e.g. adequate road networks, telecommunication networks, water and sewerage facilities, affordable housing and access to education.	90
Intellectual property (IP)	Intangible property that is the result of a person's creativity, such as patents, copyrights, trademarks, etc.	162
Inter-governmental organisations (IGOs)	Mainly global organisations whose members are nation states that uphold treaties and international law.	156
Internal deformation	A plastic-like quality caused when, under pressure, ice crystals are affected by recrystallisation.	32
Intra-plate earthquake	Occurs when there is a release in strain energy away from the plate boundaries. The causes of these events are still not fully understood.	9
Land conversion	Clearing a natural ecosystem and using the space it occupied for a different purpose.	142
Land grabbing	The acquisition of large areas of land in developing countries by domestic and transnational companies, governments and individuals. In many instances, the land is simply taken over and not paid for.	179
Life expectancy	The average number of years a person might be expected to live based on the year of their birth.	170
Littoral zone	The wider coastal zone, which includes adjacent land areas, the shore and the shallow part of the sea just offshore. It comprises four sub-zones: coast, backshore, foreshore and nearshore.	46
Management	A set of actions that facilitate the transition from one situation to another. More specifically, those actions might be aimed at solving or ameliorating a particular problem or issue.	111

Marginalisation	The social process of being made to feel apart or excluded from the rest of society. This leads to feelings of belonging to an underclass that is discriminated against.	88
Mass movement	The large-scale downward movement of material under the influence of gravity.	29
Mass movement	A collective term for the processes responsible for the downslope movement of weathered material under the influence of gravity.	53
Middle class	The global middle class can be defined as people with an annual income of over US$10,000. They are significant consumer spenders.	161
Mitigation	Any action taken to reduce or eliminate the long-term risk to human life and property from natural hazards. Those actions are largely the outcome of stage 5 in the hazard management cycle. However, they are also likely to be taken during stage 3, sometimes referred to as adjustment or adaptation.	20
Mitigation	In the present context, any actions that either reduce or eliminate the long-term risk and hazards of climate change.	145
Nationalism	The belief held by people belonging to a particular nation that their own interests are much more important than those of people belonging to other nations.	192
Natural change	The difference between births and deaths over a given period of time. Natural increase happens when the birth rate is higher than the death rate and natural decrease when the death rate is higher than the birth rate.	100
Negative multiplier	A downward cycle. Change in economic conditions can produce less spending, which results in less investment from businesses and therefore fewer employment opportunities.	87
Neo-colonialism	An indirect form of control, which means that newly independent countries are not masters of their own destinies.	152
Net migration	The difference between immigration and emigration from an area over a given period. This can be either positive or negative.	100
Ocean acidification	The decrease in the pH of the oceans caused by the uptake of carbon dioxide from the atmosphere.	142
Official development assistance (ODA)	A measure used by the OECD as an indicator of the flows of international aid. Flows are transfers of resources, either in cash or in the form of commodities and services.	178
Paleomagnetism	Results from magma locking in the Earth's magnetic polarity when it cools. Scientists can use this to reconstruct past plate movements.	11
Paraglacial	Locations are those recovering from the disturbance of glaciation.	34
Perception	The 'picture' or 'image' of reality held by a person or group of people resulting from their assessment of received information.	86
Periglacial	Locations are those where frost action and permafrost processes dominate.	34
Periglaciation	The modification of landscapes located adjacent to the margins of ice sheets and glaciers.	25
Photosynthesis	The process by which plants capture carbon dioxide from the atmosphere and then store (sequester) it as carbon in their stems and roots. Some is also stored in the soil.	134
Plant succession	The sequential development of vegetation from its initial establishment on bare ground through to the ultimate vegetation cover or climax plant community.	49
Plucking	The detachment of joint-bounded blocks by glaciers. It was thought that the ice froze onto the rock and wrenched blocks of it away. However, it is now believed that the breaking away of blocks is due to the movement of the glacier.	35
Population density	The number of people per unit area (usually per square kilometre), i.e. the total population of a given area (a place, city, region or country) divided by its area.	99
Population pyramid	A graph to show the distribution of age groups in a population. It is constructed with males on one side and females on the other, with the youngest age group at the base of the pyramid.	101
Preparedness	Educating people about what they should do in that emergency (where to seek shelter, how to assist others) as well as improving warning systems and training, and equipping rescue teams. It focuses on the emergency stage immediately following a hazard.	20
Primary effect	Occurs as a direct result of a hazard.	13

Primary energy	The forms of energy found in nature have not been subjected to any conversion or transformation process, for example water (watermills) and wind power (windmills) or the burning of gas in heating the home and by motor vehicles.	139
Profit leakage	The process whereby the profits made by a business are not retained and reinvested in the country where they were made.	179
Proglacial	Locations that lie close to the ice front of a glacier or ice sheet.	34
Quality of life	The level of social and economic well-being experienced by individuals and communities, measured by various indicators such as health, longevity, happiness and educational achievement.	83
Refugee	Defined by the UN as someone whose reasons for migrating are genuinely to do with fear of persecution or death.	187
Regelation	Occurs where basal ice is forced against a rock obstacle — it melts and then refreezes on the down-glacier side. The temporary meltwater acts as a lubricant.	32
Rendition	The practice of sending international criminal or terrorist suspects covertly to be interrogated in a country where there is less concern about the humane treatment of terrorist suspects.	180
Reservoir turnover	The rate at which carbon enters and leaves a store. It is measured by the mass of carbon in any store divided by the exchange fluxes.	134
Residence time	The average time a water molecule spends in a store or reservoir.	118
Resource nationalism	A growing tendency for state governments to take measures ensuring that domestic industries and consumers have priority access to the national resources found within their borders.	76
Rural idyll	A 'chocolate box' image of quaint villages set in beautiful countryside. A place thought to be free of most of the negative aspects associated with urban living.	105
Safe water	Water that is sufficiently clean to be fit for human consumption and use.	127
Sanctions	Political or economic decisions that aim to persuade a country back to the negotiating table without the use of military force.	158
Secondary effect	Occurs as a result of the primary effect.	13
Secondary energy	Derived from the conversion of primary energy. The most important secondary energy is electricity, which is derived from many primary sources.	139
Sequestering	The long-term storage of carbon dioxide and other forms of carbon.	134
Sink estate	Housing estate that experiences deprivation, poverty and crime, such as domestic violence, gang warfare and drugs.	88
Special economic zone (SEZ)	An industrial area, often near a coastline, where favourable conditions are created to attract international TNCs.	65
Sphere of influence	A physical region over which a country believes it has economic, military, cultural or political rights.	162
Stadials/ interstadials	Short-term fluctuations within glacial periods. Stadials are colder phases that lead to ice advances, while interstadials are slightly warmer phases during which ice sheets and glaciers retreat.	31
Stakeholder	An individual, group or organisation with a particular interest in the actions and outcomes of a project or issue-solving exercise.	94
Stakeholder	An individual, group or organisation with a particular interest in the actions and outcomes of a project or issue-solving exercise.	112
Stakeholders	Individuals, communities, organisations, businesses and government with a specific interest in a situation — in this instance, in hazard risk and hazard mitigation.	17
Studentification	Social, economic and environmental change brought about by the concentration of students in particular areas and cities, usually located close to universities.	84
Subduction zones	Broad areas where two plates are moving together, often with the thinner, more dense oceanic plate descending beneath a continental plate. Fold mountains form at the edge of the overriding plate, with associated volcanic activity. Stress between the two plates also triggers earthquakes.	11
Subglacial	Locations that lie beneath the base of a glacier or ice sheet.	34
Sublimation	The change from the solid state (ice) to gas (water vapour) with no intermediate liquid stage (water).	30

Sustainable development	Development that meets the economic, social and environmental needs of today's population without compromising the ability of future generations to meet their own needs.	173
Swash	The flow of seawater up a beach as a wave breaks.	50
Tectonic hazard profile	A technique used to try to understand the physical characteristics of different types of hazards, such as earthquakes, volcanic eruptions and tsunamis.	16
Thermohaline circulation	The global system of surface and deep-water currents within the oceans driven by differences in temperature and salinity.	135
Totalitarian government	A system of government that is centralised and dictatorial. It requires complete subservience to the state, with political control in the hands of elites.	171
Transfer pricing	A financial flow occurring when one division of a TNC charges a division of the same company based in another country for the supply of a product or service. It can lead to less corporation tax being paid.	193
Transnational corporations (TNCs)	Businesses/companies whose operations are spread across the world, and which operate in many nations as both makers and sellers of goods and services.	62
Water security	Exists when a population has sustainable access to adequate quantities of water of acceptable quality. Adequate, that is, in sustaining human wellbeing and socio-economic development and for ensuring protection against water-borne pollution and disease.	127
Weathering	The breakdown or disintegration of rock in situ.	29
Weathering	The disintegration and decomposition of rocks in situ by the combined actions of the weather, plants and animals.	53
Wetlands	Areas where the soil is frequently or permanently waterlogged by fresh, brackish or salt water. The water may be static or flowing. The vegetation may be marsh, fen or peat.	123
Xerophytes	Plants that can tolerate very dry conditions, such as those found in coastal sand dunes.	48

Now test yourself answers

Topic 1

1 At a convergent (destructive) plate boundary the two tectonic plates are moving towards each other. When a continental and oceanic plate meet, the denser oceanic plate is subducted underneath the continental plate, creating a subduction zone. As the denser oceanic plate subducts into the mantle this can lead to earthquakes and violent volcanic eruptions.

2 Tsunamis are caused by the abrupt disturbance of the ocean by an earthquake. The earthquake causes the sea to push upwards suddenly, causing waves to spread. As the waves move towards the shore, the increased friction with the ocean floor causes them to slow down and then increase in height.

3 Paleomagnetism is caused by magma locking in the Earth's magnetic polarity when it cools.

4 False.

5 The epicentre is the point directly above the hypocentre on the Earth's surface whereas the hypocentre is the point of rupture within the Earth where an earthquake starts.

6 Pyroclastic flows, because they involve hot, toxic gases and pyroclastic materials. They occur explosively without warning and move downslope at incredible speeds, giving humans little or no chance of escaping.

7 A hazard is an event that threatens both life and property. A disaster occurs when the hazard causes mass destruction and death. The difference lies in the scale of the impacts.

8 Vulnerability is the degree of exposure to the impacts of a hazard. Resilience is the ability of a society or community to cope with and recover from the effects of a hazard.

9 There are many more earthquakes during the course of a year than volcanic eruptions. Earthquakes occur in more parts of the world, including some of the most densely populated. Earthquakes occur without warning, whereas there are often tell-tale symptoms that warn of a forthcoming eruption.

10 Hazard profiles compare the physical characteristics and processes that all hazards share. Comparisons can help decision makers identify and rank the hazards that should be given priority in terms of attention and resources. Hazard profiling is more difficult when comparing across hazards, but can work well when comparing the same type of hazard event.

11 Good governance should ensure the proper use of resources in trying to minimise the impacts of a hazard. So this includes supporting a whole range of actions educating the public about what to do when a hazard strikes, planning emergency procedures, land-use planning, building design and construction, etc.

12 Simply, the scale of the impact (the damage to property, number of deaths, etc.). Tectonic mega-events are classified as high-impact, low-probability events. An example of a mega-disaster would be the Indian Ocean tsunami (2004).

13 Mitigation involves any action taken to reduce or eliminate the long-term risk to human life and property from natural hazards, whereas preparedness is educating people about what they should do in that emergency, as well as improving warning systems and training, and equipping rescue teams.

14 Modify the hazard event.
Modify both vulnerability and resilience.
Modify the potential financial loss.

15 A The likely cost of repairs or reconstruction, and C The market value of the properties to be insured.

16 Most criticism focuses on the emergency phase. Trying to put into effect even a rehearsed set of procedures is often very difficult when chaos prevails. It is difficult to co-ordinate the work of voluntary rescue teams and aid organisations, particularly if they come from overseas and lack first-hand experience of the environment in which the disaster has occurred. The destruction of infrastructure, particularly roads and airports, adds to the challenges of coping in the immediate aftermath of a disaster. Another concern is the misappropriation of funds raised by public appeals across the globe. The Haiti earthquake of 2010 provides a classic example.

Topic 2A

1 Long-term causes are explained by Milankovitch's theory and are to do with the Earth's orbit around the Sun. Short-term climate change is related to sunspot activity, volcanic activity and most recently human activity.

2 Evidence of climate change includes historical records, such as old drawings or written records. An example is frost fairs on the River Thames in 1683. Other evidence includes: tree-ring dating, ice-core analysis, shift in vegetation belts and fossil landscapes.

3 It is important, as many present-day landscapes cannot be explained without looking at past climates. For example, in northern England and Scotland there are wide, steep-sided valleys that have to have been caused by glaciers.

4 Ice cover has been shrinking since about 1850 thanks to a general warming of the global climate. Ice accumulations in mountains are generally thinner anyway because they are feeding warm-based glaciers. The lower the latitude of the high mountains, the greater the reduction, simply because of the warmer climate.

5 Aspect determines both the amount of snow falling and the accumulation. In the Northern Hemisphere, north- and northeast-facing slopes are both more sheltered and shaded. They are therefore more conducive to snow accumulation and the compaction of snow into ice.

6 Glacial areas can be covered by ice all year round whereas periglacial areas are not. Periglacial environments are underlain by permafrost and found on the fringes of polar glacial environments.

7 Weathering and mass movement contribute to upland landscapes. An example of weathering is frost shattering where water seeps into cracks within a rock and as temperatures fall the water turns to ice and expands. This process exerts pressure on the rock and, over time, chunks of rock break away, which can lead to the creation of large scree fans of broken rock.

8 Permafrost and frost-formed features such as ice-wedge polygons, patterned ground, pingos, block fields and rock glaciers. There are also landscape features resulting from wind (loess) and meltwater rivers (braided channels).

9 The balance between its gains (accumulation) and its losses (ablation). The place where accumulation and ablation balance each other out is known as the equilibrium point.

10 In the winter, temperatures start to drop, which results in more snowfall and less melting and evaporation. Accumulation exceeds ablation, which means a positive mass balance. In summer, as the temperatures rise there is less snowfall but more melting and evaporation. Ablation then exceeds accumulation equally, which means a negative mass balance.

11 Open, because there are inputs (snow, weathered material) and outputs (meltwater, moraine). Components are solar energy, snow and ice, rock debris, moraine and meltwater.

12 Positive feedback loops speed up processes, whereas negative feedback loops regulate and reduce the natural processes that restore balance.

13 Through the medium of slope. Since the basic cause of ice movement is gravity, the steeper the slope, the faster the movement.

14 It depends in which part of the world they are located.

15 Polar glacier movement is extremely slow because they are located in cold environments where there is little or no meltwater. Almost all movement is through internal deformation. Temperate glaciers move at a faster pace due to being affected by meltwater. Meltwater helps lubricate the base of the ice.

16 In three ways — supraglacial, englacial and subglacial. Supraglacial is where material is transported on top of the ice, englacial is within the ice and subglacial is below the ice.

17 Upland glaciated landforms are distinguished from lowland glaciated areas by a suite of landforms produced by erosion – for example, cirques, pyramidal peaks, U-shaped troughs, hanging valleys and truncated spurs.

18 The key feature is the cirques. As a result of headwall erosion, when two cirques lie either side of a ridge, they cut back towards each other and reduce the ridge to an arête. Where more than two cirques are converging through headwall erosion, the separating upland mass is reduced to a pyramidal peak.

19 Abrasion is an erosional process caused by the sandpapering effect of ice as rock fragments are dragged over the landscape. Scouring is where the landscape becomes smoothed and polished by rock flour.

20 (a) Terminal moraines mark the furthest point reached by a valley glacier, while recessional moraines stand in the retreat of a glacier's snout. (b) Lateral moraines are linear ridges deposited along valley sides. Medial moraines develop on the surface of a valley glacier and are formed by the joining of two separate ice streams (i.e. lateral moraines).

21 Examples include: drumlins indicate the direction of the ice flow; erratics indicate the direction of past ice flow; terminal moraine indicates the furthest point of the glacier and lateral moraine indicates the edge of the glacier.

22 It is the temperature at which ice melts at a given pressure, i.e. when ice becomes meltwater. Inside a glacier, because of the pressure, the melting point is lower than 0°C. So it is significant as the threshold between two sets of processes — ice and running water.

23 Similarities include: both are made of coarse sand and gravel, they are stratified and consist of sorted material. Differences include: eskers are created by subglacial river deposition and kames form on the ice surface, and eskers are sinuous ridges whereas kames are isolated hills.

24 They have social, economic and environmental value. Social value: glacial landscapes can provide water storage to supply different parts of a country. Economic value: tourists who visit the area bring revenue and this provides jobs in the local community. Environmental value: the glacial landscape itself is a habitat for different types of wildlife and many glacial landscapes are important for farming.

25 The threats include farming, forestry, mining and quarrying, HEP and tourism. Tourism, because tourists themselves damage the environment (trampling) and the development of tourism requires the creation of an infrastructure of hotels, cabins, ski slopes and ski lifts, access roads, etc. All these developments threaten to cause large-scale environmental damage.

26 It could impact precipitation levels, particularly a fall in snow and a change in rainfall patterns. Furthermore, there could be an increase in glacial melt due to rising temperatures and as the air warms there could be an increase in snowfall in parts of the Arctic due to the air being able to hold more moisture.

27 The two extremes are 'do nothing' (business as usual) and 'total protection'. In between, the options are sustainable exploitation of some resources and comprehensive conservation. The latter would allow some developments, such as carefully regulated ecotourism or organic eco-farming.

28 There are various stakeholders with an interest in the future of glaciated landscapes, which then brings with it different attitudes to the landscape as stakeholders can range from conservationists to oil companies, from scientists to tourist operators, from indigenous people to investors. There are a number of possible approaches to the management of cold environments. They range from 'do nothing' and 'business as usual' to 'comprehensive conservation' and 'total protection'. Which strategy is most appropriate depends on the area and the views of the players involved. In the event that unanimity will rarely prevail among stakeholders with different values and viewpoints, compromise will be the only way ahead.

Topic 2B

1 When there is a balanced state in a system where the inputs and outputs balance over time.

2 Metamorphic rocks are rocks (mainly sedimentary) that have been changed in texture, composition or structure by either intense pressure and/or heat. Igneous rocks are formed by the solidification of magma either within the Earth's crust or at the surface. Sedimentary rocks are usually deposited in layers or strata, the material being largely derived from the breakdown of other rocks (i.e. igneous or metamorphic).

3 Sedimentary rocks because of their bedding planes and often loosely consolidated nature are most prone to erosion. Igneous rocks are hard and resistant to erosion. Metamorphic rocks are somewhere in between, but where folded and fractured can be very susceptible to erosion.

4 Halophytes are plants that can tolerate salt water whereas xerophytes can tolerate very dry conditions.

5 The size and strength of a wave is dependent upon the strength of the wind, the length of time the wind blows for, water depth and wave fetch.

6 B Destructive waves have a more circular motion, and C Destructive waves have a weak swash and a strong backwash.

7 Hydraulic action is when wave energy is at its strongest and waves strike the coast with great force. Wave activity at this time is predominantly destructive rather than constructive.

8 The cave is subjected to erosion, particularly hydraulic action. The cave is gradually enlarged until it reaches a cave being excavated on the other side of the headland. When they join, an arch is formed. This is gradually weakened until the point is reached when the cave roof collapses and the outer limb of the arch becomes a stack.

9 Traction is where pebbles and boulders roll along the seabed whereas saltation happens when pebbles and boulders bounce along the seabed.

10 The two processes of mechanical weathering are freeze–thaw and salt crystallisation. Freeze–thaw occurs when the repeated freezing and thawing of water in cracks exerts pressure on the surrounding rock, causing it to break apart. Salt crystallisation occurs when salt crystals grow and exert pressure within the cracks of rocks.

11 Mass movement is the large-scale downward movement of unconsolidated material under the influence of gravity.

12 (a) Rockfalls are a rapid form of mass movement (rocks often falling vertically) whereas rotational slides are the outcome of mass movement along curved failure surfaces, and (b) screes are masses of weathered, angular rock fragments collecting below a free face to create a talus. Cliff terraces are the result of rotational sliding.

13 Barrier islands are low, sandy islands, usually in a chain-like formation parallel to the mainland. Formed by post-glacial sea-level rises, formerly continuous bars have been breached to create a string of islands.

14 Because they starve deltas of the sediment needed to maintain them. The equilibrium between deposition and erosion is changed in favour of the latter.

15 Coincidence of very low pressure and strong winds; powerful waves and swell; spring high tides; coastal morphology funnelling in the same direction as the wind.

16 Yes, but the scale and speed of coastal changes are also considerations that need to be taken into account.

17 Sediment cells are natural divisions along the coastline that function as open systems. There are 22 around the coast of England and Wales. Any change within a cell is likely to have consequences within that cell. Because of this 'wholeness', a more joined-up form of coastal management is possible.

Topic 3

1 Since 1995 the number of people using the internet has increased significantly, with over 40 per cent of individuals having access to the internet by 2018.

2 The World Bank aims to reduce poverty by working with international institutions, regional banks and national governments through grants, zero interest credits and low-interest loans and investments.

3 TNCs are important players through:
- allowing for the free flow of capital

- influencing people's political and cultural views, leading to a global culture
- allowing for the free flow of labour, as people migrate into cities or elsewhere in the world to seek better lives
- allowing for the free flow of goods and services, as they increase trade.

4 A country can remain switched off because it is landlocked or it may be heavily indebted. If indebted, it must have good relations with its neighbouring countries and if this is not a factor then it will remain landlocked, e.g. Zambia. Paying for imported goods is difficult if a country is heavily indebted, e.g. Tanzania, and leads to the country remaining switched off. Some countries are stuck in 'interdependence'. Developing nations continue to export raw materials and lose their labour force as people migrate. Places may not have the infrastructure (e.g. places are landlocked and undersea cables cannot reach certain countries), political stability or an educated enough workforce for TNCs to invest.

5 Winners: TNCS, Western countries, big business, consumers. Losers: developing countries, workers. Global shift can result in the loss of old manufacturing jobs, meaning high unemployment. When middle-class people migrate out of the rundown areas, this leaves the elderly and low-income families trapped. When old factories are abandoned it leaves the area run down and results in a rise in crime. Factories create high levels of pollution in newly industrialised countries (NICs).

6 People migrate out of impoverished states to cities as the idea of a shrinking world results in them gaining knowledge of the outside world and the opportunities available. There are many urban pull factors that attract rural dwellers to cities — for example, better education, healthcare and employment opportunities everywhere. The rural push factors — poverty, land subdivision (to make space for TNCs), crop failure and the lack of job opportunities — also play a part.

7 Elite migrants (highly influential/skilled individuals) often have multiple homes in global hub cities. Low-waged individuals migrate in huge numbers into global hubs. Both illegal and legal immigrants end up working in kitchens, as domestic cleaners and on construction sites. Internal (rural-urban) migrants contribute greatly to the growth in global hub cities.

8 Advantages: for people this can include human rights arrangements to promote education and welfare of children and those whose physical, mental or emotional problems or caring responsibilities prevent them from taking standard jobs. For the environment, this can include the enforcement of existing environmental protection laws through organisations as well as the public. Disadvantages: habitat loss and biodiversity decline on a continental scale. Around 40 per cent of the Earth's surface has been changed into productive agricultural land, e.g. aquaculture, cattle ranching — groundwater depletion and deforestation.

9 With globalisation bringing cultural imperialism, resistance is growing against those with diversity as a prevalent lifestyle choice. Those who reject globalised cultures are often the ones experiencing backlash. Economically, there are TNCs involved in globalisation that are known to exploit workers, with low wages, bad working conditions and labour-intensive work. TNCs introduce worries about environmental exploitation as their over-exploitation of the land to produce goods leads to deforestation, groundwater depletion, etc.

10 The graph indicates a decrease in the percentage of people living in extreme poverty since the 1800s. Some of the key reasons for this are the economic development of countries, and improvements to healthcare, sanitation and education systems.

11 To calculate the average of a data set you need to add the values together and divide the answer by the total number of values. 6.326/10 = 0.633

12 The migration of different groups of people into a concentrated area in a city is an impact caused by globalisation. It can create mixed societies with differing beliefs and religions, which can cause friction within communities, leading to tension. Where ideologies clash this can often lead to violence and racism.

13 Laws can limit the number of people being able to migrate. Governments limit people's access to information found online; violent and sexual imagery is often censored. Trade protectionism is also a significant factor in controlling the spread of globalisation.

14 Sourcing from local producers is beneficial to both the environment and the consumers who wish to consume locally produced goods. In the long run, this proves more sustainable and beneficial to local providers and leads to a reduction in the use of pesticides, of food miles and of individual carbon footprints.

15 Recycling can be viewed as the first step towards the more ambitious goal of a circular economy. It reduces the rate at which new natural resources are used. However, the recycling process does itself require the use of energy and water.

Topic 4A

1 Regeneration (or place making) is the long-term upgrading of urban residential, retail, industrial and commercial areas, as well as rural areas. It is important in the UK, as many places are being regenerated as we move to a post-industrial economy.

2 There has been a demise of employment in primary industries and manufacturing linked to deindustrialisation and global shift. A fifth sector has also emerged, the quinary sector, which is an important aspect of the increasing 'knowledge economy'.

3 Changes could include the decline in the 'old economy' (primary and secondary sector) and growth of the 'new economy' (tertiary and

quaternary sector). There could be an increase or decrease in the type of jobs or changes in full-time/part-time work.

4 Health: Variations in income can affect the quality of people's housing and diets. Health may suffer as a result of access to food and lifestyle choices.

Life expectancy: Longevity varies substantially between places, between regions, and both between and within settlements, especially larger cities.

Gender (biological differences between the sexes): income, occupation and education are key factors, together with associated lifestyle choices, such as diet and smoking.

Education:
- Working-class white children in poverty have lower educational achievement and are more likely to continue to underachieve.
- Boys are more likely to have low results than girls, especially those of Bangladeshi, Pakistani and black African origin.
- Only 14 per cent of variation in any individual's performance is due to the quality of the school attended.

5 There may be a change due to employment change from the old economy to the new economy. There may be changes due to an increase or decrease of wealth in your local place. Finally, there may be changes due to either a lack of investment or investment in regeneration projects.

6 Those working in the primary sector and low-level services receive lower pay than those in more skilled and professional sectors. Jobs may be seasonal and insecure compared with manufacturing and higher-level services. Prices for goods and services vary regionally.

7 This can be done by looking at land-use changes, employment trends, demographic changes and levels of deprivation.

8 Increasing connectedness is most likely linked to globalisation and can shape economic characteristics through a change in the availability of jobs or the types of jobs that are on offer. This in turn could increase the wealth in an area, which may impact local spending.

9 This question is directed at your two place studies and any relevant connections that have shaped the economic and social characteristics of those places. These could be regional and national influences, international and global influences, and recent economic and social changes.

10 The answer will vary depending on your place of study but key changes may include: employment changes from old to new economy, which can have a knock-on effect on communities; there may be inward or outward migration, which results in a change to the cultural identity of an area; furthermore, gentrification or regeneration of an area can have a big impact on the identity, as it attracts different people to the area.

11 Your mind map should be based around people's perceptions and these could include ideas such as:
- low-income households tend to seek out communities that provide lower-cost housing and have higher social welfare spending
- higher-income groups similarly cluster together and if they move into a previously lower-income location, may gentrify it.

12 The answer will depend on your chosen case study. Some reasons why places can be economically successful are the migration policy of the country, for example it could focus on bringing in well-qualified professionals. Other factors include: having a young, economically active workforce; a large proportion of jobs in the 'knowledge economy'; businesses that then attract investment from more businesses; the right buildings and infrastructure for economic success.

13 People and groups may be marginalised because of their language, religion or customs, and especially by wealth.

14 The two key factors affecting a person's lived experience and resulting level of engagement with neighbours, community groups and elections are:
- membership: a feeling of belonging, familiarity and being accepted
- influence: a sense of playing a part in a place and hence caring about it.

15 Conflicts often occur among contrasting groups in a community largely because they hold different views about the priorities and strategies for regeneration.

16 To understand the need for regeneration you need to study economic, social and environmental factors in your place of study. You need to weigh up the factors to understand whether there is a need for regeneration and also the costs and benefits of the regeneration project.

17 The UK government manages the country's economic, social and physical environments through various political decisions. Investment in infrastructure and addressing issues of accessibility are seen as major factors in maintaining economic growth. Infrastructure projects have two main characteristics — high cost and longevity — hence they require government funding.

18 Deregulating capital markets is undertaken to encourage overseas and private investment in regeneration schemes. Deregulation allows international investors to invest in the UK without government approval.

19 Local government policies can create attractive business environments for investors and workers who are highly skilled and paid, and who can choose where to work more easily. They develop local plans that designate specific areas for development.

20 This will depend on the evidence that you have collected but could focus on any of the ways in which rebranding concentrates on the attractiveness of places and specific place identities build on their historical heritage to attract national and international visitors.

21 Attempts to regenerate rural areas have included regional aid, whereby money was provided by the EU to those recognised as being less economically advantaged while the UK was still a member state, and enterprise zone schemes, which help small areas focused on attracting particular types of business. There are mixed views on enterprise zones, as they tend to be more successful in urban areas.

22 Economic:
- Employment: schemes involving an immediate job focus, other than the construction phase, such as a shopping mall or new science park, generate a greater initial rise in income compared with a refurbished or new housing scheme or new/upgraded park. Regeneration increases opportunities, but outsiders rather than locals may take new jobs.
- Income: if people's incomes have risen following a regeneration scheme it points to its success. However, if only certain groups have benefited then this may be relative. Regeneration programmes have rarely been created to tackle poverty directly.
- Poverty: getting out of the poverty trap depends in the short term on household income, but longer term on educational attainment.

Demographic:
- Improvements in life expectancy and reductions in health deprivation.

Social:
- Reductions in inequalities both between areas and within them.
- Improvements in social measures of deprivation.

Environmental:
- Reductions in pollution levels: individuals pay taxes to fund national environmental watchdogs and planners in order to control levels of pollution and overall environmental quality.
- Reductions in abandoned and derelict land: traditionally dereliction is associated with former manufacturing areas and redundant infrastructure, such as power plants.

23 This will depend on the area that is being studied.

24 Stakeholders fall into four groups:
- Providers: could be landowners, investors, contractors.
- Users or beneficiaries: those who stand to benefit (or lose out).
- Governance: local government officials, enforcers of local bye-laws and national government policy.
- Influencers: action groups, political parties.

25 The roles of different players may include government officials who may be providing investment for the project, especially if it is seen to be costly or risky; company directors provide investment where there is expected to be profit; local government officials try to ensure that local people are looked after in the project; environmental stakeholders campaign for the project to be environmentally sustainable; and local people are interested in what is in it for them and ensure local views are taken into account.

26 This will depend on the chosen project.

27 Making judgements about the success of regeneration in any place involves not just those of the actual decision makers but also those of stakeholders – the people, groups and organisations with an interest in the regeneration. In most situations, stakeholders fall into four groups:
- providers: could be landowners, investors, contractors
- users or beneficiaries: those who stand to benefit (or lose out)
- governance: local government officials, enforcers of local bye-laws and national government policy
- influencers: action groups, political parties.

From this, it follows that each stakeholder will have their own particular perception of or opinion on what constitutes 'success' and 'failure'. They will have their own vested interests and agendas. They will have their own criteria for assessing whether a particular scheme has been, or is being, managed successfully or not. The Egan Wheel, which creates an evaluative scoring system, is a useful tool when assessing the outcome of regeneration projects in rural settings.

Topic 4B

1 Ethnicity is the mix of people drawn from different cultural backgrounds. Ethnic groups may be defined on the basis of language, religion and forms of dress. Ethnicity is not an issue, provided there is a willingness on the part of minority ethnic groups to become assimilated. Isolation in tight communities is inclined to antagonise white British groups. Another issue is the unfounded fear that minority ethnic groups might be 'stealing' houses, jobs and services.

2 England is much more densely populated than Scotland, Wales or Northern Ireland. Particularly high densities can be found between Manchester and Merseyside in the North West and London and the South East. There is a noticeable concentration of the Scottish population in the Central Lowlands. Wales shows an empty heartland, with much of the population concentrated in South Wales. In Northern Ireland, the demographic focus is around Belfast.

3 This could be due to the following factors:
- the physical environment, particularly relief and climate
- the economy, agricultural or non-agricultural, with the latter generating higher densities
- population characteristics — youthful structure is likely to raise densities
- planning, by controlling the location and amount of new housing.

4 Natural change (increase or decrease) and migrational change (international migration

balance). Despite concerns about immigration, natural increase accounts for the greater part of the UK's population growth.

5 The impact of population change may include increased pressure on services such as housing, schools or care homes for the elderly, which may lead to pressure on green belt land. If the population is decreasing there may be a loss of labour or vice versa in areas where population is increasing.

6 Immigrants tend to be young adults, so they help swell the population in the 20–40-year-old age range. They may also come with their children and so help broaden the base of the pyramid. They might subsequently invite their parents to join them, in which case they might help thicken the pyramid towards the top.

7 Their significance is a practical one. They are important parameters in the planning of housing programmes. They impact the number and size of dwelling needed. It has to be said that marital status is of much less consequence today. There are many more partnerships (as opposed to married couples) than there used to be, and many have children. Also the increased incidence of single-parent households has to be taken into consideration.

8 In some areas you can get social clustering where groups of the same ethnicity live in the same area, as often there may be a feeling of shared identity and due to proximity to jobs vacancies, as migrants looking for work may live in the same area. Finally, in the long term, increasing fertility rates can increase the amount of certain ethnicities in an area.

9 Only you can answer this — but do answer it truthfully!

10 This will depend on the place that you have studied. Regional influences could include family connections or availability of work. National influences could include government policies towards migration or location of key industries, which affects the availability of jobs.

11 Proximity to central services and possibly place of work. Housing might be cheaper in older, less well-maintained areas.

12 Young, single people or recent immigrants.

13 This could be due to crime rates in the area, poor environmental quality, which leads to a lack of care taken of the city, or racist incidents in the area.

14 Young adults leave either for higher education (not to return) or for an urban-based job. This has a detrimental effect on services (fewer children needing schooling) and businesses (more difficult to recruit labour). Services and businesses close. The overall decline in prosperity and the quality of life persuades older people to move away.

15 Rural places are often more remote and can have beautiful scenery and a much quieter, more peaceful way of life. Also, they often have a stronger sense of community, all of which can lead to them being described as idyllic.

16 This question is again directed at your two place studies and any relevant evidence that supports the existence of different views on the places. The reasons are largely to do with personal perceptions and who you are. Personal factors such as age, gender, educational qualifications, type of employment and level of remuneration are among the most influential.

17 This depends on the style of media chosen and your place of study. It very much depends on what the artist is trying to show, for example photography can give a negative perception of an area if it shows derelict buildings covered in graffiti. On the other hand, a photograph of a thriving community can create a positive perception.

18 The movement of people and employment away from major cities to smaller settlements and rural locations just beyond the city or to more distant, smaller cities, towns and even remote rural areas.

19 The main source became the EU, particularly Eastern European countries like Poland, Romania and Slovakia. Being from the EU, they were allowed visa-free entry to the UK prior to Brexit.

20 This could be due to different waves of migration to the UK, for example post-colonial migration led to large-scale immigration from former British colonies, whereas more recently globalisation has attracted people from all over the world due to the UK's global links — this is particularly seen in London. International migration has led to ethnic enclaves within cities.

21 Incidence of mixed marriages or partnerships. Number of children of mixed ethnicity. Residential dispersal from original ethnic concentrations. Mixed ethnicity in particular types of employment. Degree of involvement in the wider community.

22 Often this is deliberate as migrants choose to live in areas where they feel protected and self-contained. Also, migrants are attracted by similar cultures and specialist shops or places of worship, which create a distinctiveness to that place.

23 Again, this is a question only you can answer.

24 This will depend on the place you have studied.

25 Assimilation — number of MPs from minority ethnic groups. Social progress – percentage who have a higher education qualification. Housing provision – housing association activity.

26 This will depend on the place you have studied.

27 Providers: those in the driving seat largely through ownership. Users: consumers who may be affected, either gaining or losing out. Governance: those in control, not necessarily elected. Influencers: those wishing to make an intervention or take some form of action.

28 This will depend on the place you have studied.

29 This will depend on the place you have studied.

30 This will depend on the place you have studied.

31 Again, for the last time, over to you.

Topic 5

1 Because around two-thirds of it is located in ice sheets and glaciers.
2 Vegetation can have an influence on the drainage basin system, affecting the rates of interception, overland flow, transpiration and infiltration rates.
3 Because more of the ground surface is impervious — concrete and tarmac. Water entering drains is moved more quickly to streams and rivers. Both result in a faster delivery of rainwater to the drainage network, at such a rate as to exceed the drainage capacities. Once these are exceeded, flooding ensues.
4 One factor that can affect the shape of a storm hydrograph is the presence of vegetation. If an area has a greater coverage of vegetation this will increase interception rates, reducing the speed at which water reaches the river channel, increasing the lag time. A second factor affecting a storm hydrograph is the geology of the surrounding landscape. If the area is predominantly impermeable rocks, this will reduce the amount of infiltration and percolation rates, increasing surface runoff, causing water to reach the river channel faster, reducing the lag time.
5 El Niño events involve the reversal of the normal directions of ocean currents. They involve the warming of ocean waters along the west coast of South America. La Niña is characterised by unusually cool ocean temperatures in the equatorial region of the Pacific. Both have a considerable impact on weather.
6 Because of the rather arid climate and the high standard of living, the per capita water demand is high. Particular demands include irrigation, personal showers, swimming pools, etc.
7 Wetlands are important because they provide a range of valuable goods and services as well as acting as giant filters by trapping and recycling nutrients.
8 More energy in the atmosphere means more storms and heavier precipitation. That means more water in circulation and possibly being circulated at a faster rate.
9 The amount of global warming and its impact on the climate, particularly the amount and the distribution of precipitation. Uncertainty about future levels of water demand — depends on population growth rates and rising living standards. Will the supplies be sufficient? Will the mismatch between water availability and water demand increase or decrease?
10 More frequent droughts and extension of the world's arid areas. Population growth development in those parts of the world where there is little or no water surplus. Particular factors include population growth, rising living standards, industrialisation and the extension of commercial agriculture.
11 Water is really like any other commercial commodity — its price is governed by demand and supply. Where demand exceeds supply, the price will be higher than where the situation is reversed. The amount of precipitation varies spatially, and so too does the distribution of water demand. Unfortunately the two distributions do not match. It is this mismatch that causes the price of water to vary spatially.
12 Most likely, the first outbreak will be in the Middle East where there is much unrest. Water is in short supply because of the arid climate. Human survival there means high per capita water consumption. Also there are shared rivers that are important suppliers of water. The Tigris–Euphrates and Jordan are likely flashpoints, as well as international water pipelines.
13 Environmental sustainability is about ensuring the future availability of water that is unpolluted and is deemed to be safe water. Economic sustainability is ensuring a secure water supply to all users at an affordable price.

Topic 6

1 – Atmosphere: carbon dioxide, methane
– Hydrosphere: dissolved carbon dioxide
– Lithosphere: limestone, fossil fuels
– Biosphere: living and dead organisms
2 Sequestering is the long-term storage of carbon.
3 Carbon moves within oceans through the biological and physical pump. The biological pump involves the movement of carbon into the oceans through phytoplankton photosynthesising. The carbon then moves between the organisms to the deeper levels of the ocean. The physical pump involves the diffusion of carbon into the water. In cooler waters the carbon sinks to the deeper levels whereas in warmer waters it rises and diffuses back into the atmosphere.
4 Tropical rainforest.
5 Biological carbon is carbon stored in the form of dead organic matter.
6 Fossil fuels can have implications on climate, ecosystems and the hydrological cycle. Regionally, the increased combustion of fossil fuels can lead to some areas becoming warmer or cooler, affecting liveability. In terms of ecosystems, an increase in the combustion of fossil fuels can put pressure on marine ecosystems, affecting oxygen and acidification levels.
7 Heating and/or cooling homes. More domestic appliances. Increased car ownership.
8 Fossil fuel: oil, coal, natural gas. Recyclable: nuclear. Renewable: hydro, biofuels.
9 Geothermal: primary; solar: renewable; biofuels: primary; nuclear: primary.
10 Royal Dutch Shell, Chevron, Lukoil.
11 Favourable — if relying heavily on imported fossil fuels; if likely to be low cost. Unfavourable — if there is strong public opposition; if exploitation costs are high.
12 Because the construction of the necessary infrastructure will consume energy (electricity);

transport of components to build the farms of power stations.

13 It is really the case that rising income eventually reduces environmental impacts. Not all developed countries are too concerned about the environment.

14 411.43 – 315.97 = 95.46/315.97 = 0.30 × 100 = 30%

15 In the present context, adaptation is any action that reacts and adjusts to changing climate conditions, e.g. water conservation and management, development of more resilient types of farming. Mitigation is any action that either reduces or eliminates the long-term risk and hazards of climate change, e.g. international agreement on carbon emissions, promotion of renewable energy.

Topic 7

1 Superpower status can depend on five pillars of power:
   - Economic power: a large and powerful economy gives nations the wealth to build and maintain a powerful military, exploit natural resources and develop human ones through education.
   - Military power: this is used through the threat of military action, or military force can be used to achieve geopolitical goals.
   - Political power: this is the ability to influence others through diplomacy or international organisations.
   - Cultural power: this includes how appealing a nation's way of life, values and ideology are to others.
   - Resources: these can be in the form of physical resources and human resources.

2 'Hard power' refers to the way that countries get their own way by use of force. 'Soft power' is the power of persuasion — this is often enforced through culture.

3 The reasons include:
   - postwar bankruptcy, meaning there was no money to run, or defend, colonies
   - the focus on postwar reconstruction at home meant that colonies were viewed as less important
   - anti-colonial movements, for example in India, grew increasingly strong and demands for independence could not be ignored.

4 A unipolar world is one dominated by one superpower, e.g. the British Empire or the US-dominated world of today. A bipolar world is one in which two superpowers with opposing ideologies vie for power, e.g. the USA and the USSR during the Cold War. A multipolar world is more complex: many superpowers and emerging powers compete for power in different regions.

5 Power can be maintained politically through influence over elections or government policies and military operations; through propaganda and economically through the use of aid and investment overseas; through force, for example through military zones, military alliances and nuclear power. Examples are needed to support your answer.

6 There will be less of the cost of an ageing population. There will be a youthful, vibrant labour pool available and they will be able to adapt to new skills and technologies.

7 China's strengths include its growing economy, its rapid progression in regards to global investment, FDI, size of army, level of world trade, membership of international organisations and increasing number of TNCs. This list is not complete.

8 Superpowers may change over time based on three models that need to be explained: modernisation theory, dependency theory and world systems theory.

9 The dominance of TNCs in the global economy has been caused by a number of factors:
   - Their economies of scale mean they can outcompete smaller companies and, in many cases, take them over.
   - Their bank balances and ability to borrow money to invest have allowed them to take advantage of globalisation by investing in new technology.
   - The move towards free-market capitalism and free trade has opened up new markets, allowing them to expand.

10 TNCs are able to influence the countries that they operate in from their country of origin. For example, via Western culture, cultural globalisation has led to the spread of consumerism, capitalism and a white, Anglo-Saxon culture on a global scale.

11 Because the USA has often needed to intervene militarily in overseas countries:
   - as part of a UN Security Council action
   - together with allied countries as a coalition, but outside a UN remit
   - unilaterally, that is, with no support from other countries
   - in order to maintain its place in the global pecking order and reassert its superpower status.

12 The four pillars are political, economic, social and judicial. Most important will depend on your point of view, but you will need to justify your answer.

13 This will depend upon the role in international decision making. Examples could include the role in global conflict, offering multinational aid in a humanitarian crisis or acting during world pandemics.

14 In the case of China, for example:
   - emissions per person are the same as those of an EU citizen
   - emissions could grow much higher as affluence increases
   - decisions that China makes on emissions have a disproportionate impact on global emissions because its share and potential for growth are so large.

15 It would mean a growth in the numbers of people who can afford access to health and education,

improving the quality of life for millions of people globally.

16 The growth of the middle class is a significant concern related to the growth of BRICS and other emerging powers. As there is a rising affluence in these countries, this is positive in regards to development but results in a huge strain on resources. This rising middle-class demand will affect the availability and cost of key resources, such as rare earth minerals, oil and gas, staple grains and water.

17 Examples include tensions over territory, for example between China and its neighbours over the Spratly Islands or between Russia and the former states of the USSR. Also, desirable resources can lead to tensions and conflict, for example tension over resources in the Arctic.

18 These ties could be through:
- neo-colonialism: superpowers pulling the economic and political strings of developing countries, despite not ruling them directly, as occurred during the colonial/imperial era
- unfair terms of trade: cheap commodity exports for the developing world (coffee, cocoa, oil, copper) set against expensive manufactured imports from developed countries
- the brain drain of skilled workers from developing countries to boost developed world economies
- local wealthy elites, who control imports and exports in developing countries benefiting from the neo-colonial relationship but having no interest in changing it.

19 Asia could be a very economically and politically crowded continent by 2030. China, India, Indonesia and Japan are all likely to have economies greater than US$5 trillion by that date, giving them all the means to have significant military power.

20 The Middle East has been an area of tension during the past 50 years. It has a number of characteristics that make it a frequent location of tension and conflict. The complexity of Middle Eastern politics, religions, ethnic differences and territorial disputes has led to some intractable and potentially dangerous situations. There are concerns about oil supplies, religious tensions and governance.

21 The USA spends more than US$900 billion annually on maintaining its global supremacy. This includes all military spending and intelligence services, as well as foreign aid and NASA. China spends around US$200 billion by comparison. Emerging powers have lower labour costs and lower salaries and, to some extent, can increase their military power by copying technologies that were initially developed (at huge cost) by the USA.

22 Economic restructuring in the USA meant that national debt reached US$19 trillion in 2016. Such debt is unsustainable in the long term, but the US dollar's status as the global currency of choice makes it less vulnerable to economic shocks, meaning it is unlikely that these debts will be called in suddenly. The USA has many large, innovative global TNCs, which may also continue to protect the country from challenges in the future.

23 This could include the challenges linked to the USA's economic issues, for example the financial crisis it has experienced and slowing of its economic growth compared to other emerging superpowers. The main challenge the USA faces at present is China's growing dominance.

Topic 8A

1 The HDI because it is based entirely on quantitative data. The HPI relies partly on some qualitative input.

2 Environmental quality has an impact on health, for example a polluted environment will have a detrimental impact. Health, in turn, finds expression in life expectancy. Longer life expectancy means more time to enjoy the benefits of a good-quality environment.

3 There are many aspects, including the fact that it improves the quality of the labour pool, increases aspirations for a higher standard of living and better quality of life, gives better awareness of the prerequisites of healthy living, and increases the likelihood of better governance.

4 Levels of education depend upon the level of development within a country, for example sub-Saharan Africa has some of the lowest numbers of children in education in the world. Inequalities happen within a country due to factors such as gender, ethnicity, language and religion.

5 For example, number of doctors per 100,000 people; healthy life expectancy; percentage of population vaccinated for particular diseases.

6 (a) Globally: differences in diet, healthcare and quality of life; internationally: lifestyles, diet and economic prosperity; internally: ethnicity, socio-economic class, income and housing.

(b) Variations within the developing world are explained by differential access to the basic survival needs of food, clean water, sanitation and healthcare. Variations within the developed world are more to do with differences in lifestyles, levels of poverty and deprivation, and the accessibility and quality of healthcare.

7 World Bank: poverty reduction and support for development projects; WTO: the liberalisation of trade; IMF: encouragement of free market economies; UNESCO: the promotion of education, protection of human rights and heritage; OECD: the improvement of economic and social well-being.

8 Improving the quality of life today without damaging the quality of life of future generations.

9 They are very significant in setting targets and policies, for example the MDGs and SDGs, which are driven by a desire to achieve these goals. These vary from country to country depending on spending priorities of a government.

10 The Universal Declaration of Human Rights (UDHR) has led to legally binding human rights covenants, which ensure human rights are

enforced in countries. Furthermore, the European Convention on Human Rights (ECHR) has resulted in improvements in hearing human rights cases in all EU member states and the Geneva Convention has resulted in improved protection of human rights where there is armed conflict.

11 Levels of corruption may threaten human rights if laws are not enforced appropriately. Corruption can particularly affect the judicial process and in many cases slow it down. Corruption in a country could lead to aid donors being put off.

12 Human rights can vary due to the political ideologies within a country and how much democratic freedom there is. Also, the level of corruption within a country can affect human rights and often countries actually define human rights in different ways, which can lead to certain human rights being disregarded.

13 Advantages include:
- if conflicts are not stopped they could continue for many years
- conflicts may escalate further if intervention does not occur
- development of a country can continue with the appropriate interventions
- developing countries may not have the financial ability to solve problems without the support of others.

Disadvantages include:
- military intervention could be used to mask the real reason for intervention
- if there are multiple agencies involved in the intervention then priorities could vary
- they could build a reputation for getting involved in other countries' affairs.

14 ODA as a percentage of GNI. ODA needs to be put in a common context if countries are to be compared in order to determine whether they are pulling their weight. ODA for a large and prosperous country, such as the USA, may be impressive in absolute terms but minuscule when related to its wealth. While the UK's ODA is significantly smaller than that of the USA, proportionately it is significantly larger.

15 Development programmes have been seen to cause environmental damage through oil spills and rainforest destruction. There have also been cultural issues due to traditional practices and livelihoods being affected.

16 IS is intent upon further destabilising the Middle East and North Africa. It is a threat to global oil supplies. It has a history of horrific abuse of human rights.

17 Torture is what is done, physically and mentally, to extract information. Rendition is sending someone to be tortured in a country which is lax in its enforcement of international laws that declare torture to be illegal.

18 Use examples to support your answer. In countries where there is a history of human right abuses, such as poor human rights records, there can be justification for military intervention.

19 Because human rights have an intangible quality. The best measures are negative ones, namely statistics relating to the abuse of human rights. The problem is that human rights are often subtly denied rather than overtly abused.

20 Gini coefficient.

21 The USA, the EU, China and possibly Russia.

22 It would be expensive in terms of human lives and damage to property. Military equipment is hugely costly to deploy and replace. Generally speaking, the longer the intervention, the longer the recovery period.

23 Direct military intervention is when troops and equipment are sent into a place to fight. It is a high-risk and high-cost option. Indirect military intervention is when economic or military assistance is provided without the use of troops, and involves lower costs and risks.

Topic 8B

1 Economic migrants decide to move because they want to improve their chances of employment and make more money.

2 A political refugee is someone suffering some form of persecution or discrimination to the extent of genuinely fearing for their life. An environmental refugee is someone who has been forced to move because of some environmental event, such as a natural hazard.

3 Globalisation has led to an increase in international migration, e.g. international migration is easier through free movement, demonstrated by the EU.

4 Congestion; higher costs of living, especially housing; deterioration in environmental quality; peripheries drained to the point of disaster, etc.

5 Social: to be closer to family, for a better quality of life. Economic: moving for work. Political: migrating to escape persecution and conflict; due to policy changes, e.g. free movement in the EU.

6 Advances in modern transport have greatly changed time–distance values and the costs of long-distance travel. Modern communications now allow instantaneous contact almost everywhere in the world. The significance is to draw countries closer together, creating a feeling of belonging to a global village. It has also encouraged the growth of a global economy and national interdependence.

7 For: much easier movement of people, particularly economic migrants and refugees. Against: migrants are free to converge on favoured locations, causing all manner of problems — excessive growth, overheating, social and ethnic tensions, etc. National unity can be threatened.

8 Immigration can be a key factor in elections and the political party's view may be influential in their success. Immigration can be a politically divisive issue and it will depend on the government in power.

9 They can become contested due to conflict and through countries claiming sovereignty over other states and provinces.

10 Nationalism has played a number of roles. For example, it has increased loyalty within nations and influenced migration patterns and the ethnic compositions of many nations in the modern world. It has also led to conflicts across borders and created new nations as empires disperse.

11 Because of the number of civilians and military personnel killed during the Second World War and the huge amount of reconstruction work needing to be done. Main employers were public transport and the construction industry.

12 The choice is yours! The negatives you have to weigh up are authoritarian control, corruption, economic bankruptcy and human rights abuse. Not an easy choice.

13 Global wealth has increased but this has not been evenly distributed, which has led to growing inequalities and a widening of the development gap. This uneven distribution of wealth threatens economic stability within countries with the largest wealth gaps and furthermore it threatens the economy of the poorest nations which cannot invest in their education system, which results in 'switched off' nationals who cannot benefit from the technological revolution.

14 To maintain peace between internal factions; to protect human rights; to support elected governments; to sort out failed states.

15 The main interventions made by the UN fall under the following headings:
- direct military intervention, e.g. in the Congo
- peacekeeping in states with internal conflicts, e.g. Ivory Coast
- economic sanctions against countries stepping out of line, e.g. trade embargo on Iran
- defending human rights, e.g. setting up war crime trials
- protecting refugees, e.g. in Syria
- promoting development, especially in agriculture, education and healthcare, e.g. throughout much of Africa.

16 Lending by the IMF and the WTO has helped many lagging states to develop economically. However, since 1970 tougher conditions have been attached to large-scale lending. For states experiencing severe financial difficulties, there are two forms of help available:
- structural adjustment programmes (SAPs): providing loans but with strict conditions attached
- heavily indebted poor countries (HIPC) policies: aimed at ensuring that no poor country faces a debt burden it cannot manage.

Critics of these two concessions say that they tend to increase rather than decrease poverty. The concessions also undermine the economic sovereignty of borrowing states. They are regarded by some as a neo-colonial strategy used by developed countries to maintain influence over how peripheral countries develop.

17 Ozone layer: difficult to monitor compliance; carbon emissions: compliance is a threat to the development of some countries; endangered species: difficult to police; biodiversity: poor awareness of ecosystem services.

18 IGOs have been formed to address global environmental problems, including: climate change, endangered species, ozone depletion and conservation of ecosystems.

19 Rivers that are shared by two or more countries. There are two possible forms of 'sharing': where the river forms an international boundary, e.g. the Uruguay River between Brazil, Uruguay and Argentina, and where the river's length is divided (the advantage being with the upstream partner), e.g. the Ganges shared by India and Bangladesh, the Tigris by Turkey and Iraq.

20 Spillages and discharges from ships; trampling at landing sites; disturbance of penguin colonies; increase in the pressure to provide more land-based facilities, even hotels.

21 To ensure that it is protected and that there is continued recognition of its intrinsic value.

22 For example, the growth of a materialistic consumer society; increasing multi-ethnicity; increasing exposure to a global culture.

23 This is your opinion.

24 Globalisation has had both a positive and negative impact on Western identity, and this particularly has helped the USA to shape Western identity through the growth of TNCS such as Apple and Disney. Development of communication systems has enabled Western identity to disperse around the world via television, music, media and the internet.

25 Ethnic and cultural differences; political polarisation; acute core–periphery differences.

26 Rising nationalism and disillusionment with globalisation can cause disunity within nations and a desire for independence away from globalisation, e.g. Brexit. There are positive and negative consequences of disunity, positive consequences include independent nations becoming more autonomous with responsibility for their own education, healthcare, welfare, justice and legal systems. Whereas negative consequences of disunity include challenges for businesses in terms of renegotiation of trade deals and changes to tax, regulations, and currency when a nation becomes independent; rising tensions between national divisions in emerging nations; war and conflict; and growing inequalities. Examples of areas of the world where nationalism is rising, leading to separatist movements and creating disunity include Scotland in the UK and Catalonia in Spain. Examples of where disunity has led to war and conflict include Syria.